T0251447

Sandeep Kumar, PhD
Matthias Fladung, PhD
Editors

Molecular Genetics and Breeding of Forest Trees

Pre-publication
REVIEWS,
COMMENTARIES,
EVALUATIONS . . .

"**M**olecular tools are indispensable for the development of sustainable strategies to utilize and conserve forest genetic resources. Marker technologies and the availability of transformation tools allow new insights into the genetic basis of physiological processes and genome organization of trees. Innovative breeding methods based on marker-assisted selection and genetic transformation have been developed and will gain further importance in forest genetics and tree breeding. In view of a rapidly increasing number of different marker types and transformation approaches, it is not easy to define what is 'state-of-the-art' in this field for forest trees. This gap is filled by *Molecular Genetics and Breeding of Forest Trees*, which provides concise information on the most recent and important developments in molecular forest genetics. Kumar and Fladung successfully keep the balance between instructive reviews on key aspects in molecular genetics such as functional genomics, ESTs, and SSRs, and the profound case studies addressing temperate and tropical trees."

Prof. Dr. Reiner Finkeldey
Chair of Forest Genetics
and Forest Tree Breeding,
Georg-August-University Göttingen,
Germany

More pre-publication
REVIEWS, COMMENTARIES, EVALUATIONS . . .

"**M**olecular Genetics and Breeding of Forest Trees makes it clear that the science is moving ahead impressively. The editors have carefully crafted it to include the major scientific thrusts in forest biotechnology. These span genomic maps and markers as supplements to traditional breeding through to novel means for altering the characteristics of wood and tree production, with the goal of enabling new kinds of highly domesticated, biosafe plantations to be employed.

Highlights of the book include a description of the EST databases of Genesis Research and Development Corporation, Ltd., in New Zealand; an excellent review of the state of knowledge of lignin biosynthetic genes in plants, particularly with regard to prospects for genetic engineering; and a comprehensive review of attempts to induce flowering and impart sterility in trees. A number of chapters discuss the considerable progress in the use of genomics tools to understand fundamental aspects of tree biology, including high quality chapters on pine proteomics, fungal and plant changes in gene expression during mycorrhizal association, use of microsatellite markers for population and systematic biology, and the integration of genetic maps in poplar for analysis of disease resistance.

With a focus on science, this book proves that the research progress on forest biotechnology is impressive; opportunities abound for much further scientific advance; field verifications of economic value and ecological effect are badly needed; and that the commercial and social landscape regarding public and market acceptance is complex."

Steven H. Strauss
Professor, Department of Forest Science,
Oregon State University

"**T**he knowledge and techniques of molecular biology are of great assistance in addressing diverse aspects of forest genomics. In the past decade there has been a great surge of knowledge about molecular aspects of the tree genome and its application for improving the quality and productivity of forests, which has been spread through a diverse array of journals and periodicals in need of consolidation for educational and reference material. *Molecular Genetics and Breeding of Forest Trees* is a wonderful accomplishment in this direction. The book provides a wealth of classified information dealing with tree genome analysis at the molecular level with regard to genetic diversity, quantitative trait loci, ectomycorrhizal symbiosis, physiobiochemical pathways of wood, lignin formation, and development of transgenesis and high-density linkage maps for forest trees. The renowned contributors elegantly crafted each chapter, suited alike to both classroom texts for graduate students and reference material for researchers. The language and style is simple and lucid with liberal use of illustrations. This book should be on the shelf of school and university libraries for inquisitive students and enlightened researchers."

Dr. Shamim Akhtar Ansari
Scientist "E,"
Genetics and Plant Propagation Division,
Tropical Forest Research Institute,
Jabalpur, India

Molecular Genetics and Breeding of Forest Trees

FOOD PRODUCTS PRESS®
Crop Science
Amarjit S. Basra, PhD
Senior Editor

Mineral Nutrition of Crops: Fundamental Mechanisms and Implications by Zdenko Rengel

Conservation Tillage in U.S. Agriculture: Environmental, Economic, and Policy Issues by Noel D. Uri

Cotton Fibers: Developmental Biology, Quality Improvement, and Textile Processing edited by Amarjit S. Basra

Heterosis and Hybrid Seed Production in Agronomic Crops edited by Amarjit S. Basra

Intensive Cropping: Efficient Use of Water, Nutrients, and Tillage by S. S. Prihar, P. R. Gajri, D. K. Benbi, and V. K. Arora

Physiological Bases for Maize Improvement edited by María E. Otegui and Gustavo A. Slafer

Plant Growth Regulators in Agriculture and Horticulture: Their Role and Commercial Uses edited by Amarjit S. Basra

Crop Responses and Adaptations to Temperature Stress edited by Amarjit S. Basra

Plant Viruses As Molecular Pathogens by Jawaid A. Khan and Jeanne Dijkstra

In Vitro Plant Breeding by Acram Taji, Prakash P. Kumar, and Prakash Lakshmanan

Crop Improvement: Challenges in the Twenty-First Century edited by Manjit S. Kang

Barley Science: Recent Advances from Molecular Biology to Agronomy of Yield and Quality edited by Gustavo A. Slafer, José Luis Molina-Cano, Roxana Savin, José Luis Araus, and Ignacio Romagosa

Tillage for Sustainable Cropping by P. R. Gajri, V. K. Arora, and S. S. Prihar

Bacterial Disease Resistance in Plants: Molecular Biology and Biotechnological Applications by P. Vidhyasekaran

Handbook of Formulas and Software for Plant Geneticists and Breeders edited by Manjit S. Kang

Postharvest Oxidative Stress in Horticultural Crops edited by D. M. Hodges

Encyclopedic Dictionary of Plant Breeding and Related Subjects by Rolf H. G. Schlegel

Handbook of Processes and Modeling in the Soil-Plant System edited by D. K. Benbi and R. Nieder

The Lowland Maya Area: Three Millennia at the Human-Wildland Interface edited by A. Gómez-Pompa, M. F. Allen, S. Fedick, and J. J. Jiménez-Osornio

Biodiversity and Pest Management in Agroecosystems, Second Edition by Miguel A. Altieri and Clara I. Nicholls

Plant-Derived Antimycotics: Current Trends and Future Prospects edited by Mahendra Rai and Donatella Mares

Concise Encyclopedia of Temperate Tree Fruit edited by Tara Auxt Baugher and Suman Singha

Landscape Agroecology by Paul A Wojkowski

Concise Encyclopedia of Plant Pathology by P. Vidhyaskdaran

Molecular Genetics and Breeding of Forest Trees edited by Sandeep Kumar and Matthias Fladung

Testing of Genetically Modified Organisms in Foods edited by Farid E. Ahmed

Molecular Genetics and Breeding of Forest Trees

Sandeep Kumar, PhD
Matthias Fladung, PhD
Editors

Food Products Press®
An Imprint of The Haworth Press, Inc.
New York • London • Oxford

Published by

Food Products Press®, an imprint of The Haworth Press, Inc., 10 Alice Street, Binghamton, NY 13904-1580.

Cover design by Marylouise E. Doyle.

Library of Congress Cataloging-in-Publication Data

Molecular genetics and breeding of forest trees / Sandeep Kumar, Matthias Fladung, editors.
    p. cm.
    Includes bibliographical references (p.) and index.
    ISBN 1-56022-958-6 (hardcover : alk. paper) — ISBN 1-56022-959-4 (softcover : alk. paper)
    1. Trees—Breeding. 2. Trees—Molecular genetics. 3. Forest genetics. I. Kumar, Sandeep. II. Fladung, Matthias.
SD399.5.M66 2003
634.9'56—dc21

2002156648

CONTENTS

PART III: FOREST TREE TRANSGENESIS

ABOUT THE EDITORS

Sandeep Kumar, PhD, is a tree molecular biologist with a wide range of experience in plant tissue culture, genetic transformation, and tree genomics. He completed his master's and doctorate in India and presently works at the Institute for Forest Genetics and Forest Tree Breeding in Grosshansdorf, Germany. He has been working on genetic transformation of forest trees, stable expression of transgenes, gene silencing, mechanisms of transgene integration in a tree system, and transposon/T-DNA based gene tagging in trees. His current research interests include controlled gene transfer into predetermined genomic positions and gene tagging for tree functional genomics.

Matthias Fladung, PhD, is Head of the Tree Molecular Genetics and Provenance Research Section of the Institute for Forest Genetics and Forest Tree Breeding in Grosshansdorf, Germany. He is a member of numerous scientific organizations and is the author of more than 30 original publications and 50 peer-reviewed chapters. He received his PhD in botany at the Max Planck Institute for Plant Breeding in Cologne, Germany. His research focuses on tree transgenesis, genome mapping, and functional genomics of model angiosperm and gymnosperm tree species.

CONTRIBUTORS

Nasser Bahrman, Institut National de la Recherche Agronomique (INRA), Equipe de Génétique et Amélioration des Arbres Forestiers, Cestas, France.

Wout Boerjan, Department of Plant Systems Biology, Flanders Inter-university Institute for Biotechnology (VIB), Ghent University, Gent, Belgium.

Gerd Bossinger, The University of Melbourne—School of Resource Management; Forest Science Center, Creswick, Victoria, Australia.

Penelope A. Butcher, Commonwealth Scientific and Industrial Research Organisation (CSIRO) Forestry and Forest Products, Kingston, Australia.

María Teresa Cervera, Departamento de Mejora Genética Forestal, Madrid, Spain.

Julia Charity, New Zealand Forest Research Institute Ltd, Rotorua, New Zealand.

Janice E. K. Cooke, Centre de Recherche en Biologie Forestière, Université Laval, Québec QC G1K 7P4, Canada.

Paulo Costa, Institut National de la Recherche Agronomique (INRA), Equipe de Génétique et Amélioration des Arbres Forestiers, Cestas, France.

John M. Davis, University of Florida, School of Forest Resources and Conservation, and Plant Molecular and Cellular Biology Program, Gainesville, Florida.

Lloyd Donaldson, New Zealand Forest Research Institute Ltd, Rotorua, New Zealand.

Christian Dubos, Institut National de la Recherche Agronomique (INRA), Equipe de Génétique et Amélioration des Arbres Forestiers, Cestas, France.

Sébastien Duplessis, Joint Research Unit of Université Henri Poincaré and Institut National de la Recherche Agronomique (UMR UHP-INRA) 1136, "Interactions Arbres/Micro-Organismes," INRA Centre de Nancy, Champenoux, France.

Hiroyasu Ebinuma, Pulp and Paper Research Laboratory, Nippon Paper Industries Co., Ltd., Tokyo, Japan.

Saori Endo, Pulp and Paper Research Laboratory, Nippon Paper Industries Co., Ltd., Tokyo, Japan.

Santiago Espinel, Nekazal Ikerketa eta Garapenenako Euskal Erakundea (NEIKER), Vitoria, Alava, Spain.

Patricia Faivre-Rampant, Joint Research Unit of Université Henri Poincaré and Institut National de la Recherche Agronomique (UMR UHP-INRA) Nancy, Interactions Arbres/Micro-organismes, Faculté des Sciences, Vandoeuvre lès Nancy, Cedex, France.

Jean-Michel Favre, Joint Research Unit of Université Henri Poincaré and Institut National de la Recherche Agronomique (UMR UHP-INRA) Plant-Microbes Interactions, Faculté des Sciences, Vandoeuvre lès Nancy, France.

Jean-Marc Frigerio, Institut National de la Recherche Agronomique (INRA), Equipe de Génétique et Amélioration des Arbres Forestiers, Cestas, France.

Sophie Gerber, Institut National de la Recherche Agronomique (INRA), Equipe de Génétique et Amélioration des Arbres Forestiers, Cestas, France.

Jean-Marc Gion, Institut National de la Recherche Agronomique (INRA), Equipe de Génétique et Amélioration des Arbres Forestiers, Cestas, France.

Thomas Goujon, Laboratoire de Biologie Cellulaire, Institut National de la Recherche Agronomique (INRA), Versailles Cedex, France.

Lynette Grace, New Zealand Forest Research Institute Ltd, Rotorua, New Zealand.

Kyung-Hwan Han, Department of Forestry, Michigan State University, East Lansing, Michigan.

Hu Jianjun, Associate Professor, The Research Institute of Forestry, The Chinese Academy of Forestry, Wan Shou Shan, Beijing, China.

Chandrashekhar P. Joshi, Plant Biotechnology Research Center, School of Forestry and Wood Products, Michigan Technological University, Houghton, Michigan.

Lise Jouanin, Laboratoire de Biologie Cellulaire, Institut National de la Recherche Agronomique (INRA), Versailles Cedex, France.

Jae-Heung Ko, Department of Forestry, Michigan State University, East Lansing, Michigan.

Annegret Kohler, Joint Research Unit of Université Henri Poincaré and Institut National de la Recherche Agronomique (UMR UHP-INRA) 1136, "Interactions Arbres/Micro-Organismes," INRA Centre de Nancy, Champenoux, France.

Céline Lalanne, Institut National de la Recherche Agronomique (INRA), Equipe de Génétique et Amélioration des Arbres Forestiers, Cestas, France.

Mathew A. Leitch, The University of Melbourne–School of Resource Management; Forest Science Center, Creswick, Victoria, Australia.

Juha Lemmetyinen, Department of Biology, University of Joensuu, Joensuu, Finland.

Wang Lida, Institute of Forestry Sciences, Chinese Academy of Forestry, Beijing, P.R. China.

Delphine Madur, INRA, Equipe de Génétique et Amélioration des Arbres Forestiers, Cestas, France.

Francis Martin, Joint Research Unit of Université Henri Poincaré and Institut National de la Recherche Agronomique (UMR UHP-INRA) 1136, "Interactions Arbres/Micro-Organismes," INRA Centre de Nancy, Champenoux, France.

Etsuko Matsunaga, Pulp and Paper Research Laboratory, Nippon Paper Industries Co., Ltd., Tokyo, Japan.

Armando McDonald, New Zealand Forest Research Institute Ltd, Rotorua, New Zealand.

Ralf Möller, New Zealand Forest Research Institute Ltd, Rotorua, New Zealand.

Alison M. Morse, University of Florida, School of Forest Resources and Conservation, and Plant Molecular and Cellular Biology Program, Gainesville, Florida.

Sookyung Oh, Department of Forestry, Michigan State University, East Lansing, Michigan.

Sunchung Park, Department of Forestry, Michigan State University, East Lansing, Michigan.

Cédric Pionneau, Institut National de la Recherche Agronomique (INRA), Equipe de Génétique et Amélioration des Arbres Forestiers, Cestas, France.

Christophe Plomion, Institut National de la Recherche Agronomique (INRA), Equipe de Génétique et Amélioration des Arbres Forestiers, Cestas, France.

Enrique Ritter, Nekazal Ikerketa eta Garapenenako Euskal Erakundea (NEIKER), Vitoria, Alava, Spain.

Ivan Scotti, Dipartimento di Produzione Vegetale e Tecnologie Agrarie, Università degli Studi, Udine, Italy.

Mitchell M. Sewell, Environmental Sciences Division, Oak Ridge National Laboratory, Oak Ridge, Tennessee.

Tuomas Sopanen, Department of Biology, University of Joensuu, Joensuu, Finland.

Véronique Storme, Department of Plant Systems Biology, Flanders Interuniversity Institute for Biotechnology (VIB), Ghent University, Gent, Belgium.

Timothy J. Strabala, Genesis Research & Development Corporation, Ltd., Auckland, New Zealand.

Koichi Sugita, Pulp and Paper Research Laboratory, Nippon Paper Industries Co., Ltd, Tokyo, Japan.

Denis Tagu, Joint Research Unit of Université Henri Poincaré and Institut National de la Recherche Agronomique (UMR UHP-INRA) 1136, "Interactions Arbres/Micro-Organismes," INRA Centre de Nancy, Champenoux, France.

Giovanni G. Vendramin, Istituto Miglioramento Genetico Piante Forestali, Consiglio Nazionale delle Ricerche (CNR), Firenze, Italy.

Armin Wagner, New Zealand Forest Research Institute Ltd, Rotorua, New Zealand.

Christian Walter, New Zealand Forest Research Institute Ltd, Rotorua, New Zealand.

Keiko Yamada-Watanabe, Pulp and Paper Research Laboratory, Nippon Paper Industries Co., Ltd., Tokyo, Japan.

Jaemo Yang, Department of Forestry, Michigan State University, East Lansing, Michigan.

Han Yifan, Institute of Forestry Sciences, Chinese Academy of Forestry, Beijing, P.R. China.

Birgit Ziegenhagen, Nature Conservation Division, Faculty of Biology, Philipps-University of Marburg, Marburg, Germany.

Preface

Plant molecular biology leapt from being a futuristic concept in the 1970s to an accepted part of science in the new millennium. Following the great progress in molecular bacterial genetics, the milestones in the past thirty years include the production of the first transgenic plant in the early 1980s and the development of genome-mapping methods, allowing "molecular breeding," along with the introduction of molecular marker technology. The genome of the weed plant *Arabidopsis* has been sequenced, and other plant genome-sequencing efforts are under way. These genomics technologies have also been integrated into forest tree improvements that are now producing gene structural and expressional data at an unprecedented rate.

In February 2002, the U.S. Department of Energy announced its decision to sequence the first tree genome, *Populus balsamifera* ssp. *trichocarpa*. The poplar genome sequence will assist in identifying the full suite of genes, including their promoters and related family members, all of which will guide experimental efforts to define gene function. Functional genomics in forest trees has the potential to contribute greatly to our understanding of how forest tree-specific traits are regulated and how trees differ from model plant species.

Transformation is key for the varieties of reverse and forward genetic approaches creating "knock-in" or "knock-out" mutants unraveling the functions of unknown sequences (Part I). Transformation is also very important for understanding the most fundamental biological process of forest trees: wood formation. In particular, increasing wood quality or modifying lignin and/or cellulose content are investigated in detail (Part II). A range of additional targets are of interest for genetic engineering in trees. These include increased pest and disease resistance, better growth characteristics, modification of flowering, and tolerance to abiotic stresses. Transgenesis is also needed to confirm gene function, after deductions made through comparative genomics, expression profiles, and mutation analysis. A major limitation of current plant transformation technologies is the inability to control transgene integration leading to expression variability. To overcome the problem of expression variability and gene silencing systematically, simple and reproducible transformation technologies as well as the ability to precisely modify or target defined locations within the genome are required (Part III). In addition, genetic linkage maps have significantly contributed to both the genetic dissection of complex inherited traits and positional cloning of genes of interest and have become a valuable tool in molecular breeding. The design of new mapping strategies, particularly for tree species characterized with long generation intervals and high levels of heterozygosity, together with the development of new marker technologies, has paved the way for constructing genetic maps of forest tree species (Part IV).

The aim of this book is to integrate tree transgenesis and functional and structural genomics in the context of a unified approach to forest tree molecular biology research for the benefits of students and researchers alike. It was a pleasant experience to edit this book together with many friends and colleagues. The completion of this task could not have been achieved without the cooperation of the chapter contributors and reviewers. We also wish to acknowledge the pleasure of working with the staff of The Haworth Press, Inc., including Editor in Chief (Food Products Press) Amarjit S. Basra and Senior Production Editor Peg Marr. Finally, we gratefully acknowledge the strong support of our families throughout this endeavor.

PART I:
FOREST TREE
FUNCTIONAL GENOMICS

Chapter 1

Functional Genomics in Forest Trees

Alison M. Morse
Janice E. K. Cooke
John M. Davis

INTRODUCTION

Genomic science has revolutionized how gene function is studied. Scientists have unprecedented access to gene sequence information, in some cases to the entire genome sequence. This information has motivated new perspectives and approaches to carrying out biological research on many different organisms, including forest trees. Forest tree functional genomics aims to define the roles played by all of the genes in a tree. The accomplishment of this aim will indeed be a challenge; however, along the way experiments are sure to reveal novel and unexpected aspects of tree biology. Furthermore, it seems likely that forest tree functional genomics will lead to new insights on manipulating tree genomes for practical benefits such as increasing yield or altering wood quality.

Prior to the genomics era, scientists were technologically restricted to identifying and characterizing one or a few genes at a time. Often, these genes were identified as important because they encoded proteins that were abundant or that exhibited a particular enzyme activity. Characterization was much more difficult for genes that encoded proteins of low abundance, or with transitory, ill-defined, or completely unknown activities. Consequently, many genes were not studied. For example, genes encoding proteins involved in signal transduction—the process of coordinating growth, development, and environmental responsiveness—remained poorly understood because these proteins tend to be of low abundance and have activities

This work was supported by the Florida Agricultural Experiment Station, the Department of Energy (Cooperative Agreement No. DE-FC07-97ID13529 to JMD), and the U.S. Department of Agriculture (USDA) Forest Service Southern Research Station. Cooke is the recipient of a National Sciences and Engineering Research Council of Canada (NSERC) postdoctoral fellowship.

3

that are transitory and/or poorly defined. Another difficulty associated with the "one-gene-at-a-time" approach is that it is not obvious how any one gene fits into the bigger picture of cellular and organismal processes. Today, scientists can query an organism via a genomics approach and identify the genes and proteins with potential roles in a process of interest, without any a priori knowledge of function. The comprehensive picture of an organism that is afforded by functional genomics has changed the way we approach biological research. This is because the analytical tools at our disposal to observe levels of mRNA, proteins, and metabolites and their interactions provide global phenotyping information.

Many different interpretations of the phrase "functional genomics" are currently in use in the scientific community (Hieter and Boguski, 1997). In this chapter we define functional genomics as the analysis of the roles played by all of the genes in an organism, typically involving high-throughput experiments that generate large quantities of information. A number of research areas associated with functional genomics, including proteomics and metabolomics, are beyond the scope of this chapter. Consequently, we will primarily discuss the analysis of mRNA expression abundance, i.e., analysis of the transcriptome. This chapter introduces some concepts of functional genomics, with particular reference to its use in understanding forest trees. Because the cornerstone of functional genomics is the sequence information contained in genes, we will first discuss methods that have been used to discover and sequence tree genes. We then discuss two key papers from work in yeast and *Arabidopsis thaliana* that help define the potential roles of genetic manipulation and microarrays in forest tree functional genomics. Finally, we turn to biological features of forest trees that make them unique among plants and that also affect the kinds of functional genomics approaches that are most likely to be successful.

THE FUNCTIONAL GENOMICS TOOL KIT—
TRANSCRIPT DISCOVERY AND PROFILING

The ultimate goal of transcript profiling is to identify all of the genes that are transcribed into mRNA. Because specific tissue types and developmental stages (e.g., reproductive structures, developing wood) are of particular interest in trees, a common approach is to target those genes that are expressed therein. Not only can targeting specific traits allow a gene discovery effort to focus on tree-specific processes, but it can also help guide both hypothesis generation and testing. Of course, biological processes are not regulated solely by the modulation of transcript abundance. In most cases, the proteins encoded by the genes effect changes in cellular processes. Modulation of protein activity can occur without changes in gene expression, for example by posttranslational modifications such as phosphorylation. Nonetheless, compelling evidence exists that transcriptional regulation plays a central role in regulating many biological processes. In the following sections, we review some of the gene expression analyses used to identify genes in trees.

Comparative Expressed Sequence Tag (EST) Sequencing

Partial sequencing of cDNAs selected from libraries has become a useful method for identifying genes with potential roles in the tissue or organ of interest (Adams et al., 1991; Hofte et al., 1993) and are referred to as expressed sequence tags (ESTs). Comparing tree ESTs to other plant genes already in the public sequence databases can reveal the extent to which the tree collection contains genes in common with other plants versus those genes potentially unique to trees.

The largest forest tree EST collections have been generated by randomly selecting cDNAs from libraries made from developing xylem (Allona et al., 1998; Sterky et al., 1998). Because cDNAs derived from genes that are highly expressed are more likely to be selected for sequencing than transcripts of lower abundance, the number of times a cDNA is sequenced can be thought of as a reflection of its expression level in the tissue from which it was derived. Sterky and colleagues (1998) identified a total of 5,692 ESTs from two libraries made from poplar tissues destined to form wood. Clustering of the EST sequences from each library revealed redundancies of 53 percent and 26 percent, respectively. In other words, there was a 53 percent chance that the next cDNA selected would already be present in the database for the first library. Redundancy derived from EST frequency data can be useful as a measure of differential gene expression between tissues, treatments, or species (e.g., Audic and Claverie, 1997). Whetten and colleagues (2001) used redundancy to evaluate differential gene expression between libraries made from RNA isolated from differentiating pine xylem undergoing normal or compression wood formation and identified relative transcript abundance differences in a number of genes involved in cell wall synthesis.

Suppressive Subtractive Hybridization (SSH)

One of the drawbacks to random sampling is the resequencing of highly abundant transcripts. Resequencing of previously identified cDNAs increases the cost per unique sequence that is identified. SSH techniques are used to increase the abundance of rare transcripts in libraries relative to the abundance of common transcripts, thus reducing the total number of sequencing reactions that have to be performed to identify a given number of unique sequences. Essentially, these techniques remove transcripts in common between two experimental samples while maintaining those transcripts that are unique (Hesse et al., 1995). Not only does this reduce the number of highly abundant cDNAs in the library, it also maximizes the opportunity to identify genes of low abundance that may play critical roles in the experimental sample of interest. Covert and colleagues (2001) used SSH to identify pine genes that were regulated by, and thus may play important roles in, the fusiform rust disease state. SSH was also used to create libraries from *Eucalyptus globulus–Pisolithus tinctorius* ectomycorrhiza to provide insights into ectomycorrhizal symbioses (Voiblet et al., 2001).

Differential Display

Differential display is a method that allows researchers to detect and clone either up- or down-regulated genes in a single experiment (Liang and Pardee, 1992). Although this technique is not as high throughput as others, it has the advantage of allowing researchers to target cloning efforts to genes that are expressed differentially between two or more samples. Those cDNAs that differentially amplify between the samples of interest can be eluted from the gel, cloned, and sequenced. Theoretically, all differentially expressed genes in a tissue that are expressed at sufficient levels for detection can be identified if enough random primer combinations are used. Although this technique allows differentially expressed genes to be quickly identified and cloned, the sequences are often truncated, which requires further cloning efforts if full-length sequences are desired. This technique has been successfully used to identify a number of differentially expressed tree genes, including poplar genes expressed in association with nitrogen availability (Cooke and Davis, 2001), pine genes expressed in association with pathogen defense responses (Mason and Davis, 1997; Davis et al., 2002), and pine genes expressed in association with specific stages of zygotic and somatic embryo development (Xu et al., 1997; Cairney et al., 2000).

Serial Analysis of Gene Expression (SAGE)

SAGE can be used to quantify the expression of both known and unknown genes in a particular sample (Velculescu et al., 1995). In brief, short 10 to 12 base pair DNA sequences, or "tags," are identified in mRNAs isolated from the samples of interest. Each tag, generated by standard molecular biological techniques followed by sequencing, represents the expression profile of a particular gene in the sample from which it was derived. The outcome of SAGE analysis is a list of the different tags and their associated count values or expression profiles. An important consideration in SAGE analysis is the availability of an EST collection that can be used for assigning genes to SAGE tags. SAGE analysis has been used to quantify changes in gene expression underlying wood quality differences between the crown and base of a loblolly pine tree (Lorenz and Dean, 2002). The authors identified a total of 150,855 tags representing a maximum of 42,641 different genes expressed along the vertical developmental gradient in the tree stem.

Expression Arrays

Either cDNAs themselves or sequence information derived from the cDNAs are used to generate gene expression arrays. The precise number of genes whose expression can be monitored under various experimental conditions or treatments depends on the total number of genes available within a particular species of interest and the array method of choice. There are three major types of array technologies currently in use: nylon filter arrays (macroarrays; Pietu et al., 1996; Desprez et al., 1998), glass slide cDNA arrays (microarrays; Schena et al., 1995), and oligonucleotide-based arrays created by photolithography (e.g., the GeneChip, Affymetrix, Santa Clara, California; Fodor et al., 1991; Lockhart et al., 1996). Regardless of the particular system used, all three of these techniques have some basic factors in common. The genes or DNA templates that are to be tested are fixed to a solid support. The array is then interrogated with RNA-derived probes isolated from the samples to be tested. Genes that are expressed in the sample of interest are identified based on the signal emitted from the array.

Nylon filter-based arrays are gridded with DNA fragments, usually derived from cDNAs. PCR-amplified cDNAs are spotted either manually or robotically in ten to hundred nanoliter volumes. Usually fewer than 5,000 cDNAs are screened on a nylon membrane. Because the gridding devices and solid supports for macroarrays are based on standard molecular biology methods and materials, the platform is highly flexible, meaning that the cDNAs and their arrangement can be altered at will. Probes are commonly labeled with a radioisotope followed by serial hybridization so that each membrane reveals transcript abundance in the single RNA population from which the probe was derived. The amount of hybridization to each cDNA is recovered using a phosphorimager and associated software. Signal intensities are generated from the hybridization signals for each gene, compared among membranes.

Slide-based microarrays are also gridded with DNA fragments. Polymerase chain reaction (PCR)-amplified and purified products are spotted robotically onto treated glass slides in densities of upward of 25,000 spots per slide. Of necessity, spot volumes are in the few nanoliters per spot range. The platform is not particularly flexible since array production runs tend to be large, with 100 or more slides printed per run. Unlike nylon filter arrays, hybridization is not sequential; RNA from reference and test samples are labeled with different fluorescent dyes and mixed together for hybridization to a single slide. The fluorescence emission intensities for each gene are measured after appropriate excitation of each fluorophore. The ratio of the fluorescence intensities for each gene on the array yields a measure of its relative differential expression between the two samples. Despite the relatively high costs associated with DNA microarrays, many universities now have laboratories and core facilities using this technology.

Oligonucleotide-based arrays are created by a photolithographic process that etches relatively short DNA sequences directly onto a solid support. Although more technically difficult than nylon and slide-based arrays, the use of oligonucleotides can offer a finer discrimination between gene family members than a larger piece of DNA because sequences can be selected for synthesis that minimize the potential for cross hybridization between gene family members. The oligonucleotides are synthesized at extremely high density, such that the entire *Arabidopsis* transcriptome can be assayed on a single chip. The platform is rigid, in that chips are typically manufactured in large production runs for organisms of interest to large communities of researchers. As with nylon filter-based arrays, each sample to be tested for gene expression is hybridized to an individual chip. RNA isolated from the samples is converted into biotin-labeled cRNA molecules that are hybridized to individual cassettes followed by staining with a fluorophore-tagged streptavidin. A confocal microscope is then used to detect fluorescence from the hybridized cRNA. Currently these chips are not manufactured for forest tree species and as such their suitability for tree research is limited.

There are a number of resources available that contain references for more information about DNA array technologies. GRID IT (<http://www.bsi.vt.edu/ralscher/gridit/>) is a Web site jointly maintained by Virginia Tech and the Forest Biotechnology Group at North Carolina State University with information and links that include process overviews, array fabrication, protocols, and sources for equipment and analysis software. The Stanford Microarray Database (<http://genome-www5.stanford.edu/MicroArray/SMD/>) contains software applications for image capture and analysis along with publicly accessible data. Microarrays.org (<http://www.microarrays.org>) contains an overview of the microarray process along with protocols and software.

Bioinformatics—From Numbers to Knowledge

Scientists with large array data sets are confronted with a variety of challenges that include issues related to data management, analysis, presentation, and distribution for easy access and searching by other researchers. Unlike genome or EST sequencing data that have relatively well-established presentation methods and analysis tools (e.g., GenBank, <http://www.ncbi. nlm.nih.gov>; ExPASy, <http://www.expasy.ch/>), data generated through DNA array technologies do not currently have a scientifically agreed upon structure.

Many, although not all, published reports using microarray technologies contain a reference, most often the author's Web site, for the reader to access DNA array data presented in the article. Depending on the array platform and the investigators, the type and arrangement of data available can vary widely (raw, background subtracted, or fully scaled and normalized ratios). These differences can pose significant challenges for other scientists reproducing the experiments or utilizing the data for further analyses. The lack of consistency in the ways DNA array data are presented has resulted in an effort to standardize what information should be included in microarray data presentations. The Microarray Gene Expression Data Society (<http:// www.mged.org>) was founded with the purpose of promoting standardization of DNA microarray data for presentation and exchange. The group has recently proposed the "minimum information about a microarray experiment" (MIAME; Brazma et al., 2001) as a start toward ensuring that interpretation and reproducibility can be independently verified. The inclusion of such experimental variables as biological growth and treatment conditions can assist other investigators in designing their own experiments such that independent experiments can be meaningfully integrated (Brazma, 2001). GenBank has added the "gene expression omnibus" (GEO; <http:// www.ncbi.nlm.nih.gov/geo/>) repository for gene expression and array hybridization data. Scientists can now search and retrieve publicly available data that includes DNA and protein arrays as well as SAGE data.

More crucial issues facing bioinformaticians are related to the methods of data analysis for obtaining biologically relevant results. Much of the difficulty arises from the variability that exists across biological samples and arrays. Reproducibility at both the biological and technical levels must be incorporated into array analyses. Prior to any analysis for biological significance, array data must be tested for obvious errors such as spot misalignments or hybridization failures, followed by scaling and normalization steps to minimize the effects of systematic errors. Appropriate statistical analyses must be matched to the experimental design (e.g., Kerr and Churchill, 2001;

Terry Speed's Microarray Data Analysis Group, <http://stat-www.berkeley.edu/users/terry/zarray/html/>). Because a large number of gene expression patterns can be monitored using microarray technology, a variety of different clustering algorithms have been employed to visualize how genes group together based on their expression profiles (Quackenbush, 2001; Sherlock, 2000). These types of analyses can identify genes predicted to have common functions and mechanisms of transcriptional control. Clustering can also assign putative functions to unknown genes based on the functions of the known genes in the cluster.

FUNCTIONAL GENOMICS IN ACTION

In many ways the use of microarrays for functional genomic analysis of models—yeast and *Arabidopsis*—can help guide efforts for functional genomic analysis of forest trees. Here we briefly discuss the experimental approach and findings of two significant papers (Spellman et al., 1998; Maleck et al., 2000) in which microarray analysis played a pivotal role in increasing the understanding of the regulation of a complex biological process. One common feature of these papers is their use of transgenic lines and/or mutants in combination with microarrays to define genetic regulons (genes that are regulated together). In both papers the authors had access to enough genomic sequence information to identify promoter elements that may link cellular signaling with microarray outputs.

The main focus of Spellman and colleagues (1998) was to understand the regulatory mechanisms that coordinate the cell cycle in yeast. The authors used a variety of techniques to synchronize yeast cell populations at particular stages of the cell cycle. Once synchronized, the cells were harvested at various times after release to monitor the expression of each yeast gene (approximately 6,000 genes total). Monitoring gene expression changes over a time course would allow identification of genes whose transcripts cycled in association with other well-known physiological/anatomical features of the cell cycle. Importantly, these synchronization methods included transgenic overexpression of regulatory genes, temperature-sensitive mutants in regulatory genes, mechanical separation of discrete cell types, and addition of a pharmacological agent to the growth medium. The use of all these different treatments is an effective way to identify the genes that are intrinsically regulated during the cell cycle, because each method individually carries along its own unique artifacts that are not produced by the other methods. For example, temperature-sensitive mutants are first blocked at a specific stage of the cell cycle by growth at elevated temperatures. However, elevated temperature also induces the expression of heat shock genes whose transcript abundance could confound experiments in which temperature-sensitive mutants were the sole condition. Thus, using multiple treatments in combination serves to remove artifacts due solely to a single treatment. Using microarray analyses, the authors identified 800 genes as cell cycle regulated out of 6,200 tested, more than three times the number previously postulated (Price et al., 1991). The results from these experiments revealed potentially new and surprising gene expression patterns that would have remained hidden in a "one-gene-at-a-time" analysis.

Spellman and colleagues (1998) analyzed the microarray data by categorizing genes into clusters, or regulons, based on the similarity of their transcript profiles. Clustering of genes suggests that a common mechanism controls the regulation of that particular group of genes, implying a related function. The simplest explanation for coregulation is the presence of shared *cis* elements in the promoter regions of the genes in a cluster. These *cis* elements would be predicted binding sites for one or more transcription factors in common between the genes within the cluster. The yeast promoters indeed shared *cis* elements, some of which were previously identified in the literature as having a role in cell cycle regulation. *cis* element discovery is important because it provides a link between microarray output data (transcript abundance) and the signal transduction networks that regulate the gene expression patterns (transcription factor binding sites).

Plant researchers working on the model plant *Arabidopsis thaliana* are closest to the yeast benchmark with respect to functional genomics approaches to understanding biological processes. Maleck and colleagues (2000) used functional genomics to analyze the transcriptome of *Arabidopsis thaliana* induced during systemic acquired resistance to disease (SAR), an inducible defense response that provides enhanced resistance against a variety of pathogens. A total of 16 different conditions or treatments were utilized including transgenic lines, mutant lines, pharmacological treatments, and pathogen challenges, which allowed the authors to identify genes central to SAR. Two different clustering methods for analyzing gene expression profiles were applied to the expression ratios of the approximately 7,000 genes in the data set. These analyses identified a "PR-1 regulon" that appears to be intrinsic to SAR.

The expression of the *PR-1* gene is a well-established marker for SAR, and the array analysis was able to distinguish genes in the PR-1 regulon from genes that are pathogen regulated but not intrinsic to SAR. Importantly, promoters of genes in the PR-1 regulon contained promoter *cis* elements that were previously identified as binding sites for the WRKY class of plant transcription factors which are known to be involved in regulating defense and stress-responsive genes (Eulgem et al., 2000). These *cis* elements are now useful reagents to functionally test the roles of WRKY transcription factors in SAR. Like Spellman and colleagues (1998), this paper showed the importance of using multiple conditions to analyze complex processes or pathways and showed how promoter *cis* elements can be identified as a first step toward unraveling signal transduction networks.

In summary, both Spellman and colleagues (1998) and Maleck and colleagues (2000) used a diverse set of conditions to minimize artifacts, used clustering techniques to identify putative roles for genes with unknown function, and took full advantage of genomic sequence information to identify how expression patterns of interest might be controlled in signal transduction networks.

We must keep in mind that yeast and *Arabidopsis* are model systems in which directed genetic manipulations have become standard practice. In contrast, no forest tree has the full suite of characters that make an ideal genetic model system. Consequently, we should not expect to conduct identical experiments in yeast, *Arabidopsis,* and forest trees. Rather, the forest tree research community should look to the model systems for guidance in how functional genomics might be implemented in forest trees, given that sequence information from tree genomes is accumulating at a rapid pace. We may need to choose alternate methodologies in some cases, since cornerstone techniques such as directed genetic manipulations are not routine in many forest tree species.

EN ROUTE TO FUNCTIONAL GENOMICS IN FOREST TREES

The previously discussed papers illustrate that transcriptional profiling of a tissue or process is the first step toward identifying the underlying regulatory mechanisms. An excellent example of how forest tree scientists are beginning to use functional genomics approaches to decipher mechanisms underlying complex systems is provided by Hertzberg and colleagues (Hertzberg, Aspeborg, et al., 2001; Hertzberg, Sievertzon, et al., 2001). The question of how cambial activity is regulated to give rise to terminally differentiated cell types in trees was approached using an elegant procedure that isolated RNA from a few cell layers for microarray analysis. Cell types were highly differentiated with respect to their transcript profiles, and clustering was used to group genes into regulons that shared similar transcript profiles. For genes with no known function, expression profiling across developmental gradients can be a significant step toward assigning function. Likewise, the coupling of transgenic and/or pharmacological manipulation with microarray analysis of xylogenesis will be an important step toward using a functional genomics approach to understand how wood formation is controlled.

Transgenic manipulation of putative regulators can be used in combination with arrays to define the downstream effects of those regulators (Spellman et al., 1998; Maleck et al., 2000) and to begin the establishment of gene expression networks. *Agrobacterium*-mediated transformation is central to a variety of forward genetic approaches involving insertional mutagenesis (Azpiroz-Leehan and Feldmann, 1997); forward genetics involves a phenotypic screen that is followed by a search for the mutant gene that causes it (phenotype to gene). Insertional mutagenesis via T-DNA tagging or transposon mutagenesis would be in the category of forward genetic manipulation. *Agrobacterium*-mediated transformation is also central to a variety of reverse genetic approaches; a reverse genetic approach seeks to assign a function to a gene already in hand by creating a transgenic plant in which the expression of the gene is altered in some way (gene to phenotype). These types of constructs often result in dominant "mutations" that alter the phenotype of the plant without the need for crossing lines. Poplars with ectopic overexpression, antisense, or RNAi transgenes are included in the category of reverse genetic manipulation.

In poplar, *Agrobacterium*-mediated transformation is now routine (Leple et al., 1992), suggesting that both forward and reverse genetic approaches are technically feasible. However, in practice, the application of some genetic approaches is made difficult by the dioecious nature of poplar trees. Dioecy is a property of poplar trees in which male and female reproductive structures are borne on different trees. Consequently, the normal practice of selfing to generate homozygous plants that reveal phenotypes controlled by recessive alleles cannot be performed in poplar, even if the prolonged time to flowering (six to eight years) was not a deterrent. As a result, many forward genetic approaches are limited in their applicability because these methods tend to generate recessive mutant alleles. Still, certain forward genetic approaches, such as activation tagging, do generate dominant, gain-of-function phenotypes that could be used to generate mutant lines useful in functional genomic analysis of poplars (cf. Bradshaw et al., 2000). Ma and colleagues (2001) reported the identification of over 600 T-DNA tagged poplar lines, four of which exhibited overt phenotypic changes, indicating the potential usefulness of this forward genetic approach. Many of the traits modified in forest trees through genetic manipulation have utilized reverse genetic approaches (Pena and Seguin, 2001). The types of forward and reverse genetic experiments that are available to tree researchers are narrowed to those that can generate a phenotype—either morphological or molecular—in a primary transformant. In summary, we expect to see reverse genetic approaches (transgenic trees) to be the predominant approach applied in functional genomics analysis of poplar.

In February 2002, the U.S. Department of Energy announced its decision to sequence the first tree genome, *Populus balsamifera* ssp. *trichocarpa.* Sequencing will be a joint international venture with participants in the project to include scientists from the Oak Ridge National Laboratory, the University of Washington, the Joint Genome Institute, the Canadian Genome Sequence Center, and the Swedish University of Agricultural Sciences. Once sequenced, it will be possible to put every poplar gene onto an individual microarray for identifying gene expression patterns. Knowing the poplar genome sequence in its entirety will define the full number of genes and gene families, which may enable more precise up- or down-regulation of individual gene family members in transgenic trees. This will enable scientists to tailor gene expression array analyses to examine individual members of a gene family that may be regulated differentially under various experimental conditions.

As coregulated tree genes are identified by array analysis, forest scientists will be able to start assigning putative functions to unknown genes based on the functions of the known genes in the cluster. These coregulated genes will also open avenues into understanding promoter function. Rather than having to identify regulatory motifs in individual promoters of genes by promoter deletion analyses, putative *cis* elements can be identified based on shared sequences with coregulated genes and will be enhanced in poplar by the availability of the genome sequence.

All forest tree species are not as tractable as poplar for functional genomic analysis. Gymnosperm species have large genomes, and loblolly pine (*Pinus taeda* L.) is the most economically important species in this group. The likelihood of a genomic sequencing project for pine is low in the foreseeable future because of the vastness of the genome (Lev-Yadun and Sederoff, 2001). The pine genome is approximately seven times that of the human genome (Venter et al., 1998) and 160 times that of *Arabidopsis thaliana* (Somerville and Somerville, 1999). In certain ways, however, pines are good models for genetic analysis. Many genetic markers have been mapped in the pine genome, due in large part to the high degree of genetic diversity. Unlike poplars, pines can be selfed to generate homozygous recessive offspring, and in fact, many visible mutations—including potential null alleles—are readily apparent in seed lots produced from selfing (Remington and O'Malley, 2000). The pine megagametophyte may offer a unique research tool for identifying potentially interesting phenotypes because it is a maternally derived, relatively simple haploid tissue in which to uncover mutants. Loblolly pine megagametophyte tissue has been used to identify a mutant loblolly pine with severely reduced cinnamyl alcohol dehydrogenase (CAD) activity, an enzyme in the lignin biosynthetic pathway (MacKay et al., 1997; Ralph et al., 1997). Although these aspects of pine may foster genetic analysis, a major challenge in performing functional genomic analysis in pines is demonstrating gene function, since transformation is difficult. The development of efficient stable transformation procedures will greatly assist progress in pine functional genomics in the future.

CONCLUSIONS

Functional genomics in forest trees has the potential to contribute greatly to our understanding of how forest tree-specific traits are regulated and how trees differ from model species. Gene discovery projects have provided many of the raw materials (gene sequences) required for large-scale surveys of tree expression profiles, and the next challenge is to unravel the regulatory networks that coordinate gene expression. Forest tree functional genomics is at the point where we can begin to test hypotheses regarding tree-specific processes. Transgenic trees will most likely play a crucial role in accomplishing this goal. The poplar genome sequence will assist in identifying the full suite of genes, including their promoters and related family members, all of which will guide experimental efforts to define gene function. *Arabidopsis* and yeast provide models for forest trees in that the combination of genetic approaches with array analysis is a powerful way to carry out functional genomics.

The practical applications of functional genomics research in forest trees may include generation of trees with enhanced wood quality, increased pest and disease resistance, or better growth characteristics. These potential benefits notwithstanding, it is difficult to underestimate how important transgenic trees will be in understanding the fundamental biology of forest trees. Reverse genetic approaches via transgenics would appear to be the best option for defining the functions of tree genes, since the long generation interval of trees precludes most forward genetic strategies for defining gene function. Due to their importance in tree biology, regulatory networks that govern wood properties will probably be the first to be functionally dissected through genomics approaches.

REFERENCES

Adams, M.D., Delley, J.M., Gocayne, J.D., Dubnick, M., Polymeropoulos, M.H., Xiao, J., Merril, C.R., Wu, A., Olde, B., and Moreno, R.F. (1991). Complementary DNA sequencing: Expressed sequence tags and human genome project. *Science* 252: 1651-1656.

Allona, I., Quinn, M., Shoop, E., Swope, K., St. Cyr., S., Carlis, J., Riedl, J., Retzel, E., Campbell, M.M., Sederoff, R., and Whetten, R. (1998). Analysis of xylem formation in pine by cDNA sequencing. *Proc. Natl. Acad. Sci. USA* 95: 9693-9698.

Audic, S. and Claverie, J.M. (1997). The significance of digital gene expression profiles. *Genome Research* 7: 986-995.

Azpiroz-Leehan, R. and Feldmann, K.A. (1997). T-DNA insertion mutagenesis in *Arabidopsis:* Going back and forth. *Trends Gen.* 13: 152-156.

Bradshaw, H.D. Jr., Ceulemans, R., Davis, J., and Stettler, R. (2000). Emerging model systems in plant biology: Poplar *(Populus)* as a model forest tree. *J. Plant Growth Regul.* 19: 306-313.

Brazma, A. (2001). On the importance of standardization in life sciences. *Bioinformatics* 17: 113-114.

Brazma, A., Hingamp, P., Quackenbush, J., Sherlock, G., Spellman, P., Stoeckert, C., Aach, J., Ansorge, W., Ball, C., Causton, H., et al. (2001). Minimum information about a microarray experiment (MIAME)—Toward standardization for microarray data. *Nature* 29: 365-371.

Cairney, J., Xu, N.F., MacKay, J., and Pullman, J. (2000). Transcript profiling: A tool to assess the development of conifer embryos. *In Vitro Cell Devel. Biol.-Plant.* 36: 55-162.

Cooke, J.E.K. and Davis, J.M. (2001). Using a targeted genomics approach to study nitrogen responses in poplar. International Union of Forest Research Organizations (IUFRO) Molecular Biology of Forest Trees, Stevenson, Washington, July 22-27.

Covert, S.F., Warren, J.M., Holliday, A.B., and Long, T. (2001). Molecular analysis of the fusiform rust disease interaction. IUFRO Conference, Stevenson, Washington, July 22-27.

Davis, J.M., Wu, H.G., Cooke, J.E.K., Reed, J.M., Luce, K.S., and Michler, C.H. (2002). Pathogen challenge, salicylic acid, and jasmonic acid regulate expression of chitinase gene homologs in pine. *Mol. Plant-Microbe Interact.* 5: 380-387.

Desprez, T., Amselem, J., Caboche, M., and Hofte, H. (1998). Differential gene expression in *Arabidopsis* monitored using cDNA arrays. *Plant J.* 14: 643-652.

Eulgem, T., Rushton, P.J., Robatzek, S., and Sjomssich, I.E. (2000). The WRKY superfamily of plant transcription factors. *Trends Plant Sci.* 5: 199-206.

Fodor, S.P., Read, J.L., Pirrung, M.C., Stryer, L., Lu, A.T., and Solas, D. (1991). Light-directed, spatially addressable parallel chemical synthesis. *Science* 251: 767-773.

Hertzberg, M., Aspeborg, H., Schrader, J., Andersson, A., Erlandsson, R., Blomqvist, K., Bhalerao, R., Uhlen, M., Teeri, T.T., Lundeberg, J., Sundberg, B., and Nilsson, P. (2001). A transcriptional roadmap to wood formation. *Proc. Natl. Acad. Sci. USA* 98: 14732-14737.

Hertzberg, M., Sievertzon, M., Aspeborg, H., Nilsson, P., Sandberg, G., and Lundeberg, J. (2001). cDNA microarray analysis of small plant tissue samples using a cDNA tag target amplification protocol. *Plant J.* 25: 585-591.

Hesse, H., Frommer, W.B., and Willmitzer, L. (1995). An improved method for generating subtracted cDNA libraries using phage-lambda vectors. *Nucleic Acids Res.* 23: 3355-3356.

Hieter, P. and Boguski, M. (1997). Functional genomics: It's all how you read it. *Science* 278: 601-602.

Hofte, U.J., Deprez, T., Amselem, J., Chiapello, H., Caboche, M., Moisan, A., Jourjon, M.F., Carpenteau, J.L., Berthomieu, P., and Guerrier, D. (1993). An inventory of 1152 expressed sequence tags obtained by partial sequencing of cDNAs from *Arabidopsis thaliana. Plant J.* 4: 1051-1061.

Kerr, M.K. and Churchill, G. (2001). Statistical design and the analysis of gene expression microarray data. *Genet. Rev. Camb.* 77: 123-128.

Leple, J.C., Brasileiro, A.C., Michel, M.F., Delmotte, F., and Jouanin, L. (1992). Transgenic poplar: Experiments of chimeric genes using four different constructs. *Plant Cell Rep.* 11: 137-141.

Lev-Yadun, S. and Sederoff, R. (2001). Pines as model gymnosperms to study evolution, wood formation, and perennial growth. *J. Plant Growth Regul.* 19: 290-305.

Liang, P. and Pardee, A.B. (1992). Differential display of eukaryotic messenger RNA by means of the polymerase chain reaction. *Science* 257: 967-971.

Lockhart, D.J., Dong, H., Byrne, M.C., Follettie, M.T., Gallo, M.V., Chee, M.S., Mittmann, M., Wang, C., Kobayashi, M., Horton, H., and Brown, E.L. (1996). Expression monitoring by hybridization to high-density oligonucleotide arrays. *Nat. Biotechnol.* 14: 1675-1680.

Lorenz, W.W. and Dean, J.F.D. (2002). SAGE profiling and demonstration of differential gene expression along the axial developmental gradient of lignifying xylem in loblolly pine *(Pinus taeda). Tree Physiol.* 22: 301-310.

Ma, C., Meilan, R., Brunner, A., Carson, J., Li, J., and Strauss, S. (2001). Activation tagging in poplar: Frequency of morphological mutants. IUFRO Conference, Stevenson, Washington, July 22-27.

MacKay, J.J., O'Malley, D.M., Presnell, T., Booker, F.L., Campbell, M.M., Whetten, R.W., and Sederoff, R.R. (1997). Inheritance, gene expression and lignin characterization in a mutant pine deficient in cinnamyl alcohol dehydrogenase. *Proc. Natl. Acad. Sci. USA* 94: 8255-8260.

Maleck, K., Levine, A., Eulgem, T., Morgan, A., Schmidt, J., Lawton, K.A., Dangl, J.L., and Dietrich, R.A. (2000). The transcriptome of *Arabidopsis thaliana* during systemic acquired resistance. *Nature* 26: 403-410.

Mason, M.E. and Davis, J.M. (1997). Defense responses in slash pine: Chitosan treatment alters the abundance of specific mRNAs. *Mol. Plant-Microbe Interact.* 10: 135-137.

Pena, L. and Seguin, A. (2001). Recent advances in the genetic transformation of trees. *Trends Biotech.* 19: 500-506.

Pietu, G., Alibert, O., Guichard, V., Larny, B., Bois, F., Leroy, E., Mariage-Samsom, R., Houlgatte, R., Soularue, P., and Auffray, C. (1996). Novel gene transcripts preferentially expressed in human muscles revealed by quantitative hybridization of a high density cDNA array. *Genome Research* 6: 492-503.

Price, C., Nasmyth, K., and Schuster, T. (1991). A general approach to the isolation of cell cycle-regulated genes in the budding yeast, *Saccharomyces cerevisiae. J. Mol. Biol.* 218: 543-556.

Quackenbush, J. (2001). Computational analysis of microarray data. *Nat. Rev. Genet.* 2: 418-427.

Ralph, J., MacKay, J.J., Hatfield, R.D., O'Malley, D.M., Whetten, R.W., and Sederoff, R.R. (1997). Abnormal lignin in a loblolly pine mutant. *Science* 277: 235-239.

Remington, D.L. and O'Malley, D.M. (2000). Whole-genome characterization of embryonic stage of inbreeding depression in a selfed loblolly pine family. *Genetics* 155: 337-348.

Schena, M., Shalon, D., Davis, R.W., and Brown, P.O. (1995). Quantitative monitoring of gene expression patterns with a complementary DNA microarray. *Science* 270: 467-470.

Sherlock, G. (2000). Analysis of large-scale gene expression data. *Curr. Opin. Immunol.* 12: 201-205.

Somerville, C. and Somerville, S. (1999). Plant functional genomics. *Science* 285: 380-383.

Spellman, P.T., Sherlock, G., Zhang, M.Q., Lyer, V.R., Anders, K., Eisen, M.B., Brown, P.O., Botstein, D., and Futcher, B. (1998). Comprehensive identification of cell cycle-regulated genes of the yeast *Saccharomyces cerevisiae* by microarray hybridization. *Mol. Biol. Cell* 9: 3273-3297.

Sterky, F., Regan, S., Karlsson, J., Hertzberg, M., Rohde, A., Holmberg, A., Amini, B., Bhalerao, R., Larsson, M., Villarroel, R., et al. (1998). Gene discovery in the wood-forming tissues of poplar: Analysis of 5,692 expressed sequence tags. *Proc. Natl. Acad. Sci. USA* 95: 13330-13335.

Velculescu, V.E., Zhang, L., Vogelstein, B., and Kinzler, K.W. (1995). Serial analysis of gene expression. *Science* 270: 484-487.

Venter, J.C., Adams, M.D., Sutton, G.G., Kerlevage, A.R., Smith, H.O., and Hunkapiller, M. (1998). Shotgun sequencing of the human genome. *Science* 280: 1540-1542.

Voiblet, C., Duplessis, S., Encelot, N., and Martin, F. (2001). Identification of symbiosis-regulated genes in *Eucalyptus globulus-Pisolithus tinctorius* ectomycorrhiza by differential hybridization of arrayed cDNAs. *Plant J.* 25: 181-191.

Whetten, R.W., Sun, Y.H., Zhang, Y., and Sederoff, R. (2001). Functional genomics and cell wall biosynthesis in loblolly pine. *Plant Mol. Biol.* 47: 275-291.

Xu, N., Johns, B., Pullman, G., and Cairney, J. (1997). Rapid and reliable differential display from minute amounts of tissue: Mass cloning and characterization of differentially expressed genes from loblolly pine embryos. *Plant Mol. Biol Rep.* 15: 377-391.

Chapter 2

Expressed Sequence Tag Databases from Forestry Tree Species

Timothy J. Strabala

INTRODUCTION

Eukaryotic genome DNA sequence databases, such as those from *Arabidopsis thaliana* (Arabidopsis Genome Initiative, 2000), *Caenorhabditis elegans* (*C. elegans* Sequencing Consortium, 1998), *Drosophila melanogaster* (Adams et al., 2000), *Homo sapiens* (International Human Genome Sequencing Consortium, 2001), and *Saccharomyces cerevisiae* (Mewes et al., 1997), have received much attention in the recent literature. Such databases have told us much about phylogenetic relationships among higher eukaryotes (*C. elegans* Sequencing Consortium, 1998; International Human

This manuscript discusses the cumulative efforts of many scientists over the past six years. It is impossible to individually name everyone who has contributed to the Genesis databases. However, without the initiation of these EST projects by Dr. Jim Watson of Genesis, in collaboration with Fletcher Challenge Forests, none of the work described here could have taken place. Many forestry team staff members at Genesis contributed to the data discussed in this chapter. Clare Eagleton (slot blots), Marion Wood (subtracted libraries), as well as Barry Flinn and Christina Balmori (random-primed libraries) deserve particular mention for their work that I have discussed here. Andy Shenk, Niels Nieuwenhuizen, Annette Lasham, Lenny Bloksberg, Steve Rice, Ranjan Perera, Angela De Ath, Sandra Fitzgerald, Susan Wheeler, Richard Forster, Paul Sanders, Jonathan Phillips, Sathish Puthigae, and Sean Simpson have all made significant contributions to the project as well. The sequencing staff at Genesis, headed originally by Alastair Grierson and now by Matt Glenn, with Jonathon Fraser and Coralie Hansen, has done an excellent job in consistently delivering high-quality sequence data. The bioinformatics staff at Genesis, including, but not exclusive to, Ilkka Havukkala, Dingyi Xu, Paul Bickerstaff, Mehran Aghaei, and Annette McGrath, has likewise done a superb job of providing the laboratory scientists with useful information for analyzing and interpreting the huge amounts of data from this project. Our ArborGen partners, Rubicon, International Paper, and MeadWestvaco, have helped guide the science ongoing in the sequencing efforts. Finally, I would like to thank Anne-Marie Smit and Andy Shenk for helpful discussions and critical reading of the manuscript.

Genome Sequencing Consortium, 2001), synteny among animal species (International Human Genome Sequencing Consortium, 2001), and the number of genes predicted to exist in these species (Adams et al., 2000; International Human Genome Sequencing Consortium, 2001). Expressed sequence tag (EST) databases for most of these organisms also exist (Delseney et al., 1997; FlyBase Consortium, 1999; Gerhold and Caskey, 1996; Stein et al., 2001). In the case of yeast, transcribed sequences (the "transcriptome") have been examined not primarily via ESTs, but rather by serial analysis of gene expression (SAGE) (Velculescu et al., 1997). Genomic and EST databases are complementary to one another, as genomic sequence data can demonstrate by linkage that two otherwise separate EST contigs in fact belong to the same gene. Conversely, the ability to predict the transcriptome from a genomic database is limited (Gopal et al., 2001; Reymond et al., 2002), and EST databases serve as a means to rapidly access the expressed portion of the genome. Such access to the transcriptome provided by EST databases is of particular use in organisms with genomes too large or redundant to make whole-genome sequencing practical, such as maize (Gai et al., 2000) and pine species (Kinlaw and Neale, 1997). Genetic research is also enhanced by EST databases because they provide microsatellite (Williams et al., 2001; Moriguchi et al., 2003) data useful for gene mapping in such species.

The ready availability of gene sequences from EST and genomic sequencing allows direct access to plasmids containing a given DNA molecule of interest, obviating the need for much physical screening for genes in cDNA libraries. This has resulted in the scale of experimentation moving away from analysis of single genes to parallel experiments using thousands of genes. For example, researchers are using microarray analysis founded on EST sequencing work to analyze global differences in gene expression in mutant lines (Furlong et al., 2001). In forestry tree species, microarrays have been used to examine differential expression in compression wood versus normal wood (Whetten et al., 2001) and differential gene expression at a number of developmental stages during xylogenesis (Hertzberg et al., 2001) and somatic embryogenesis (van Zyl et al., 2003).

In recent years, there have been EST sequencing projects in pine (Kinlaw et al., 1996; Allona et al., 1998) and other forestry tree species (Sterky et al., 1998; Ujino-Ihara et al., 2000), directed primarily toward examining vascular development. Concurrent with these published forestry tree EST projects, *Pinus radiata* and *Eucalyptus grandis* EST sequencing projects were begun at Genesis Research and Development in collaboration with Fletcher Challenge Forests in New Zealand in late 1995. This collaboration became a joint venture in 1996, which expanded to include the U.S. forestry companies International Paper and Westvaco (now MeadWestvaco) in 2000. To facilitate the utilization of the Genesis EST databases, the activities of this joint venture are carried out through the company ArborGen LLC. To date, over 500,000 pine and eucalyptus ESTs have been generated, with more than 150,000,000 nucleotides contained in the databases. More than 60 different cDNA libraries representing different organs, cell types, and treatments have been constructed and examined, collectively comprising the world's most comprehensive storehouse of EST data from forestry tree species. Here, I will briefly review the results from the four published forestry EST sequencing projects and then discuss in more detail the approaches and efforts ongoing in the EST projects at Genesis.

FORESTRY TREE EST DATABASES

The bulk of the EST data in published forestry tree EST sequencing projects has come from cDNA libraries prepared from developing vascular tissue, with an emphasis on developing xylem in economically important tree species. Xylogenesis is of great importance to the forestry and forest products industries, primarily because this developmental process results in lignin deposition in the trunk and branches of the tree. Lignin contributes a great deal of structural strength to the tree and is therefore important in creating sturdy timber for construction. On the other hand, lignin is the primary obstacle encountered in pulp extraction in the paper industry. Therefore, the reduction of total lignin content, or its modification to a more easily extracted form in trees grown for pulping, would be highly desirable.

The molecular mechanism of xylogenesis as a developmental process is an area under intensive study, aided by the use of the zinnia in vitro tracheary element (TE) induction system (Fukuda and Komamine, 1980). This system is presently being used as a tool in a number of laboratories for the identification and cloning of genes that are likely involved in various aspects of xylogenesis (Milioni et al., 2001; Demura et al., 2002; Milioni et al., 2002; Grigor et al., 2003). This cell-based method is complemented by EST data derived from tree vascular tissue cDNA libraries. The size of trees serves as an advantage rather than an impediment in such experiments due to the ease with which large quantities of developing xylem, phloem, and vascular cambium tissue, essentially uncontaminated by other cell types, may be obtained. Like the zinnia TE system, xylem tissue EST databases have led to the identification of many genes transcribed during xylem development (Allona et al., 1998; Kinlaw et al., 1996; Sterky et al., 1998; Ujino-Ihara et al., 2000; Genesis, unpublished results).

cDNA Libraries

Although the forestry tree EST projects have emphasized developing xylem, other developmental and physiological phenomena have also been addressed, as summarized in Table 2.1. Specifically, the *Pinus taeda* seedling and developing phloem cDNA libraries were sequenced to identify expressed genes from this species (Kinlaw et al., 1996). The *Populus* project examined a vascular cambium cDNA library from *P. tremula* × *P. tremuloides* and a developing xylem library from *P. trichocarpa* (Sterky et al., 1998). The purposes of this project were to obtain a survey of transcripts present in the vascular cambial region and developing xylem, and to compare the two databases to examine differences in EST profiles to identify genes that might be involved in xylogenesis. The *P. taeda* xylem EST project also took a comparative approach to the expression of genes in vascular development, with four different libraries, compression wood and side wood primary libraries and subtracted libraries from compression wood and side wood mRNA, used to generate ESTs (Table 2.1). The two subtracted libraries were created to enrich for cDNAs specific to compression wood and side wood formation. Finally, the *Cryptomeria japonica* effort constructed a library from vascular cambium tissue harvested from the tree two days after it was felled, presumably enriching it in cDNAs induced by drought and wounding (Ujino-Ihara et al., 2000). Although there are obvious species and physiological differences among all these libraries (Table 2.1), a comparison of the results of the various projects is informative nonetheless.

Characteristics of Published Forestry Tree EST Databases

The largest of the published forestry tree EST projects, at least in terms of the number of sequences, is that of *Populus* spp., with a database of 5,692 ESTs from its two vasculature libraries (Table 2.1). The *P. taeda* seedling and developing phloem databases comprised a total of 330 sequences from two libraries. The *P. taeda* xylem project was composed of four libraries with a total of 1,097 sequences, and the *C. japonica* vascular cambium effort sequenced 2,231 cDNAs. Nearly all of the libraries were oligo-dT-primed and directionally cloned (Table 2.1), with the exception being the random-primed seedling library.

Despite the differences in source species and tissues, many of the characteristics of the databases were remarkably similar. The average "read lengths" (the number of nucleotides successfully sequenced from a given cDNA clone) of the libraries were generally 400 to 500 nucleotides (Table 2.1).

TABLE 2.1. Characteristics of published forestry tree cDNA libraries and EST databases

Species	Tissue/cell type (treatment)	cDNA library type	Sequencing depth	Average read length (bp)	Redundancy rate[a]	Percentage of ESTs with putative function	Reference
Pinus taeda	whole seedlings	random-primed	460[b]	200	1.27	43	Kinlaw et al. (1996)
Pinus taeda	developing phloem	oligo-dT-primed, directionally cloned	100	ND[c]	1.00	19	Kinlaw et al. (1996)
Pinus taeda	developing xylem (compression wood)	oligo-dT-primed, directionally cloned	577	510[d]	1.49[d]	59[d]	Allona et al. (1998)
Pinus taeda	developing xylem (compression wood-enriched)	subtracted, directionally cloned	72	510[d]	1.49[d]	59[d]	Allona et al. (1998)
Pinus taeda	developing xylem (side wood)	oligo-dT-primed, directionally cloned	324	510[d]	1.49[d]	59[d]	Allona et al. (1998)
Pinus taeda	developing xylem (side wood-enriched)	subtracted, directionally cloned	124	510[d]	1.49[d]	59[d]	Allona et al. (1998)
Populus trichocarpa	developing xylem	oligo-dT-primed, directionally cloned	883	446	1.35	54	Sterky et al. (1998)
Populus tremula × tremuloides	vascular cambium	oligo-dT-primed	4,809	391	2.15	63	Sterky et al. (1998)
Cryptomeria japonica	vascular cambium (two days postfelling)	oligo-dT-primed, directionally cloned	3,746[e]	450	2.13	62	Ujino-Ihara et al. (2000)

[a]Calculated as total ESTs ÷ singleton ESTs
[b]230 clones were sequenced from both 5' and 3' ends
[c]No data available
[d]Collective figure for all four libraries
[e]1,515 of 2,231 clones were sequenced from both 5' and 3' ends.

23

Again, the exception was the *P. taeda* random-primed cDNA library, with a read length of approximately 200 bp. Although the authors did not state it directly, the phloem library from this study presumably had a similar average read length (Kinlaw et al., 1996). A high portion of ESTs from all four projects were singletons, as is typical for smaller databases. Specifically, in the *Populus* vascular cambium library, 47 percent of the 4,809 ESTs were singletons, corresponding to a "redundancy rate" of 2.15 EST sequences required to obtain a new (singleton) EST. In contrast, 74 percent of the 883 *Populus* developing xylem database ESTs were singletons (Table 2.1). A total redundancy rate of 1.49 for the 1,097 ESTs from all four *P. taeda* xylem libraries was observed, and approximately 47 percent of the 2,231 ESTs were represented only once in the *Cryptomeria* database (Table 2.1). Consistent with the notion that smaller databases have lower redundancy rates, the *P. taeda* seedling database had 79 percent singletons, and all 100 sequences in the developing phloem database were unique, corresponding to a redundancy rate of 1.00 (Table 2.1).

The percentage of ESTs to which the various groups could attribute a putative function among the various databases was again quite similar, between 54 percent and 63 percent, regardless of the source species or originating tissue with the exceptions of the *P. taeda* seedling and developing phloem libraries (Table 2.1). The ESTs from the random-primed seedling library and the oligo-dT-primed phloem library were found to be BLASTX "hits" (Altschul et al., 1997) to sequences in public databases at frequencies of 43 percent and 19 percent, respectively. These results are probably due to the fact that the average read length of the seedling library at least is significantly shorter than that for the other EST databases (Table 2.1), which would make attribution of function more difficult simply due to less sequence overlap in the BLAST comparisons. That the phloem library has only 19 percent hits to known genes is possibly due to relatively short (200 nt) read lengths resulting in ESTs from the oligo-dT-primed phloem library largely containing 3'-untranslated region (3'-UTR) and 5'-untranslated region (5'-UTR) sequences, which tend to be poorly conserved among divergent species. Like other databases, all of the forestry tree databases show that a very large fraction of ESTs have no assignable function. Clearly, many aspects of forestry tree biology are still awaiting elucidation.

Identified ESTs Reflect the Physiology of the Source Tissue

Closer examination of the genes for which function can be predicted in the various EST databases largely confirms the physiological and developmental roles of the source tissues. For example, although the most abundant EST in the *P. taeda* seedling library had no predictable function, the next most abundant ESTs were hits to the Rubisco (ribulose bisphosphate carboxylose/oxygenase) small subunit and light-harvesting complex mRNAs. Together, these three hits constituted nearly 10 percent of the total ESTs from the seedling library (Kinlaw et al., 1996). Thus, genes related to photosynthesis are heavily represented in light-germinated seedlings.

In a similar vein, the xylem databases from the *Populus* and the *P. taeda* xylem EST projects contained abundant ESTs involved in cell wall biosynthesis. In the *P. taeda* study, approximately 10 percent of ESTs showed similarity to known cell wall biosynthesis enzymes, including genes encoding enzymes involved in lignin biosynthesis which comprised 1.5 percent of the database (Allona et al., 1998). All but one of the lignin biosynthesis-related ESTs were sequenced from the compression wood primary or subtracted libraries, an observation consistent with compression wood's higher lignin content than side wood or vertical wood (Kramer and Kozlowski, 1979). Also heavily represented in the *P. taeda* database were ESTs coding for amino acid metabolic enzymes and proteins involved in cellular signaling and transcription. Northern blot analysis of several ESTs confirmed that mRNAs corresponding to the selected ESTs are expressed in xylem, but none of the selected ESTs were xylem specific. Surprisingly, ESTs for enzymes expected to be more xylem specific, such as lignin biosynthesis genes, were not also examined in this manner. As with *P. taeda,* the genes attributed to cell wall biosynthesis were approximately twice as abundant on a percentage basis in the *Populus* xylem database, relative to the vascular cambium database from this study (Sterky et al., 1998). Thus it appears that the *Populus* developing xylem library is in fact enriched in cDNAs involved in xylogenesis relative to the vascular cambium library.

Other differences were found in the representation of broad categories of ESTs between the two *Populus* libraries, with twice the proportion of genes involved in protein synthesis in the cambial region library relative to the developing xylem library (Sterky et al., 1998). The abundance of protein synthesis transcripts in the vascular cambium would seem to be in preparation for the synthesis of the various enzymes involved in the cell wall deposition required of protoxylem cells later in development.

In contrast to the *Populus* vascular cambium EST database, of which 3 percent of total ESTs were related to stress responses, 12 percent of the *Cryptomeria* vascular cambium database consisted of stress related ESTs, a fourfold higher abundance than *Populus*. In fact, six of the fourteen most abundant ESTs in the *Cryptomeria* database were stress related. Such a high proportion of stress-responsive ESTs relative to *Populus* must almost certainly be due to the fact that the *Cryptomeria* cambial tissue was not isolated from the source tree until two days after it was felled. It is of course possible that the *Cryptomeria* tree in this study was under stress prior to felling. It would therefore be of interest to determine the representation of stress-related ESTs in a cambial region library from a control tree, using tissue harvested immediately after felling. It appears that the increased percentage of stress-related genes in the *Cryptomeria* database comes primarily at the expense of hits to protein synthesis gene ESTs, which comprise 6 percent of the *Cryptomeria* database in contrast to 15 percent of the *Populus* cambial database. The proportions of other directly comparable categories of genes between the *Cryptomeria* and *Populus* databases were roughly equal. Further emphasizing the differences between the *Populus* and *Cryptomeria* cambial tissue databases, no ESTs are found in the *Cryptomeria* database that are orthologous to ESTs in the *Populus* database, and vice versa.

In summary, it appears that the three published forestry tree vascular tissue sequencing projects for which comparable data are available have yielded roughly similar results, in that cell wall enzyme ESTs are well represented in vascular cambium and developing xylem libraries. The redundancy rates of the databases are low, as expected for smaller databases from high-quality cDNA libraries. The *P. taeda* xylem and vascular cambium sequencing project databases (Allona et al., 1998) showed the utility of deriving a library from tissues that had been stimulated to induce the transcription of the genes of interest (i.e., lignin biosynthesis). In the same vein, the *Populus* project demonstrated a form of enrichment for cell wall synthesis genes by the use of xylem tissue as opposed to cambial tissue. Cell wall synthesis genes were also found in *Cryptomeria* cambial tissue ESTs. The data from the *Cryptomeria* project also suggest that the method of making the library from tissues harvested well after the plant had been felled was highly effective in enriching for stress-related genes, which were increased fourfold on a percentage basis as compared to an equivalent *Populus* library. Furthermore, these genes were apparently strongly induced, with half of the most frequently sequenced ESTs in the *Cryptomeria* database showing similarity to known stress-responsive genes.

THE GENESIS PINE AND EUCALYPTUS EST PROJECTS

The Genesis EST project's approach has been broad from its inception, both in terms of the sources of the libraries for sequencing and in the depth of sequencing (numbers of ESTs sequenced from a given library). As can be seen in Table 2.2, although more emphasis is placed on certain tissues and organs such as the vasculature in general, EST data from virtually every available tissue and organ have been generated. At the time these data were

TABLE 2.2. The Genesis *Eucalyptus grandis* and *Pinus radiata* cDNA libraries

Library name	Tissue	Sequencing depth
A) *Eucalyptus grandis*		
EGBA	mature shoot buds	10,267
EGCA	lower trunk cambium	9,123
EGCB	upper branch cambium/phloem	4,858
EGFA	phloem	1,481
EGFB	earlywood phloem	16,773
EGIA	floral mixed developmental stage	25,342
EGIB	floral	3,744
EGL0	mature leaf	1,598
EGLA	expanding leaf	4,858
EGRA	secondary roots	22,875
EGRB	primary roots	23,358
EGSA	fruit/developing seed	11,670
EGX0	xylem	7,386
EGXA	earlywood xylem	7,113
EGXB	trunk xylem	11,876
EGXC	lower trunk xylem	11,643
EGXD	xylem	5,218
EGXE	control xylem	1,376
EGXF	tension wood day 4 xylem	1,256
EGXP	lower trunk xylem PCR amplification	1,167
B) *Pinus radiata*		
PRBA	shoot buds	6,808
PRCA	suspension-cultured cells	8,692
PRCB	juvenile trunk cambium/phloem	590
PRFB	earlywood phloem	17,188
PRGA	female strobilus	5,018
PRHB	epibrassinolide-treated 23-day-old seedlings	22,165
PRHG	GA_3-treated 23-day-old seedlings	9,908
PRHI	IAA-treated 23-day-old seedlings	840
PRHJ	jasmonic acid treated 23-day-old seedlings	20,177
PRHK	control 23-day-old seedlings	20,940
PRHS	salicylic acid-treated 23-day-old seedlings	17,742
PRHT	ACC-treated 23-day-old seedlings	842
PRHZ	zeatin-treated 23-day-old seedlings	17,601
PRMA	fascicle meristem	17,721
PRPA	male strobilus	6,968
PRPB	early development male strobilus	13,166

TABLE 2.2 *(continued)*

PRPC	17-year mature wood—lower trunk phloem	2,228
PRPD	long fiber extreme wood type, phloem	6,497
PRPE	short fiber extreme wood type, phloem	5,601
PRPF	density and growth ex wood, phloem	6,345
PRRB	root	5,971
PRRC	secondary roots	23,945
PRRD	primary roots	26,652
PRSA	female strobilus	8,572
PRSB	early development female strobilus	6,280
PRSC	receptive female strobilus	5,016
PRXD	early xylem (late subtracted from early)	1,699
PRXE	late xylem (early subtracted from late)	1,207
PRXF	random-primed xylem	6,431
PRXG	juvenile trunk xylem	5,732
PRXH	17-year tree—mature lower trunk xylem	4,627
PRXI	long fiber extreme wood xylem tissue	6,520
PRXJ	short fiber extreme wood xylem	5,422
PRXK	high density and growth extreme wood, xylem	4,971
PRXL	17-year tree—xylem	5,286
PRXM	17-year tree—transition zone 19m ht. xylem fiber	1,021
PRXY	xylem	5,768
PRYA	young seedling—arial and fascicle	1,782
PRYB	106-day whole seedlings, arial tissues	5,704
PRYC	176-day whole seedlings, arial tissues	4,636
Total *E. grandis*		182,982
Total *P. radiata*		344,279
Other libraries		6,227
Total ESTs		533,488

examined (August 2001), sixty cDNA libraries from *Pinus radiata* or *Eucalyptus grandis* had been cloned and sequenced (Table 2.2). In excess of 519,000 ESTs from these two species, representing more than 168,000 nonredundant ESTs, now reside in the forest tree databases at Genesis (Table 2.2). Most of the cDNA libraries were derived from field-grown trees. Additionally, pine seedling libraries have been sequenced, both for genes expressed early in the growth of *P. radiata* and for the survey of hormone induction in a gymnosperm species. Likewise, a suspension-cultured cell library has been sequenced to survey genes involved in dedifferentiated cell growth in culture. As unamplified oligo-dT-primed libraries, useful comparisons and contrasts can be drawn from in "*in silico* northern" analyses to the survey libraries of field-grown trees.

Redundancy rates of Genesis libraries are significantly higher than those in the published cambial region and xylem sequencing efforts (Table 2.3). Redundancy rates can be influenced by many factors. However, there are likely two main reasons for the higher redundancy rates in the Genesis database as compared to the other forestry tree databases. First, many of the Genesis libraries have been sequenced to a greater depth than has been done for the other forestry tree EST sequencing projects (compare Tables 2.1 and 2.2). Second, Genesis has sequenced from a large number of libraries, and in order to qualify as novel, an EST must not have a contig partner in any of the other libraries in the database. The redundancy rates of several libraries in the Genesis database are higher than average, despite lower sequencing depths than many of its lower redundancy libraries. In particular, the EGL0, EGXB, PRCA, PRHG, PRHT, and PRXY libraries all have redundancy rates above 8, with sequencing depths below 12,000. These libraries generally derive from experiments designed to assess ESTs under specific circumstances or from highly metabolically specialized tissues. The PRHG and PRHT libraries derive from seedlings induced with gibberellin (GA_3) and 1-aminocyclopropane-1-carboxylic acid (ACC), the biosynthetic precursor to ethylene, respectively (Table 2.2). The PRCA library is from pine suspension-cultured cells, whereas EGL0 was derived from eucalyptus leaves (Table 2.2), and ESTs from EGL0 are heavily skewed toward photosynthesis-related genes (data not shown). The PRXY and EGXB libraries may simply be of slightly higher redundancy due to artifacts arising from their construction, although some aspect of the source plants' physiology at the time of their harvest could also be responsible.

Comparison of the Genesis EST Databases
to Public Sequence Databases

Figure 2.1 summarizes the similarity of ESTs in the Genesis pine and eucalyptus databases against the public sequence databases. ESTs are ranked into several categories based on the expectation (E) value delivered by the best match to the EST resulting from a BLASTX search (Altschul et al., 1997) against the SwissProt-TrEMBL databases. The two categories with the highest representation are at the ends of the spectrum, either a known gene or a gene with no apparent homologue in the SwissProt-TrEMBL databases. The high proportion of unknowns in these databases can be partially explained by the fact that cDNAs in directionally cloned libraries are sequenced at Genesis exclusively from their 5' ends. Therefore, 5'-UTR sequences from

TABLE 2.3. Redundancy rates for Genesis *Eucalyptus grandis* and *Pinus radiata* EST libraries

Library	Total ESTs	In a contig	Singleton	Redundancy rate[a]
A) *Eucalyptus grandis*				
EGBA	9,936	7,592	2,344	4.23
EGCA	9,214	7,491	1,723	5.34
EGCB	4,832	3,962	870	5.55
EGFA	1,481	1,149	332	4.46
EGFB	16,221	13,431	2,790	5.81
EGIA	25,082	21,918	3,164	7.92
EGIB	3,655	2,973	682	5.35
EGL0	1,592	1,421	171	9.3
EGLA	4,811	3,375	1,436	3.35
EGRA	22,038	16,880	5,158	4.27
EGRB	22,535	17,977	4,558	4.94
EGSA	11,337	9,031	2,306	4.91
EGX0	7,239	6,012	1,227	5.89
EGXA	6,785	5,571	1,214	5.58
EGXB	11,655	10,383	1,272	9.16
EGXC	11,751	9,710	2,041	5.75
EGXD	4,775	3,801	974	4.9
EGXE	1,376	1,132	244	5.63
EGXF	1,071	868	203	5.27
EGXP	1,159	708	451	2.56
B) *Pinus radiata*				
PRBA	6,614	5,683	931	7.1
PRCA	8,555	7,643	912	9.38
PRCB	590	491	99	5.95
PRFB	16,899	14,353	2,546	6.63
PRGA	5,289	325	4,964	1.06
PRHB	21,916	18,976	2,940	7.45
PRHG	9,814	8,851	963	10.19
PRHI	765	676	89	8.59
PRHJ	20,144	17,383	2,761	7.29
PRHK	20,779	18,032	2,747	7.56
PRHS	17,533	15,081	2,452	7.15
PRHT	756	678	78	9.69
PRHZ	17,505	15,521	1,984	8.82
PRMA	17,529	15,269	2,260	7.75
PRPA	6,723	5,587	1,136	5.91

PRPB	13,121	11,274	1,847	7.1
PRPC	2,228	1,946	282	7.9
PRPD	6,552	4,705	1,847	3.54
PRPE	5,573	3,923	1,650	3.37
PRPF	6,290	5,373	917	6.85
PRRB	5,893	5,128	765	7.7
PRRC	23,608	18,979	4,629	5.1
PRRD	26,156	22,923	3,233	8.09
PRSA	8,317	7,028	1,289	6.45
PRSB	6,159	5,222	937	6.57
PRSC	4,983	4,265	718	6.94
PRXD	1,661	1,382	279	5.95
PRXE	1,198	924	274	4.37
PRXF	6,372	5,517	855	7.45
PRXG	5,744	4,702	1,042	5.51
PRXH	4,597	3,867	730	6.29
PRXI	6,332	5,586	746	8.48
PRXJ	5,378	4,284	1,094	4.91
PRXK	4,838	4,281	557	8.68
PRXL	5,449	243	5,206	1.04
PRXM	1,021	41	980	1.04
PRXY	5,647	5,162	485	11.64
PRYA	1,780	1,371	409	4.35
PRYB	5,643	4,584	1,059	5.32
PRYC	4,698	3,446	1,252	3.75
Total *E. grandis*	178,545	145,385	33,160	5.38
Total *P. radiata*	340,649	280,705	59,944	5.68
All libraries	519,194	426,090	93,104	5.58

[a]Calculated as total ESTs ÷ singleton ESTs

full-length clone sequences would not be represented in this protein sequence database. Supporting this point, a combined search against SwissProt-TrEMBL (using the BLASTX algorithm; Altschul et al., 1997) and the EMBL nonredundant (using the BLASTN algorithm; Altschul et al., 1997) databases results in a general shift of E values toward known sequences, presumably due to the inclusion of noncoding sequences in the compared sequences (Figure 2.1).

Even this result must be interpreted with caution, because as public database sizes increase, there is a higher likelihood of an EST being a strong hit

versus SwissProt-TrEMBL

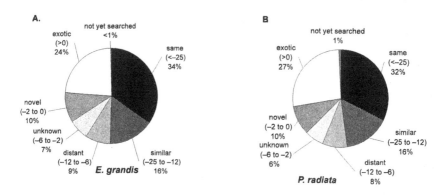

versus SwissProt-TrEMBL + EMBL

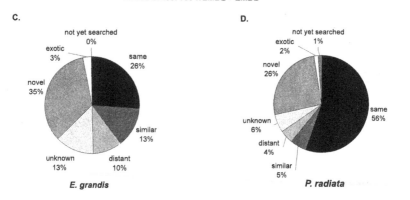

FIGURE 2.1. Degree of relationship of Genesis forestry sequences to sequences in SwissProt-TrEMBL or the combined SP-TrEMBL + EMBL databases. The total *E. grandis* and *P. radiata* databases were compared to either SwissProt-TrEMBL (A and B, respectively) or to the combined SwissProt-TrEMBL + EMBL (C and D, respectively) databases using the BLAST algorithm (Altschul et al., 1997). In the case of C and D, the best E value from either SwissProt-TrEMBL or EMBL database was used to identify the EST. A and B show the logs of the E values used to categorize the degree of similarity to a sequence from the SwissProt-TrEMBL and EMBL databases. The same ranges of E values are used in C and D.

to a sequence which itself has no known function. Therefore, the strength of the hit should not necessarily be interpreted to suggest that the function of the gene that a given EST encodes is known. For example, a higher proportion of ESTs from the *P. radiata* libraries is found in the "same" category as compared to *E. grandis* (Figure 2.1C and D). Although the Genesis *E. grandis* BLASTN and BLASTX searches result in comparison to relatively few eucalyptus sequences in the EMBL-TrEMBL databases, the *P. radiata* ESTs are being compared to nearly 40,000 ESTs from the *P. taeda* sequencing effort contained in the EMBL database. Due to the near identity of *P. radiata* and *P. taeda* sequences, a high percentage of ESTs from *P. radiata* are therefore found in the "same" category. When viewed in this light, it is still impressive that approximately 25 percent of the *P. radiata* and *E. grandis* ESTs are not significant hits to any other sequence in the EMBL/ SwissProt-TrEMBL databases (Figure 2.1). Therefore, it appears likely that a significant fraction of ESTs in both the *P. radiata* and *E. grandis* databases are strongly species specific, or more likely, genus specific, given the similarities between *P. taeda* and *P. radiata* at the nucleotide level.

Globally Abundant ESTs in the Genesis Databases

The most abundant genes in the Genesis databases are shown in Table 2.4. Many of the ESTs are hits to so-called "housekeeping" genes, specifically, elongation factor 1α (EF-1α), tubulin α chain, and polyubiquitin. Consistent with the hypothesis that highly abundant (1 to 2 percent of total mRNA in a cell) mRNAs are generally tissue specific (Bishop et al., 1974), it should be noted that none of the most abundant genes are found in the complete databases at a frequency higher than 0.43 percent. There is little overlap between the *P. radiata* and *E. grandis* databases in terms of the most abundant ESTs, with only EF-1α and *S*-adenosyl methionine synthetase (AdoMet synthetase) appearing in both databases. Three genes related to *S*-adenosyl methionine (AdoMet) metabolism appear in the *E. grandis* list of most abundant ESTs. AdoMet synthetase and *S*-adenosyl homocysteinase catalyze the anabolic and catabolic reactions on either side of methyl transfer reactions such as ethylene biosynthesis. Although it is tempting to speculate that such reactions are contributing to ethylene production in the tree from which the tissues were harvested, no direct evidence supports such a hypothesis. The third gene is a hit to AdoMet decarboxylase, which creates the precursor to propylamine donor reactions, as in spermine and spermidine synthesis. In plants, polyamines often play a role in stress responses or possibly in signal transduction (Galston and Sawhney, 1990). It is unclear which, if either, of these phenomena might be occurring in this case. Finally,

TABLE 2.4. The ten most abundant ESTs in the *Eucalyptus grandis* and *Pinus radiata* databases

EST hit identifier[a]	Number of hits in database	Frequency in database (%)
A) *Eucalyptus grandis*		
Tubulin α chain	789	0.43
S-Adenosylhomocysteinase	544	0.30
Isoflavone reductase-like protein	519	0.28
Putative xyloglucan endotransglycosylase	336	0.18
S-adenosyl methionine synthase	323	0.18
Polyubiquitin	296	0.16
ADP ribosylation factor-like protein	274	0.15
γ tonoplast intrinsic protein	266	0.15
Elongation factor 1α	250	0.14
S-Adenosylmethionine decarboxylase proenzyme	243	0.13
B) *Pinus radiata*		
Elongation factor 1α	1,316	0.38
S-adenosyl methionine synthase	889	0.26
Hypothetical protein (SB36)	881	0.26
Heat shock protein 82	844	0.25
Calcium-binding protein	836	0.24
Glycine-rich protein homologue	797	0.23
Hypothetical 58.3 kDa protein	753	0.22
F15K9.16 (putative extracellular dermal glycoprotein precursor)	748	0.22
No significant hit (protein)	746	0.22
BE187478 NXNV_98_G04_F Nsf Xylem Normal Wood Vertical *Pinus taeda*	742	0.22
5-Methyltetrahydropteroyltriglutamate - homocysteine methyltransferase	739	0.21

[a]Maximum E value for ESTs with putative assigned functions = 1×10^{-1}

four of the most abundant ESTs in *P. radiata* are either hits to hypothetical proteins or an anonymous *P. taeda* EST. The divergence between angiosperms and gymnosperms occurred approximately 275 million to 290 million years ago (Savard et al., 1994). It is possible that *P. radiata*'s evolutionary distance from the angiosperms that make up the bulk of plant sequence data in the public databases explains the preponderance of highly abundant anonymous ESTs in *P. radiata* relative to *E. grandis* (Table 2.4). These anonymous ESTs may therefore represent conifer-specific or gymnosperm-specific genes.

Genesis Xylem ESTs and Metabolic Specialization

Like the other forestry tree EST sequencing projects, the Genesis project has devoted a significant amount of effort to the sequencing of xylem and vascular cambium libraries. At the time these data were examined, 47,035 *E. grandis* and 48,684 *P. radiata* xylem ESTs, constituting 26 percent and 14 percent of the total *E. grandis* and *P. radiata* databases, respectively, had been generated. Many of the globally abundant ESTs discussed in the previous section are also found to be abundant in the xylem data set, particularly in *E. grandis,* with the four most abundant ESTs in global and xylem data sets unchanged except for relative ranking to one another (Tables 2.4 and 2.5). In contrast to the whole-plant view of abundant ESTs, when one examines a specific tissue such as developing xylem, genes involved in metabolic specialization of the tissue become more evident. Again included among these xylem-abundant ESTs are genes encoding AdoMet synthetase (Table 2.5). Sterky and colleagues (1998) also noted that AdoMet synthetase was one of the most abundant transcripts in their *Populus* developing xylem library. AdoMet is the methyl group donor for *O*-methylation of monolignol precursors, which would explain the AdoMet synthetase EST abundance in their database. The Genesis xylem databases are in general agreement with this observation. AdoMet synthetase abundance is strongly enriched in *E. grandis* xylem relative to its global abundance, with two different AdoMet synthetase genes together comprising 0.65 percent of the xylem database (Table 2.5), whereas its global abundance in *E. grandis* is 0.18 percent (Table 2.4). In contrast to this observation, AdoMet synthetase abundance is only slightly higher in *P. radiata* xylem as compared to its global abundance (0.31 percent and 0.26 percent, respectively; Tables 2.4 and 2.5). This difference presumably reflects the need for additional AdoMet in angiosperm developing xylem for synthesis of doubly methylated syringyl monolignols relative to gymnosperms, which primarily synthesize singly methylated guaiacyl monolignols (Freudenberg, 1965).

The Genesis xylem EST databases are also enriched in other genes involved in cell wall biosynthesis (Table 2.5). Whereas no cell wall-related ESTs appear in the ten most abundant ESTs at the whole-plant level, both xylem data sets show three cell wall-related hits in their twenty most abundant ESTs (Table 2.5). Interestingly, while *E. grandis* has two abundant lignin biosynthesis-related ESTs, *P. radiata* shows no lignin hits among its

TABLE 2.5. The twenty most abundant genes in the *Eucalyptus grandis* and *Pinus radiata* xylem libraries

Xylem EST hit identifier[a]	Number of hits in xylem database	Frequency in xylem database (%)
A) *Eucalyptus grandis*		
Tubulin α chain	646	1.4
Isoflavone reductase-like protein	321	0.68
S-Adenosyl-homocysteinase	286	0.61
S-Adenosyl methionine synthetase	195	0.41
Caffeic acid 3-O-methyltransferase	177	0.38
Xyloglucan endotransglycosylase	155	0.33
Arabinogalactan protein POP14A9	142	0.30
No significant hit (protein or nucleic acid)	128	0.27
Putative aquaporin	119	0.25
DTDP-glucose 4-6-dehydratase	118	0.25
F14J9.26 protein	117	0.25
β tubulin	116	0.25
Serine hydroxymethyltransferase	114	0.24
S-Adenosyl methionine synthetase I	113	0.24
No significant hit (protein or nucleic acid)	113	0.24
Aluminum-induced protein	99	0.21
Putative myosin heavy chain	98	0.21
Phenylalanine ammonia lyase	90	0.19
Putative nitrilase-associated protein	85	0.18
B) *Pinus radiata*		
Elongation factor 1 α	353	0.75
Photoassimilate-responsive protein PAR-like protein	244	0.52
Actin	231	0.49
Tubulin α chain	137	0.29
Methyltetrahydropteroyl triglutamate-homocysteine methyltransferase	191	0.41
Adenosyl homocysteinase	166	0.35
OSR4OG2 gene (fragment)	166	0.35
α expansin OSEXP7	162	0.34
S-Adenosyl methionine synthetase	144	0.31
HSP82	144	0.31
BE187478 NXNV_98_G04_F Nsf Xylem Normal wood Vertical *Pinus taeda*	117	0.25

Pinus taeda clone p3H6/CDM8 arabinogalactan-like protein	107	0.23
GB|AAD25141.1	104	0.22
DTDP-glucose-4-6-dehydratase	96	0.20
Ascorbate oxidase-related protein precursor	95	0.20
ADP, ATP carrier-like protein	94	0.20
Tubulin α chain	83	0.18
Endo-1-4 α glucanase	75	0.16
Heat shock cognate 70 kDa protein	74	0.16
Elicitor-inducible chitinase	74	0.16

[a]Maximum E value for ESTs with putative assigned functions = 1×10^{-1}

twenty most abundant ESTs. Although developing xylem is not of a single cell type, the preponderance of ESTs that relate to specialization of function in *E. grandis* and *P. radiata* developing xylem libraries is similar to that of EST databases from mammalian cell lines (Nelson et al., 1998; Sleeman et al., 2000).

Strengths and Weaknesses of EST-Based Transcript Profiling As a Predictive Tool

The abundance of ESTs in oligo-dT-primed unamplified cDNA libraries should serve as directly comparable predictors of the expression of a given gene between two or more tissue types, particularly with abundantly expressed genes. Because of a common reference point, analyses from such libraries provide data not only pertaining to location of expression but also to relative abundance of the gene among cell and tissue types (Nelson et al., 1998). This type of analysis is frequently referred to as "transcript profiling," "digital northern analysis," or "*in silico* northern analysis," and it allows the rapid prediction of the location of expression of large numbers of genes. Additionally, transcript profiling allows the formulation of testable hypotheses as to promoter function as well as possible metabolic or regulatory roles for many genes essentially simultaneously, based on their tissue and organ distributions. Statistical tools have been developed to assess the confidence with which one may predict tissue specificity of expression (Audic and Claverie, 1997).

Figure 2.2 illustrates the strengths and weaknesses of transcript profiling, with three examples of comparisons between predicted *in silico* northerns and RNA slot blots. The genes examined were an *E. grandis* gene bearing

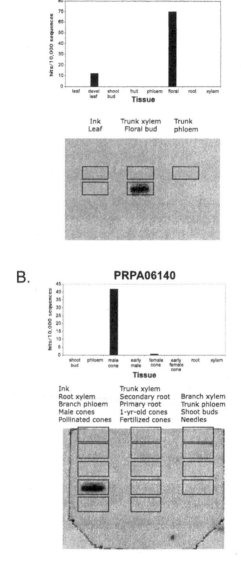

FIGURE 2.2. Comparison of *in silico* northern prediction of tissue distributions versus slot RNA blot hybridizations. The figure shows the EST names at the top, the *in silico* northern distributions of EST hits (A), the slot blot maps (B), and the slot blot autoradiograms (C).

C.   **PRBA01670**

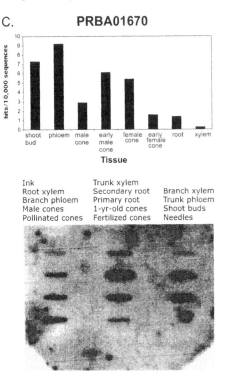

FIGURE 2.2 *(continued)*

similarity to pollen coat protein genes and abscisic acid (ABA)-inducible genes (Figure 2.2A), a *P. radiata* gene of unknown function (Figure 2.2B), and a *P. radiata* presumed flavanone-3-hydroxylase gene (SwissProt-TrEMBL E value: 5×10^{-10}; Figure 2.2C).

The *E. grandis* putative pollen coat protein gene (EGIA01147) is predicted by its transcript profile to be expressed in a highly organ-specific manner, with ESTs from the EGIA01147 contig appearing at a frequency of 203 ESTs in a total of 29,086 sequences in two *E. grandis* floral libraries (Figure 2.2A). Interestingly, EGIA01147 also appears in a developing leaf library (EGLA, Figure 2.2A), a tissue not represented on the slot blot in Figure 2.2A because this library had not been sequenced at the time that this slot blot was done. Based on the *in silico* northern, the band in developing leaf slot would be one-sixth as intense as the floral band on the slot blot. Despite this finding, the significance test of Audic and Claverie (1997) shows that this gene is differentially expressed in flowers relative to developing leaves (or any other eucalyptus library in the database) with greater than 99.9 percent confidence. However, the number of ESTs corresponding to this gene in the developing leaf library, six, statistically differs from zero with greater than 98 percent confidence (Audic and Claverie, 1997). Differential expression of EGIA01147 in developing leaves is also predicted by a G-test (Sokal and Rohlf, 1969) with greater than 96 percent confidence. These statistical analyses suggest that the expression of EGIA01147 in the developing leaf is not coincidental and that the gene could be playing some role in leaf development as well as reproduction. It is possible that the presence of EGIA01147 in the developing leaf library is related to its similarity to ABA-inducible genes (Figure 2.2A). As expected, the expression of this same gene between the two floral libraries EGIA and EGIB is not predicted to be differential, with a confidence level of only 70 percent for differential expression.

Male strobilus-specific expression of the *P. radiata* unknown function gene (PRPA06140) was predicted by its transcript profile, with 29 hits in a library of 6,968 ESTs. Like the *E. grandis* floral-specific gene, PRPA06140 shows highly tissue-specific expression on a slot blot of various tissues and organs. Furthermore, the *in silico* data also predict with greater than 99.9 percent confidence that the expression of PRPA06140 is temporally regulated in male strobili, with no ESTs of this gene found at an earlier developmental stage (PRPB library, 13,166 sequences; Figure 2.2B). There is one hit to PRPA06140 in a female strobilus library, PRSA, which consists of 8,572 sequences (Figure 2.2B). Differential expression between PRPA and PRSA is predicted with greater than 99.9 percent confidence. In contrast, the expression in female strobili is not significantly predicted to be differential above any of the other pine libraries, with about 80 percent confidence for differential expression even when compared to the largest library in the pine database, PRRD, consisting of 26,652 sequences (Table 2.2). This prediction is confirmed in the slot blot analysis, which reveals no expression of this gene in any tissue or organ other than male strobili (Figure 2.2B).

Finally, there is generally good agreement between the predicted expression of the putative flavanone-3-hydroxylase gene (PRBA01670) and its observed expression on a slot blot (Figure 2.2C). The gene was predicted by the *in silico* data to be most highly expressed in phloem, shoot buds, and male and female reproductive organs. As can be seen in the slot blot, the gene is in fact broadly expressed, with detectable expression in virtually all examined tissues. As predicted, the gene appears to be expressed in phloem, reproductive tissues, and needles (fascicles), and poorly expressed in developing xylem. A prediction for differential expression is made by Audic and Claverie's (1997) test between xylem and other tissues. It appears from the slot blot that the strongest expression of this gene is in root tissues, which is not predicted by the *in silico* analysis. It is possible that this discrepancy is attributable to random sampling, since there is a finite probability that such a lower than expected number of EST clones (based on the slot blot) harboring genes corresponding to PRBA01670 could occur. However, the size of the EST libraries makes random sampling unlikely to be accountable for this discrepancy, since the overall number of ESTs among all the root libraries number greater than 56,000. Alternatively, there may a closely related gene preferentially expressed in root that would cross hybridize on the slot blot. Because it would not align perfectly with the PRBA01670 contig, due to sequence differences between gene family members, it would not be found in the transcript profile. Indeed, a BLASTN search of the Genesis database nearly doubles the number of root clones which would likely cross hybridize with PRBA01670 but would not be identified by transcript profiling. Still, even the addition of these ESTs in an *in silico* northern analysis cannot completely make up the discrepancy between the slot blot analysis and the *in silico* prediction. Another possible explanation is that yet another gene was preferentially expressed in roots to which the probe hybridized that has not yet been found in the EST database. This could occur if a mRNA has an exceptionally long 5'-UTR, such that its open reading frame (ORF) might never be sequenced from a library of full-length cDNA clones. Such caveats are reasons why it is unlikely that EST data will ever substitute completely for physical means of mRNA abundance determination.

Despite the limitations of transcript profiling discussed here, its predictions can still be quite powerful, as demonstrated in Figure 2.2A and B. Scientists at Genesis have used this predictive power as the basis for the cloning of the promoter of the most highly expressed *P. radiata* ubiquitin gene in the Genesis EST collection, based on 5'-UTR sequences. In addition to the high level of predicted expression of this gene, its promoter was predicted by transcript profiling of the *P. radiata* database to strongly drive transcription in all tissues. Studies of this promoter at Genesis have shown it to be considerably stronger and more uniform in expression than the "constitutive" promoters in common use in transgenic plants (manuscript in preparation).

SPECIAL PURPOSE LIBRARIES

Although the oligo-dT-primed, unamplified "survey" libraries are highly useful for most purposes at Genesis, there are limitations to the type and amount of data that they provide. Special purpose libraries such as subtracted and random-primed libraries are therefore important because they provide a means by which researchers can gain additional EST data that are not readily available from the oligo-dT-primed libraries comprising the bulk of the Genesis EST database. For example, we have used subtracted libraries to enrich for cDNAs abundant in earlywood and latewood xylem development. We have identified a number of ESTs presumably involved in some aspect of earlywood and latewood physiology from these libraries. In addition, we have used a random-primed xylem cDNA library to provide ESTs from portions of the transcriptome that might not otherwise be sequenced in directionally cloned full-length oligo-dT-primed libraries. This means that random-primed libraries also allow the sequencing of genes that would not otherwise be identified due to long 5'-UTRs and allow the joining of previously separate contigs. These efforts are described in more detail in the following sections.

Earlywood- and Latewood-Enriched Subtracted Libraries

A means of searching for transcripts differentially expressed in a given tissue, developmental stage, or physiological state is the construction and generation of ESTs from subtracted cDNA libraries. We chose to examine differences in latewood versus earlywood development, since cell wall biosynthesis changes markedly during the growing season, with heavy lignification occurring preferentially in latewood (Kramer and Kozlowski, 1979). Latewood cDNA was therefore subtracted from earlywood cDNA, yielding an earlywood-enriched library (PRXD) and vice versa (PRXE). We profiled ESTs that appeared in PRXD or PRXE exclusively, as well as genes enriched in PRXD and PRXE but found in other libraries. Table 2.6A and C shows the ten most abundant ESTs found exclusively in PRXD or PRXE. This stringent selection criterion presumably reveals very rare transcripts that are enriched in earlywood and latewood. Six of the twenty ESTs exclusive to PRXD or PRXE in this table align with anonymous *P. taeda* ESTs. Furthermore, of these twenty ESTs, only two have significantly identifiable protein sequences. It is tempting to speculate that these ESTs represent rare novel transcripts involved in earlywood and latewood biogenesis for which there are no homologues present in the public sequence databases. However, it is also possible that method used to create the subtracted libraries, which generates short (approximately 200 bp) cDNA inserts, is responsible for the inability to identify these ESTs. This short sequence length likely decreases the ability of sequence comparison software to make an unambiguous identification with sequences in the public databases. Thus, although there are unique sequences identified from the subtracted libraries, further work is required to determine whether these sequences are novel or are part of a known gene.

TABLE 2.6. Genes unique to or abundant in the PRXD (earlywood EST) and PRXE (latewood EST) libraries

		Number of hits in library		
EST hit identifier	E value	PRXD	PRXE	All other
A) The ten most abundant ESTs unique to PRXD (earlywood) library				
No significant hit	NA	16	0	0
AI856113 sc31g10.x1 Gm-c1014 Glycine max cDNA clone GENOME SYST...	2e-07	15	0	0
Q9SHG2 F20D23.27 PROTEIN	0.11	13	0	0
Q9LVH6 ALDOSE 1-EPIMERASE-LIKE PROTEIN.	1e-22	10	0	0
Q9SW97 IONOTROPIC GLUTAMATE RECEPTOR ORTHOLOG GLR6.	3e-06	8	0	0
No significant hit	NA	6	0	0
AI919822 1366 Pine Lambda Zap Xyem library *Pinus taeda* cDNA cl...	5e-57	5	0	0
No significant hit	NA	4	0	0
No significant hit	NA	4	0	0
No significant hit	NA	4	0	0
B) The ten most abundant ESTs in the PRXD (earlywood) library				
BG040132 NXSI_106_F10_F NXSI (Nsf Xylem Side wood Inclined) Pinus	0.0	52	2	42
No significant hit	NA	18	0	5
No significant hit	NA	16	0	0
AI856113 sc31g10.x1 Gm-c1014 Glycine max cDNA clone GENOME SYST...	2e-07	15	0	0
No significant hit	NA	15	0	20
NXCI_034_A08_FNXCI (Nsf Xylem compression wood inclined) Pinus ta...	1e-104	14	0	7
ACTIN	1e-08	14	0	43
Q9SHG2 F20D23.27 PROTEIN	0.11	13	0	0
UBIQUITIN-LIKE PROTEIN	5e-6	12	0	25
NXNV_071_F08_F Nsf Xylem Normal wood Vertical Pinus ta...	9e-18	12	0	71

TABLE 2.6 *(continued)*

C) The ten most abundant EST's unique to PRXE (latewood) library				
Q9XEL3 PUTATIVE DEHYDRIN	4e-34	0	23	0
AI725185 1084 PtIFG2 *Pinus taeda* cDNA clone 9108r, mRNA sequence.	1e-41	0	22	0
AI725339 1205 PtIFG2 *Pinus taeda* cDNA clone 9297r, mRNA sequence.	8e-43	0	3	0
BE049814 NXNV_144_F04_F Nsf Xylem Normal wood Vertical Pinus ta...	1e-33	0	3	0
No significant hit	NA	0	3	0
No significant hit	NA	0	3	0
BE451936 NXCI_006_E06_F NXCI (Nsf Xylem Compression wood Inclin...	1e-62	0	2	0
No significant hit	NA	0	2	0
BAB09205 26S PROTEASOME SUBUNIT-LIKE PROTEIN	2e-14	0	1	0
No significant hit	NA	0	1	0
D) The ten most abundant ESTs in the PRXE (latewood) library				
O49149 ABSCISIC ACID- AND STRESS-INDUCIBLE PROTEIN	2e-18	0	230	43
Q9ZTT5 CAFFEOYL-COA O-METHYLTRANSFERASE	6e-38	0	34	28
Q9XEL3 PUTATIVE DEHYDRIN	4e-34	0	23	0
AI725185 1084 PtIFG2 *Pinus taeda* cDNA clone 9108r, mRNA sequence.	1e-41	0	22	0
TUBULIN BETA-2 CHAIN (BETA-2 TUBULIN)	4e-34	3	22	234
1084 PtIFG2 *Pinus taeda* cDNA clone 9108r, mRNA sequence	3e-42	0	16	1
Q41096 WATER DEFICIT STRESS INDUCIBLE PROTEIN LP3-2 (FRAGM...	7e-05	0	16	5
COBALAMIN-INDEPENDENT METHIONINE SYNTHASE	7e-29	2	9	803
100 Loblolly pine N *Pinus taeda* cDNA clone 3N10A, mRNA...	3e-81	0	7	120
579 PtIFG2 *Pinus taeda* cDNA clone 8929M 3', mRNA seque...	5e-48	4	7	9

When one omits the criterion for exclusivity to either PRXD or PRXE, then ESTs of higher abundance in these libraries are seen (Table 2.6B and D). Although some overlap occurs among ESTs found in Table 2.6A and 2.6B as well as 2.6C and 2.6D, these ESTs generally represent more abundant transcripts than those that appear exclusively in the PRXD or PRXE portions of the database. The more interesting of the two subtracted libraries from the perspective of identifiable abundant ESTs is the PRXE (latewood) library. Of the most abundant clones observed, three appear to be water stress and/or ABA-inducible. In fact, at 230 ESTs, a putative ortholog to the ABA-inducible and stress-inducible gene O49149 is the most abundant EST in the PRXE portion of the database, constituting approximately 20 percent of its ESTs. When compared to the percentage representation of the most abundant ESTs in the xylem databases which are all under 1 percent of the total (Table 2.6), this is a stunning enrichment. Although this EST is observed elsewhere in the *P. radiata* database, the extreme abundance of such an EST in PRXE suggests that it is highly abundant in the latewood tissue relative to earlywood xylem and that it is likely present in response to drought. The expression of drought-stress genes would be consistent with the latewood xylem having been collected in the dry late summer. The presence of abundant putative lignin biosynthetic pathway gene in the PRXE library, caffeoyl-CoA-O-methyltransferase, is also consistent with latewood, which is made up of small cells with heavily lignified thick cell walls (Kramer and Kozlowski, 1979).

In contrast to PRXE, the PRXD-abundant genes are rather unremarkable, with many ESTs yielding no signficant hits or hits to *P. taeda* ESTs of unknown function. However, based on the examples of physiological relevance of the identified genes to latewood in PRXE library, these as yet unidentified ESTs likely have importance for the physiology of earlywood.

Random-Primed Libraries

As noted by Nelson and colleagues (1998), in order to directly compare frequencies of ESTs from primary libraries in transcript profiles, a "common anchoring reference point" is required. The polyA tails of cDNA clones generally serve as this common reference point in EST sequencing projects. Therefore, oligo-dT-primed cDNA libraries comprise the bulk of the Genesis primary libraries. Unfortunately, this leads to less than complete identification of the coding sequences of many ESTs, particularly if sequencing is carried out from the 3' ends of such cDNAs. Because of their essentially equal representation of the entire transcriptome, random-primed libraries offer a means of quickly adding novel data to an EST database for a given cell type, organ, or tissue.

A random-primed xylem library, designated PRXF, has been constructed at Genesis. Like the other libraries in the Genesis collection, PRXF is an unamplified library for EST sequencing purposes. There are two critical differences in the construction of this library compared to the bulk of the primary libraries in the Genesis collection that make it useful for bridging gaps in contigs and generating novel ESTs. The foremost difference between PRXF and the other libraries is the use of random hexameric oligonucleotides, rather than oligo-dT oligonucleotides, to prime first strand cDNA synthesis. As mentioned previously, representation of essentially the entire xylem transcriptome results from this random priming of first strand cDNA synthesis, since any portion of a mRNA has an essentially equal chance of being reverse transcribed into cDNA. The second major difference between PRXF and the oligo-dT-primed libraries lies in the fact that the PRXF library is not directionally cloned. That is, a given cDNA insert has an equally likely chance of being cloned in either of two possible orientations relative to the cloning vector. Therefore, sequencing with a given universal primer means that there is a 50 percent probability that a clone in the sense orientation will be sequenced, versus a 100 percent theoretical probability for a directionally cloned library, assuming the correct choice of sequencing primer. This essentially doubles the potential information content available in PRXF from a single universal sequencing primer relative to directionally cloned libraries.

Several predictions can be made as to how EST sequencing of a random-primed library would enhance an EST database. First, the PRXF ESTs would be predicted to extend contigs of ESTs from other libraries at their 5' and 3' ends. An assessment of the pine contigs in the database shows that 57 contigs are extended at either their 5' or 3' ends by at least 200 bp by a PRXF EST. An apparent bias toward extension of 3' ends of contigs is a further confirmation of the overall quality of Genesis oligo-dT-primed libraries, as it suggests that there are relatively few clones that are not full length. Second, it would be expected that otherwise separate contigs would be bridged by PRXF ESTs, due to the random access to sequence data provided by the random-primed library. Indeed, 21 contigs are bridged uniquely by PRXF ESTs, adding further informatic value to contigs being closer to full-length sequences and drawing together sequences for which a relationship might not otherwise be readily deduced. Third, due to the random access to the transcriptome provided by PRXF, a high number of contigs in the database should consist only of PRXF ESTs. As seen in Table 2.7, this is indeed the case, with 177 (12.9 percent) of PRXF contigs consisting of only PRXF members. In comparison, the xylem library with the next most single library

TABLE 2.7. Xylem single library contigs in the *P. radiata* database

Library	Number of single library contigs	Percentage of contigs as single library
PRXD	55	8.4
PRXE	7	1.6
PRXF	177	12.9
PRXG	19	0.7
PRXH	34	1.6
PRXI	79	3.7
PRXJ	32	1.3
PRXK	72	3.6
PRXL	51	1.9
PRXM	16	3.7
PRXY	118	6.3

EST count is PRXY, with 118 (6.3 percent) of contigs consisting only of PRXY library ESTs (Table 2.7). Thus, these PRXF contigs provide additional novel bioinformatic data that would not otherwise be present in our database, emphasizing the less-biased access to the transcriptome that random priming provides. The construction and sequencing of random-primed libraries from other tissues should provide similar enhancements of the Genesis databases.

CONCLUSION

High-throughput genomics studies, including EST sequencing, are increasingly proving to be useful tools for the molecular biologist. These techniques have led to the large-scale examination of tree species at the molecular level. The ability to isolate large quantities of vascular tissue from tree species has led to high-quality libraries from various tree species that are being used to study not only lignification (Hertzberg et al., 2001; Whetten et al., 2001) but also the involvement of other cell wall proteins in secondary xylem development (Zhang et al., 2000). Forestry xylem and cambial EST databases tag a wealth of genes involved in cell wall synthesis.

A result of the forestry and other EST projects has been the identification of a huge number of genes that have no known function. Even these bioinformatic data can be used as predictive tools for gene function or for the grouping of genes in a metabolic or signaling pathway. Among the methods used for such predictions are extrapolation from physically linked genes in an organism (Pellegrini et al., 1999) or analysis of coordinate expression of ESTs in a cell or organ type (Marcotte et al., 1999). Further complementing the genome and transcriptome sequence databases are large-scale functional genomics projects, such as genome-wide gene knockout programs in yeast (Smith et al., 1995) and arabidopsis (Sussman et al., 2000) as well as proteomics projects (Zhu et al., 2001). Although some of these methods cannot be applied to forestry tree species, data from such efforts in herbaceous plants will enhance the ability to predict gene function based on homology from forestry tree databases. This will lead to future projects like that of Zhang and colleagues (2000) to examine other aspects of secondary xylem formation, such as cellulose synthesis. Such research will eventually allow biotechnologists to address problems that have nagged the forestry industries, such as spiral grain, as well as juvenile/mature and earlywood/latewood quality differences.

EST databases are proving to be powerful predictive tools for gene expression studies as well as for mapping genes into metabolic pathways. As EST databases increase in size, so do their predictive power. The Genesis EST databases and others like them are of a sufficient size to essentially be predictive for the expression of abundantly expressed and moderately expressed genes, particularly with regard to tissue-specific expression. These predictions from the EST databases are driving the Genesis promoter cloning effort. Constitutive, developmentally regulated, organ-specific, and cell type-specific promoters are all being cloned with the assistance of the *P. radiata* and *E. grandis* transcript profile predictions. The ongoing promoter cloning will eventually result in an extensive collection of promoters to allow engineered control of transgene expression at the whole-plant, tissue-specific, and cell-specific levels in commercial forestry trees and other plant species. This combination of genes and promoters, when combined with conventional breeding, will enhance forestry tree improvement initiatives in the future.

REFERENCES

Adams, M. D., Celniker, S. E., Holt, R. A., Evans, C. A., Gocayne, J. D., Amanatides, P. G., Scherer, S. E., Li, P. W., Hoskins, R. A., Galle, R. F., et al. (2000). The genome sequence of *Drosophila melanogaster*. *Science* 287: 2185-2195.

Allona, I., Quinn, M., Shoop, E., Swope, K., St. Cyr, S., Carlis, J., Riedl, J., Retzel, E., Campbell, M. M., Sederoff, R., and Whetten, R. W. (1998). Analysis of xylem formation in pine by cDNA sequencing. *Proc. Natl. Acad. Sci. USA* 95: 9693-9698.

Altschul, S. F., Madden, T. L., Schaffer, A. A., Zhang, J., Zhang, Z., Miller, W., and Lipman, D. J. (1997). Gapped BLAST and PSI-BLAST: A new generation of protein database search programs. *Nucl. Acids Res.* 25: 3389-3402.

The Arabidopsis Genome Initiative (2000). Analysis of the genome sequence of the flowering plant *Arabidopsis thaliana*. *Nature* 408: 796-815.

Audic, S. and Claverie, J.-M. (1997) The significance of digital gene expression profiles. *Genome Res.* 7: 986-995.

Bishop, J. O., Morton, J. G., Rosbash, M., and Richardson, M. (1974). Three abundance classes in HeLa cell messenger RNA. *Nature* 250: 199-204.

The *C. elegans* Sequencing Consortium (1998). Genome sequence of the nematode *C. elegans:* A platform for investigating biology. *Science* 282: 2012-2018.

Delseney, M., Cooke, R., Raynal, M., and Grellet, F. (1997). The *Arabidopsis thaliana* cDNA sequencing projects. *FEBS Lett.* 405: 129-132.

Demura, T., Tashiro, G., Horiguchi, G., Kishimoto, N., Kubo, M., Matsuoka, N., Minami, A., Nagata-Hiwatashi, M., Nakamura, K., Okamura, Y., et al. (2002). Visualization by comprehensive microarray analysis of gene expression programs during transdifferentiation of mesophyll cells into xylem cells. *Proc. Natl. Acad. Sci. USA* 99: 15794-15799.

The FlyBase Consortium (1999). The FlyBase database of the Drosophila genome projects and community literature. *Nucl. Acids Res.* 27: 85-88.

Freudenberg, K. (1965). Lignin: Its constitution and formation from p-hydroxycinnamyl alcohols. *Science* 148: 595-600.

Fukuda, H. and Komamine, A. (1980). Establishment of an experimental system for the study of tracheary element differentiation from single cells isolated from the mesophyll of *Zinnia elegans. Plant Physiology* 65: 57-60.

Furlong, E. E. M., Andersen, E. C., Null, B., White, K. P., and Scott, M. P. (2001). Patterns of gene expression during *Drosophila* mesoderm development. *Science* 293: 1629-1633.

Gai, X., Lal, S., Xing, L., Brendel, V., and Walbot, V. (2000). Gene discovery using the maize genome database ZmDB. *Nucl. Acids Res.* 28: 94-96.

Galston, A. W. and Sawhney, R. K. (1990). Polyamines in plant physiology. *Plant Physiology* 94: 406-410.

Gerhold, D. and Caskey, C. T. (1996). It's the genes! EST access to human genome content. *Bioessays* 18: 973-981.

Gopal, S., Schroeder, M., Pieper, U., Sczyrba, A., Aytekin-Kurban, G., Bekiranov, S., Fajardo, J. E., Eswar, N., Sanchez, R., Sali, A., and Gaasterland, T. (2001). Homology-based annotation yields 1,042 new candidate genes in the *Drosophila melanogaster* genome. *Nature Genet.* 27: 337-340.

Grigor, M. R., Phillips, J., Puthigae, S., and Strabala, T. J. (2003). From ESTs to gene function: A pipeline for gene discovery in forestry. *NZ Biosci.* 12: 5-7.

Hertzberg, M., Aspeborg, H., Schrader, J., Andersson, A., Erlandsson, R., Blomqvist, K., Bhalerao, R., Uhlén, M., Teeri, T. T., Lundeberg, J., Sundberg, B., Nilsson, P., and Sandberg, G. (2001). A transcriptional roadmap to wood formation. *Proc. Natl. Acad. Sci. USA* 98: 14732-14737.

International Human Genome Sequencing Consortium (2001). Initial sequencing and analysis of the human genome. *Nature* 409: 860-921.

Kinlaw, C. S., Ho, T., Gerttula, S. M., Gladstone, E., Harry, D. E., Quintana, L., and Baysdorfer, C. (1996). Gene discovery in loblolly pine through cDNA sequencing. In M. R. Ahuja, W. Boerjan, and D. B. Neale (eds.), *Somatic Cell Genetics* and *Molecular Genetics of Trees* (pp. 175-182). Dordrecht, The Netherlands: Kluwer Academic Publishers.

Kinlaw, C. S. and Neale, D. B. (1997). Complex gene families in pine genomes. *Trends Plant Sci.* 2: 356-359.

Kramer, P. J. and Kozlowski, T. T. (1979). *Physiology of Woody Plants.* Orlando, FL: Academic Press.

Marcotte, E. M., Pellegrini, M., Ng, H.-L., Rice, D. W., Yeates, T. O., and Eisenberg, D. (1999). Detecting protein function and protein-protein interactions from genome sequences. *Science* 285: 751-753.

Mewes, H. W., Albermann, K., Bähr, M., Frishman, D., Gleissner, A., Hani, J., Heumann, K., Kleine, K., Maierl, A., Oliver, S. G., et al. (1997). Overview of the yeast genome. *Nature* 387 (supplement): 7-8.

Milioni, D., Sado, P.-E., Stacey, N. J., Domingo, C., Roberts, K., and McCann, M. C. (2001). Differential expression of cell-wall-related genes during the formation of tracheary elements in the *Zinnia* mesophyll cell system. *Plant Mol. Biol.* 47: 221-238.

Milioni, D., Sado, P.-E., Stacey, N. J., Roberts, K., and McCann, M. C. (2002). Early gene expression associated with the commitment and differentiation of a plant tracheary element is revealed by cDNA-amplified fragment length polymorphism analysis. *Plant Cell* 14: 2813-2824.

Moriguchi, Y., Iwata, H., Ujino-Ihara, T., Yoshimura, K., Taira, H., and Tsumura, Y. (2003). Development and characterization of microsatellite markers for *Cryptomeria japonica* D.Don. *Theor. Appl. Genet.* 106: 751-758.

Nelson, P. S., Ng, W. L., Schummer, M., True, L. D., Liu, A. Y., Bumgarner, R. E., Ferguson, C., Dimak, A., and Hood, L. (1998). An expressed-sequence-tag database of the human prostate: Sequence analysis of 1168 cDNA clones. *Genomics* 47: 12-25.

Pellegrini, M., Marcotte, E. M., Thompson, M. J., Eisenberg, D., and Yeates, T. O. (1999). Assigning protein functions by comparative genome analysis: Protein phylogenetic profiles. *Proc. Natl. Acad. Sci. USA* 96: 4285-4288.

Reymond, A., Camargo, A. A., Deutsch, S., Stevenson, B. J., Parmigiani, R. B., Ucla, C., Bettoni, F., Rossier, C., Lyle, R., Guipponi, M., et al. (2002). Nineteen additional unpredicted transcripts from human chromosome 21. *Genomics* 79: 824-832.

Savard, L., Li, P., Strauss, S. H., Chase, M. W., Michaud, M., and Bousquet, J. (1994). Chloroplast and nuclear gene sequences indicate late Pennsylvanian time for the last common ancestor of extant seed plants. *Proc. Natl. Acad. Sci. USA* 91: 5163-5167.

Sleeman, M. A., Murison, J. G., Strachan, L., Kumble, K., Glenn, M. P., McGrath, A., Grierson, A., Havukkala, I., Tan, P. L. J., and Watson, J. (2000). Gene expression in rat dermal papilla cells: Analysis of 2529 ESTs. *Genomics* 69: 214-224.

Smith, V., Botstein, D., and Brown, P. O. (1995). Genetic footprinting: A genomic strategy for determining a gene's function given its sequence. *Proc. Natl. Acad. Sci. USA* 92: 6479-6483.

Sokal, R. R. and Rohlf, F. J. (1969). *Biometry: The Principles and Practice of Statistics in Biological Research*. San Francisco, CA: W.H. Freeman and Company.

Stein, L., Sternberg, P., Durbin, R., Thierry-Mieg, J., and Spieth, J. (2001). WormBase: Network access to the genome and biology of *Caenorhabditis elegans*. *Nucl. Acids. Res.* 29: 82-86.

Sterky, F., Regan, S., Karlsson, J., Hertzberg, M., Rohde, A., Holmberg, A., Amini, B., Bhalerao, R., Larsson, M., Villarroel, R., et al. (1998). Gene discovery in the wood-forming tissues of poplar: Analysis of 5,692 expressed sequence tags. *Proc. Natl. Acad. Sci. USA* 95: 13330-13335.

Sussman, M. R., Amasino, R. M., Young, J. C., Krysan, P. J., and Austin-Phillips, S. (2000). The arabidopsis knockout facility at the University of Wisconsin-Madison. *Plant Physiology* 124: 1465-1467.

Ujino-Ihara, T., Yoshimura, K., Ugawa, Y., Yoshimaru, H., Nagasaka, K., and Tsumura, Y. (2000). Expression analysis of ESTs derived from the inner bark of *Cryptomeria japonica*. *Plant Mol. Biol.* 43: 451-457.

van Zyl, L., Bozhkov, P. V., Clapham, D. H., Sederoff, R. R., and von Arnold, S. (2003). Up, down and up again is a signature global gene expression pattern at the beginning of gymnosperm embryogenesis. *Gene Expr. Patterns* 3: 83-91.

Velculescu, V. E., Zhang, L., Zhou, W., Vogelstein, J., Basrai, M. A., Bassett, D. E. J., Hieter, P., Vogelstein, B., and Kinzler, K. W. (1997). Characterization of the yeast transcriptome. *Cell* 88: 243-251.

Whetten, R., Sun, Y.-H., Zhang, Y., and Sederoff, R. (2001). Functional genomics and cell wall biosynthesis in loblolly pine. *Plant Mol. Biol.* 47: 275-291.

Williams, C. G., Zhou, Y., and Hall, S. E. (2001). A chromosomal region promoting outcrossing in a conifer. *Genetics* 159: 1283-1289.

Zhang, Y., Sederoff, R. R., and Allona, I. (2000). Differential expression of genes encoding cell wall proteins in vascular tissues from vertical and bent loblolly pine trees. *Tree Physiology* 20: 457-466.

Zhu, H., Bilgin, M., Bangham, R., Hall, D., Casamayor, A., Bertone, P., Lan, N., Jansen, R., Bidlingmaier, S., Houfek, T., et al. (2001). Global analysis of protein activities using proteome chips. *Science* 293: 2101-2105.

Chapter 3

Proteomics for Genetic and Physiological Studies in Forest Trees: Application in Maritime Pine

Christophe Plomion
Nasser Bahrman
Paulo Costa
Christian Dubos
Jean-Marc Frigerio
Jean-Marc Gion
Céline Lalanne
Delphine Madur
Cédric Pionneau
Sophie Gerber

INTRODUCTION

Integrated approaches combining the systematic sequencing of expressed genes (e.g., Sterky et al., 1998; Allona et al., 1998) and the monitoring of mRNA expression levels for large numbers of genes (e.g., Voiblet et al., 2001) are now considered to be the strategy of choice for tracking genes of interest. However, although the *Arabidopsis* genome project provides an overview of the genes involved, it is becoming increasingly clear that this is only a very fragmentary beginning to understand their role and function. The problems start with the observation that the old paradigm of one gene = one protein is incorrect, and that in eukaryotic cells the situation is more accurately six to eight proteins per gene. Thus, although there may be only 25,000 genes in a plant species (*Arabidopsis* Genome Initiative, 2000), there are many more resultant proteins once splice variants and essential

Current works on maritime pine proteomics are funded by Grants from France (INRA-LIGNOME, Région Aquitaine) and the European Union (QLK5-CT-1999-00942). We thank H. Thiellement and P. Gere for their critical review.

posttranscriptional modifications are included. Genome information is not able to define all of these protein components. It is also clear that for a substantial number of these proteins, the study of mRNA quantities provides only a partial view of gene product expression (Gygi et al., 1999; Anderson and Seilhammer, 1997) due to (1) large differences between mRNA and protein turnovers (a protein can still be abundant while the mRNA is no longer detectable because its synthesis has stopped), (2) posttranslational modifications, such as removal of signal peptides, phosphorylation, and glycosylation, that play important roles in their activity and subcellular localization, and (3) complex interactions with other proteins, processes that cannot be deduced from microarray data. The problems become even more complicated because complex networks of proteins can be very divergent in different organs or developmental phases of the same organism, despite the same genomic information.

Given that the genetic information is indicative only of the cell's potential and does not reflect the actual state in a given cell at a given time, the concept of "proteome" (for PROTEin complement expressed by a genOME; Wilkins et al., 1995) has emerged to provide complementary and critical information for understanding biological processes in the postgenome era. Proteomics is now considered as a priority by many universities and research institutes and is beginning to be widely applied to the model plant arabidopsis as well as to other important crop species (reviewed by Thiellement et al., 1999; Zivy and de Vienne, 2000; Rossignol, 2001; Roberts, 2002; Kersten et al., 2002). However, forest tree proteomics still remains largely embryonic, in spite of the fact that a large collection of expressed sequence tags (ESTs) is available and genome sequencing for poplar will be completed very soon.

The proteome concept relies on two-dimensional polyacrylamide gel electrophoresis (2D-PAGE; O'Farrell, 1975; Klose, 1975) and recent innovations in mass spectrometry (Yates, 1998). 2DE separates polypeptides according to two independent criteria: their charges (isoelectric point, pI) in the first dimension (isoelectrofocusing) and their apparent molecular masses (M_r) in the second dimension (SDS[sodium dodecyl sulfate]-PAGE). Although it was published more than 25 years ago, 2DE is still the main technique used to analyze mixtures of denatured proteins. Recent improvements in realization of 2DE allowed considerable progress in terms of resolution and reproducibility (Görg et al., 2000), allowing thousands of protein spots to be individually displayed. The use of dedicated software for image analysis has made it possible to quantify protein abundance and compare a large number of spots in as many samples. The value of mass spectrometry (MS) to biologists has been established by its effectiveness in routinely characterizing proteins at the femtomoles to picomoles levels, as well as in evaluating posttranslational modifications and in studying macro-molecular complexes. In addition, dramatic gains in analytical sensitivity already allow proteomic analyses to be carried out on small number of cells, largely obviating the need for molecular amplification strategies. The accessibility to mass spectrometers is therefore considered a critical point in postgenomic projects.

In this chapter, we will show that 2DE is a valuable tool in forest tree genomics to map the expressed genome; to evaluate the modifications of protein expression in respect to genetic, environmental, and developmental factors; and to determine the extent to which quantitative phenotypic variance is mirrored by quantitative variation of proteins. Examples will be taken from an economically and ecologically important pine species (*Pinus pinaster* Ait.) in southwestern Europe, for which proteomics has been developed during the last ten years for population genetics studies, genetic mapping, and genome expression analysis.

CHARACTERIZATION OF THE PINE PROTEOME

Resolving a Mixture of Denatured Proteins

The 2DE procedure that was developed in our laboratory (reviewed in Bahrman et al., 1997) includes: (1) acetone-TCA (trichloroacetic acid) protein extraction (Damerval et al., 1986); (2) separation of denatured poly-peptides in a pI ranging from 4 to 7 (wider immobilized pH gradients [IPG] strips will be tested in the near future), and a M_r range of about 14 to 110 kDa; and (3) silver nitrate, Sypro Ruby, amido black, or Coomassie brillant blue G-250 staining (Figure 3.1A). A single protocol was optimized to reveal a maximum of protein spots in all pine tissues/organs (phloem, xylem, needles, root, pollen, bud, megagametophytes) analyzed and was also shown to work well for poplar, eucalyptus, and oak (Barreneche et al., 1996). However, if these proteomes correspond to the most abundant and soluble proteins, in no case are the cell proteins exhaustively separated and visualized. In particular, the nuclear and hydrophobic membrane-associated proteins are not extracted. In addition, the low abundance proteins (e.g., transcription factors, protein kinases) are not visible on standard 2DE gel protein maps. In maritime pine, depending on the tissue, from 300 to more than 1,000 spots can be observed on silver-stained gels. The recovery of more proteins can be achieved by several ways, mainly the following:

- The use of extensive immobilized pH gradients (IPG) and acrylamide concentration gels (Wasinger et al., 1997; Urquhart et al., 1997).
- The analysis of different organs and tissues (e.g., roots, Chang et al., 2000; xylem, Plomion et al., 2000).
- The analysis of proteins after subfractionation according to cell type and subcellular compartments, as described for the plasma membrane (Rouquié et al., 1997; Santoni et al., 1998, 1999), the chloroplast (Ferro et al., 2000; van Wijk, 2000; Peltier et al., 2000), the mitochondria (Prime et al., 2000; Harvey Millar et al., 2001; Kruft et al., 2001), and the peribacteroid membrane of root nodules (Panter et al., 2000).
- Extensive proteomic approach also relies on the handling of hydrophobic proteins, which are rarely detectable in 2DE gels (Adessi et al., 1997). Recent results obtained with new detergents in combination with thiourea seem promising for the recovery of such proteins on 2DE gels (Chevallet et al., 1998).
- More recently, multidimensional liquid chromatography (Washburn et al., 2001) was developed to resolve low abundance proteins, proteins with extremes in pI and M_r, and integral membrane proteins.

The availability of standardized and reproducible procedures for the analysis of proteins by 2DE opened the way to large-scale investigations. The gel-to-gel comparison of proteins from different tissues, developmental stages, or environmental conditions (analytical 2DE, Figure 3.1B) are made by checking the spot positions (sometimes with the help of coelectrophoresis of extracts) using advanced image analysis systems (Pleissner et al., 2000). For example, using the same genotypes of maritime pine, comparison of needle, xylem, and root 2DE patterns indicates that 35 percent, 18 percent, and 23 percent of the spots were specific to each tissue, respectively (Figure 3.2). The other spots had identical M_r and pI, indicating that they likely correspond to identical gene products as discussed by Costa and colleagues (1999). The finding that most spots were common to different tissues confirmed earlier results by Bahrman and Petit (1995) who compared needle, bud, and pollen proteins. They interpreted the presence of the same proteins in well-differentiated organs as "house-keeping proteins" (e.g., actin, HSP70, alpha- and beta-tubulin, glutamine synthetase, etc.). Conversely, proteins accumulating only in particular tissues appear to be present in specialized cells and to have specific functions, such as the needle proteins involved in photosynthesis. Moreover, comparison of 2DE maps from the same tissue (e.g., developing xylem, Figure 3.3) across different species (*Pinus pinaster, Eucalyptus gunnii,* hybrids between *Populus tremula* × *P. alba*) also demonstrates the variability of tissue-specific proteomes and the necessity for

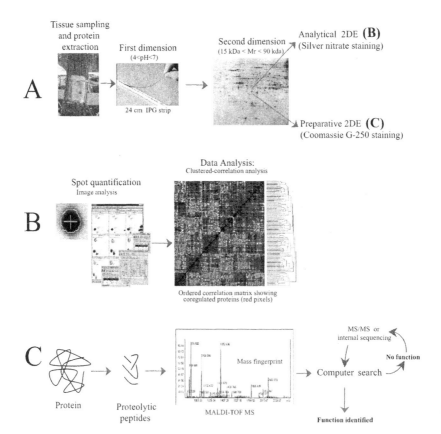

FIGURE 3.1. Schematic illustration of standard proteome analysis by two-dimensional gel electrophoresis (2DE) and mass spectrometry (MS). (A) Protein are extracted and separated by 2DE and stained. (B) Silver nitrate staining is used for analytical 2DE. Images are scanned and spot intensity analyzed using a computer-assisted system. Data are then submitted to statistical analysis (clustered· correlation, analysis of variance, principal components analysis, etc.). (C) Sypro Ruby or Coomassie brilliant blue G-250 are used for preparative 2DE. Spots are excised from the gel, subjected to in-gel digestion with trypsin, and the resulting peptides are analyzed by MS. (See also color photo section.)

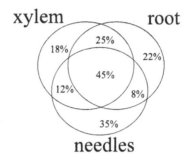

FIGURE 3.2. Tissue-specific and common proteins between xylem, root, and needles of maritime pine

in-depth characterization of several systems before comparative proteomics can be performed.

Protein Identification

The typical strategy for rapid identification of large numbers of proteins has been reviewed by Li and colleagues (1997) and is summarized in Figure 3.1C. A protein spot excised from the gel is digested with a site-specific protease (e.g., trypsin), resulting in a set of peptides. Robotics can be used for spot excision, enzymatic digestion, and loading samples in a mass spectrometer. Two complementary approaches are then used to obtain functional information of the protein.

Peptide Mass Fingerprinting or Peptide Mapping

In this case, the masses of the proteolytic peptides are measured by MALDI-TOF MS (matrix-assisted laser-desorption ionization time-of-flight mass spectrometry) and compared using dedicated search engines (reviewed by Fenyö, 2000; Rowley et al., 2000), with the predicted tryptic masses of each entry in the nucleotide (including ESTs) and protein databases. This approach requires a 100 percent match between the "query" and the homologous peptide sequences and is a powerful method for protein identification in species for which complete (e.g., *Arabidopsis*) or nearly complete genome information (e.g., rice) is available, because the full sequence of all genes is accessible. This approach could easily be extended to small-genome size forest tree species, such as *Populus* and *Eucalyptus*, as soon as their genome sequencing is completed.

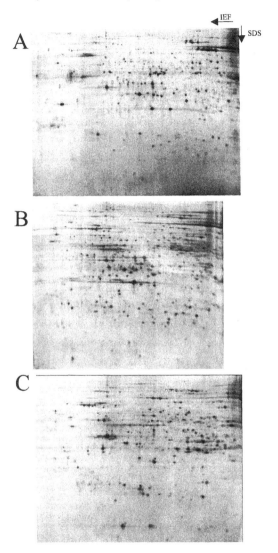

FIGURE 3.3. Example of 2DE gels of differentiating xylem proteins extracted from *Populus* (A), *Eucalyptus* (B), and *Pinus* (C). A detailed protocol for protein extraction, protein quantification (modified Bradford method), isoelectrofocusing (first dimension) using either homemade ampholyte-based gels or 24 cm IPG strips, SDS-PAGE (second dimension), and staining is available at <http://www. pierroton.inra.fr/genetics/2D/Proteomevert/Protocoles/protocole.pierroton.html>.

Peptide Sequence Tags

When a protein cannot be unambiguously identified by the first method, amino acid sequences can be obtained by microsequencing (Edman sequencing) or interpreted from MALDI-TOF postsource decay MS (PSD/ MS) or electrospray ionization tandem MS (ESI-MS/MS) mass profiles (peptide fragmentation). In the later case, individual peptides are screened in the first section of the tandem mass spectrometer, and selected peptides are subsequently fragmented along the protein backbone by collision with inert gas molecules. The masses of the peptide fragments are then measured and compared to theoretical mass spectra calculated from the protein sequences in the databases. Amino acid sequences can finally be deduced from the mass differences between fragments ranked in order size (nicely sketched by Li and Assmann, 2000) and classical homology-based searching software (e.g., BLAST, FastA) are used to search against EST and annotated genomic sequence data. Although ESI-MS/MS is not yet a high-throughput technique, it is more appropriate than MALDI-TOF for the identification of proteins from species for which the genome sequence is not available but for which large collections of partial cDNAs are being obtained (e.g., *Pinus taeda, Pinus pinaster, Robinia pseudoacacia, Cryptomeria japonica*). Küster and colleagues (2001) showed that in the case of the *Arabidopsis* genome, data (masses and sequences) from any two peptides was always sufficient for an unambiguous identification of the protein. This is because it is extremely unlikely that any of the few retrieved sequences, even from short peptides, happen to occur in the same gene by chance.

It should be pointed out that for organisms such as forest trees that have not yet been sequenced, MS-based approaches (MALDI-TOF or ESI-MS/MS) are strongly dependent on the quality and the size of the EST databases.

Quality. EST sequences are often biased toward medium- and high-abundance genes and do not represent all tissues or cell types. In genomic databases, on the other hand, the coding sequences of all proteins are present. To maximize the chance of getting most of the genes in EST databases, sequencing programs need to take all tissues/organs, different developmental stages, and environmental conditions into account. The accession to low-abundance ESTs also requires the sequencing from normalized libraries.

Size. In both strategies (peptide mapping and peptide fragmentation), because ESTs are usually too short to obtain significant protein coverage and sufficient number of matching peptides, correctly assembled ESTs and correctly predicted ORFs are required to maximize the matching probability.

Because very few ESTs were available until recently, microsequencing and ESI-MS/MS were first used to identify the putative function of maritime pine proteins (Costa et al., 1999). Partial internal sequences (6 to 15 amino acids) were obtained for 35 and 28 proteins collected on 2DE gels from photosynthetic (needles) and wood forming (developing secondary xylem) tissues, respectively. Among these 63 proteins, 90 percent could be easily identified, of which 42 percent were already described in conifers. Among the 57 proteins identified, 62 percent corresponded to enzymes and 38 percent to structural proteins (e.g., actin, heat shock protein [HSP], protease), including those involved in carbon fixation, lignification, oxygen radical scavenging, defence, and nitrogen metabolism, but also water stress-inducible proteins and chaperones. The accessibility to pine ESTs (60,226 *Pinus taeda* ESTs, and 8,989 *Pinus pinaster* ESTs are available as of April 2003) will make it possible to characterize proteins via direct searches with peptide masses (peptide mass fingerprints or peptide fragmentation), a less-expensive method to identify protein function compared to Edman sequencing. In a pilot study, the mass profiles of 200 maritime pine xylem proteins are being analyzed.

Protein Database

A proteomic database for maritime pine is accessible on the Internet (http://www.pierroton.inra.fr/genetics/2D/). This database includes scanned gels, clickable images with characterized protein spots highlighted by hyperlinked symbols. Individual protein entries are linked to other protein databases such as OWL (http://www.bioinf.man.ac.uk/dbbrowser/OWL/) by active cross-references. This database contains information in addition to sequence data, including (1) 2DE patterns of various organs (root, needles, bud, pollen) and tissues (megagametophyte, xylem, phloem); (2) location of polymorphic protein markers on a genetic linkage map; and (3) the identification of drought stress-responsive needle proteins, as well as their seasonal variation, and the identification of reaction wood-responsive proteins. In the near future, the protein database will be linked with the EST database with the goal of adding functional annotations to both known and unknown ESTs, such as expression patterns.

GENETIC VARIABILITY OF QUALITATIVE AND QUANTITATIVE PROTEIN VARIANTS

The 1990s brought a dramatic increase in the applications of molecular markers in population genetics studies (Wang and Szmidt, 2001) as well as for generating genetic linkage maps (Cervera et al., 2000) and detecting some of the underlying "genetic factors" controlling quantitative trait variation (namely the QTLs, quantitative trait loci) (Sewell and Neale, 2000). Although DNA-based markers (e.g., amplified fragment length polymorphisms [AFLPs], simple sequence repeats [SSRs]) have provided knowledge about the localization of QTLs and estimating the magnitude of QTLs' effects on quantitative traits, they appear to have a limited value for assessing the relationship between phenotypic variation and known-function genes. Alternatively, proteins revealed by 2DE provide relevant markers for mapping the expressed genome but also have the great advantage of providing candidate proteins to characterize the genes that could correspond to QTLs.

Qualitative Variants

2DE Provides Physiologically Relevant Markers to Map the Expressed Genome

Genetic mapping of protein markers has been reported in very few plant species (reviewed by de Vienne et al., 1996). Compared to other molecular techniques, 2DE is expensive and requires a long and arduous apprenticeship. In addition, the interpretation of the gels requires a tremendous amount of experience, since all the markers remain on the same gel. Two types of polymorphism are observed on 2DE gels: either position shift (PS) variants or presence/absence (P/A) variations. Both were shown to be under monogenic control. PS variants are those cases in which two allelic spots are located relatively close to each other on 2DE gels (e.g., having close pI and/or M_r). They correspond to an allelic difference at the locus (structural gene) coding for the protein, resulting in modification of its primary structure. P/A variants are most likely the result of a polymorphism in a major regulatory element (in *cis* or *trans*) of the structural gene as discussed in Thiellement and colleagues (1999, 2001). Such markers are physiologically relevant as they reveal loci whose transcripts are translated in the organ analyzed.

In maritime pine, Bahrman and Damerval (1989) were the first to report a linkage analysis for 119 protein loci using 56 megagametophytes (a nutritive haploid tissue surrounding the embryo of conifer seeds) of a single tree. Extending this approach, Gerber and colleagues (1993) reported a 65 loci linkage map covering one-fourth of the pine genome, using 18 maritime pine trees with an average of 12 megagametophytes per tree. A more conventional pedigree (inbred F2) was used to map 61 proteins using haploid (Plomion et al., 1995) and diploid (Plomion et al., 1997; Costa et al., 2000) tissues of the same seedlings. In the latter case, protein loci were found on each chromosome (Thiellement et al., 2001; see also <http://www.pierroton. inra.fr/genetics/pinus/>), interspersed among other markers (random amplification of polymorphic DNA [RAPD] and amplified fragment-length polymorphism [AFLP]). Even if protein markers are limited in number, their number could be substantially increased by analyzing some physiologically contrasted organs or tissues of the same individuals, until full coverage of the genome is obtained. Proteins can be partially sequenced and may be recognizable by sequence similarity to other proteins published in sequence databases, therefore providing functional markers expressed in the tissues analyzed. Such maps of expressed genes will be of invaluable help for the candidate gene/protein strategy of QTL characterization.

2DE Provides Interesting Markers to Study the Intraspecific Variability of Proteins

The structure of genetic variability in natural populations has always been a subject of great interest to population geneticists, evolutionists, and plant breeders. It is generally accepted that the choice of a molecular screening technology for analyzing the extent and distribution of genetic diversity in natural populations will depend on many factors (http://webdoc.gwdg.de/ebook/y/1999/whichmarker/). Because each type of marker presents advantages and limitations, a variety of techniques such as 2DE are needed. The strength of the 2DE technique is that protein loci sample the genome differently than most polymerase chain reaction (PCR)-based techniques—2DE reveals the genetic variability of only expressed genes—and therefore provides a different level of information with respect to the diversity of questions being addressed. Intraspecific variability of maritime pine proteins detected by 2DE was demonstrated by Bahrman and colleagues (1994) and Petit and colleagues (1995) who studied the level and structure of diversity of several populations from the natural range of maritime pine. Taking advantage of the possibility to distinguish between allelic forms of protein loci in the megagametophyte, they found that most of the spots revealed by 2DE were variable. The intrapopulation and interpopulation variability levels were of similar magnitude. In addition, proteins were found to display a similar level of genetic differentiation among populations compared to other biochemical markers (terpenes and isozymes).

Applications of Qualitative Protein Variation
in Quantitative Genetics Studies

The use of marker-assisted selection (MAS) in breeding programs relies on the presence of linkage disequilibrium between marker loci and quantitative trait loci. Because linkage disequilibrium decreases in each generation due to recombination, the efficiency of MAS will quickly decline unless markers are found that are physically linked to the QTLs, or in the extreme case, being the QTLs themselves (Zhang and Smith, 1992). The possibility to study the genetic variation (in pedigrees and in natural populations) at the protein level may in this respect be extremely useful. Proteins act directly on biochemical processes, and thus must be closer to the "build up" of the phenotype, compared to DNA-based markers. Therefore, 2DE appears as a very interesting technique to understand the variability in trait expression. In this context, proteins certainly constitute more informative markers compared to DNA markers. In maritime pine, Gerber and colleagues (1997) demonstrated the rationale of this approach. For several traits including seed weight and growth-related traits, they detected significant "protein-trait" associations among the 84 protein loci genotyped on 18 unrelated trees, suggesting some of these proteins to be responsible for the trait variation itself.

Quantitative Variants

The use of dedicated software allowing spot intensity to be quantified and compared between 2DE gels has made it possible to precisely measure the extent of protein accumulation variability and its genetic determinism.

Inheritance of Amounts of Individual Proteins

The inheritance of protein amounts was studied in a three-generation F2 inbred pedigree of maritime pine (Costa and Plomion, 1999). A significant difference of protein amount was found among both parents and their hybrid for 31 percent of the studied proteins, from which 78 percent followed a nonadditive mode of inheritance. For 50 percent of the polypeptides with nonadditive inheritance, the hybrid spot exhibited an intensity similar either to the lowest or to the highest intensity of one of the parental spots. For the other 50 percent of proteins, the hybrid spot showed a greater intensity than the most intense parental spot, in almost all cases. "Dominance" or "positive overdominance" inheritance is likely to reflect interacting complementary regulatory factors inherited from both parents, resulting in a similar or higher F1 hybrid protein synthesis level or increased stability. The low level of additivity in maritime pine (22 percent of the spots) contrasts with the extent of additivity observed in pea (96 percent, de Vienne et al., 1988). As discussed by Costa and Plomion (1999), such a divergence could be explained by a higher genetic load in allogamous forest tree species and a greater opportunity for epistasis between regulatory elements and the gene encoding the structural proteins in pine due to a higher degree of polymorphism.

Genetic Determinism of Quantitative Variation of Proteins

Damerval and colleagues (1994) were the first to investigate the genetic determinism of quantitative variation of proteins separated by 2DE using a QTL detection strategy. They used a linkage map constructed with restriction fragment length polymorphisms (RFLPs) and PS loci segregating in a F2 progeny of maize to locate by interval mapping (Lander and Botstein, 1989) the regulatory factors or PQL (protein quantity loci) that explain part of the spot intensity variation. For the 72 proteins analyzed, 70 PQLs were detected for 42 proteins, 20 proteins having more than one PQL. PQLs were found to be distributed all over the genome. Recently, PQLs controlling the accumulation of maritime pine needle proteins were detected in a F2 progeny (Costa and Plomion, 1999) and the same conclusions were drawn.

The question arises now as to whether or not the variability of genetic expression and its consequences in terms of protein quantity variation may also play a role in the phenotypic variability. Correlations reported between protein quantity and hybrid performances in maize might suggest that PQL could indeed correspond to QTL of agronomical traits (Leonardi et al., 1991). Similar results were observed in maritime pine. In addition, more significant correlations between quantity variation of protein and quantitative traits were observed than between qualitative variations and the same traits (Gerber et al., 1997).

PQLS As a Tool for Characterizing QTLS

Positional candidate genes (CG) whose genetic map locations coincide with QTLs have been described for some plant species (recently reviewed by Pflieger et al., 2001) including forest trees (Gion, 2001) and fruit trees (Etienne et al., 2002); however, very few of them have been isolated and characterized at the molecular level (Frary et al., 2000; Fridman et al., 2000). Indeed, although QTLs can be localized on genetic maps (see review by Paterson, 1995), they provide only an approximate position of the underlying genes, confidence interval of a QTL being large, usually ranging from 10 to 40 cM (Mangin et al., 1994). Considering that the average physical equivalent of 1 cM distance can be large—e.g., 230, 460, 2,500, 3,500, and 12,000 kb in arabidopsis, eucalyptus, maize, wheat, and pine, respectively—it becomes obvious that a colocation between a CG and a QTL can be purely fortuitous. Validation of positional CG will therefore be necessary before such information is implemented in breeding programs. There is usually no direct strategy to demonstrate that the polymorphism of a positional CG actually explains a part of the variation of a trait. Among others, validation procedures include positional cloning (Rommens et al., 1989), complementation test (Doebley et al., 1995), genetic transformation and detailed functional analysis of the genetically modified plants (Ahuja, 2000), and association studies in natural populations (Rothschild and Soller, 1997; Goldstein and Weale, 2001).

An alternative strategy involves the use of proteins revealed by 2DE. Because proteins define the business end of gene expression pathways, a CG can hardly be retained if its product does not display genetic variability of quantity or activity. This means that a necessary (but not sufficient) condition for retaining a CG is that a QTL for the quantity (PQL) and/or the activity (AQL, activity quantity loci) of its product is detected in the chromosomal region exhibiting the apparent colocation between the CG and the trait's QTL (Figure 3.4A). An application of the PQL approach to unravel the biological characterization of agronomical trait's QTL was recently given in maize (de Vienne et al., 1999). A colocation between a CG for drought response (the ASR1 gene, an ABA/water stress/ripening related protein) was colocalized on chromosome 10, with QTLs for xylem sap, ABA content, leaf senescence and anther-silking interval (a symptomatic trait of drought effect), and a major PQL controlling the presence/absence variation of the ASR1 protein under drought condition.

An extension of this approach was also proposed by de Vienne and colleagues (1999). Proteins whose PQL(s) or AQL(s) are colocated with QTL(s), while the CG does not, can identify "candidate proteins" (Figure 3.4B). Again, given the large confidence interval of PQL/QTL or AQL/QTL positions, colocation in a single chromosomal region could just be due to fortuitous linkage. But for the same trait, more than one association observed in at least two different genomic regions would increase the chance of revealing physiological link between the protein and the trait. In this

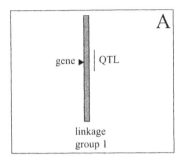 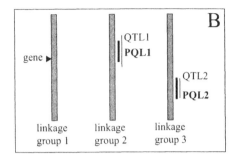

FIGURE 3.4. Illustration of the "candidate gene" (A) and "candidate protein" (B) approaches (*Source:* Adapted from de Vienne et al., 1999)

connection, de Vienne and colleagues (1999) showed that three PQLs controlling the quantity of a single leaf maize protein and three QTLs of growth colocalized in three different chromosomes. As a triple co-location is very unlikely by chance, this result strongly suggests a physiological link between the protein and the trait.

The PQL Approach Applied to Glutamine Synthetase and Juvenile Growth in Maritime Pine

The PQL strategy was applied for the cytosolic glutamine synthetase (GS, EC 6.3.1.2) in maritime pine. In plants, ammonium is assimilated into organic nitrogen mainly through the reaction catalyzed by GS. The amide group from the product of the GS reaction, glutamine, is then transferred to glutamate by the action of the glutamate synthase (GOGAT, EC 1.4.7.1 and 1.4.1.14). This metabolic pathway is of crucial importance, since glutamine and glutamate are the donors for the biosynthesis of major nitrogen-containing compounds, including amino acids, nucleotides, chlorophylls, polyamines, and alkaloids. Single-strand conformation polymorphism (SSCP) mapping allowed to map one locus of GS on linkage group #2 (Plomion et al., 1999) of a saturated maritime pine genetic map (Plomion et al., 1995). Whereas, two isoforms of the proteins (GS1a and GS1b) were identified on 2DE gels from needles by microsequencing (Costa et al., 1999). For both isoforms, a position shift of the protein was observed between the two parents of an F2 inbred pedigree that segregated 1:2:1 in the F2s. Interestingly, both protein loci exactly cosegregate, i.e., no recombinant could be found when 68 F2 individuals were analyzed by 2DE (Avila et al., 2000). In addition, both GS proteins were found to cosegregate with the GS gene mapped using the

PCR-based SSCP approach. A QTL for early height growth was previously located in the same genomic region of linkage group #2 (Plomion et al., 1996), making GS a "positional" CG for juvenile growth. To validate this positional CG, the amount of GS was quantified in the 68 individuals of the progeny, and one major PQL was detected, the most likely position of which corresponded to the expressed locus on linkage group #2 (Plomion et al., unpublished). The apparent colocation between the gene, the protein, the QTL, and the PQL constitutes a strong indication that GS is actually involved in the genetic control of juvenile growth variation in maritime pine. This hypothesis received additional support from a genetic engineering experiment. Compared to control plants, genetically modified poplar clones overexpressing the pine GS showed an average height increase of 76 percent and 21.3 percent at two and six months, respectively (Gallardo et al., 1999). This example demonstrates the interest and explicative potential of the PQL/candidate protein approach.

VARIABILITY OF PROTEOME EXPRESSION IN PHYSIOLOGICAL STUDIES

Differential genome expression between organs, developmental or physiological stages, and contrasted abiotic or biotic conditions is the subject of many studies in plant genomics (e.g., Plomion et al., 2000). Parallel to mRNA level studies, the identification of key proteins has emerged as a requirement for any physiological understanding of the patterns of gene expression and regulation. In this section we will review the work that is being conducted in maritime pine, aiming at identifying and characterizing drought stress-responsive proteins and determining those that are implicated in drought tolerance.

Breeding Trees for Drought Stress Adaptation

Compared to annual crops, genetic improvement of forest trees is a very long and time consuming process. From the selection to the harvesting of improved varieties, via seed production in seed orchards, it may take more than half a century for fast-growing conifers. On this time scale, climatologists have predicted a global climate change with an increase in temperature (Hughes, 2000). The maintenance of sufficient growth and productivity under drought stress conditions will therefore require the availability of varieties or natural resources that are adapted to present and future climatic conditions if major losses are to be avoided. At the intraspecific level, the genetic improvement of such characteristics has not yet been considered in forest tree breeding programs, primarily due to the complex nature of environmental stress tolerance and the time lag required for a breeding cycle to take place. However, application of genomic technology to forest trees would be very helpful in speeding up breeding and facilitating the creation of tolerant varieties, either by marker-assisted selection or genetic engineering.

Identification of Drought Stress-Responsive Proteins in Maritime Pine

If changes in gene/protein expression leading to the synthesis and activation of proteins is well documented in model organisms such as *Arabidopsis thaliana* and *Craterostigma plantagineum* (Cushman and Bohnert, 2000), very little is known about the biochemical and molecular mechanisms involved in drought response in forest trees. Given the characteristics of these species (long life span, size of such organisms), it is possible that the problems aroused by water-deficit stress have been solved by strengthening specific biochemical pathways compared to those of annual plants. If the same genes are likely to be expressed in annual and perennial organisms, it is possible that they differ in the way in which they are regulated. We therefore think that the genetic improvement of tree growth and productivity in drought-stressed environments cannot be derived solely from studies of model organisms but requires an understanding of the molecular mechanisms involved in stress responses and stress adaptation in trees. In this context, and similar to the work carried out at the transcriptome level (random EST sequencing and differential display of cDNAs; Dubos and Plomion, 2001, 2002), 2DE is used to identify which biochemical pathways are changed via observed alterations in sets of drought stress-responsive proteins.

A first experiment was conducted on eight maritime pine genotypes to identify drought stress-responsive proteins during a progressive water deprivation on two-year-old plants raised on the ground (Costa et al., 1998). Stress was applied during several weeks by withholding water during vegetative growth. The experiment was designed to separate the drought stress-responsive proteins to those being underexpressed or overexpressed during the season: needles were sampled before (predawn water potential of –0.5 MPa), during (down to –2 MPa), and after (back to –0.5 MPa) the stress. Out of approximately 1,000 needle spots that were quantified, 38 presented highly significant intensity variations during stress. Some proteins were expressed de novo while others accumulated or were suppressed. The internal microsequences of some water-deficit responsive polypeptides allowed the characterization of their function and their putative role in the stress. A modification of polypeptide abundance was observed for a diverse set of proteins, including those involved in photosynthesis (Rubisco activase), cell elongation (actin), antioxidative systems (superoxide dismutase, glutathione peroxidase), and lignification (CoA-*O*-methyltransferase). Although we did not sequence all the proteins varying during the stress, none of the sequenced proteins showed sequence homology with genes that have been commonly found to be implicated in desiccation stress, e.g., dehydrin, late embryogenesis abundant proteins *(lea)* or abscisic acid responsive *(rab)* genes.

Based on this first screening, a second experiment (Costa, 1999) was set up to study fewer proteins (163 spots chosen among the previous set and likely to be drought responsive) in more genotypes (120). Needles were sampled before the stress and after six weeks of water deprivation. Seventy-three percent (120 spots) of the studied spots showed significant differences of protein accumulation between the control and stress treatment, showing that the applied stress had a strong effect on needle protein expression. We observed that 62.5 percent (75 spots) were down-regulated, whereas 37.5 percent (45) were up-regulated. Clustered-correlation analysis (as exemplified in Figure 3.1B) was used to classify proteins according to their expression profiles, revealing gene products which can be under the same control mechanism (Costa, 1999). Based on known function proteins (one-fourth of the proteins were of known function), we identified biologically meaningful clusters, such as (1) proteins corresponding to specific gene products (gene family), (2) proteins characterized by a similar function (same metabolic pathway), (3) proteins located in the same cellular compartment, and (4) proteins involved in different metabolic pathways but showing similar patterns of regulation in the two tested environmental conditions. The possibility of obtaining regulatory correlations and anticorrelations between proteins provides us with a new category of homology ("regulatory homology") in tracing functional relationships and possibly inferring protein function of unknown proteins and identifying partner proteins.

A third experiment was conducted in a growth chamber on young seedlings raised in hydroponic solution to facilitate the accessibility to the root system (Dubos, 2000). Osmotic stress was applied on three week-old seedlings for a duration of three weeks by lowering the osmotic potential of the nutrient solution using high molecular weight polyethylene glycol. Protein accumulation was quantified on 2DE gels from roots, stems, and needles in the control (–0.08 MPa) and stress (–0.45 MPa) conditions. A total of 1,171 protein spots were studied in the three tissues, from which 11.8 percent showed significant differences in quantity between both treatments (Table 3.1). In most cases, the accumulation of proteins presented a minimum two-

TABLE 3.1. Number of studied spots in the three tissues and water deficit-responsive proteins classified according to their behavior in the stress condition (quantitative variation, spot appearance, or spot disappearance)

Tissues	Number of spots	Number of water deficit-responsive proteins	Behavior			
			Increase		Decrease	
			Quantitative variation	Spot appearance*	Quantitative variation	Spot disappearance*
Needle	424	34	16	11	18	5
Stem	361	51	32	8	19	5
Root	386	53	32	20	21	4

*Does not indicate that the protein expression is turned on or off, but rather that the quantity of the protein is low or is no longer detected by silver staining.

fold increase or decrease between both conditions (Figure 3.5). We are now focusing on the characterization of some of these proteins by tandem MS.

These studies provide us with better insights into the nature of the molecular machinery involved in drought stress response in pine species. Whether the drought-responsive proteins that were identified actually play important roles in drought tolerance remains to be determined. Our strategy establishes a causal relationship between protein expression and tolerance to drought using a "forward genetics" approach (Costa, 1999). The idea is that if the variability of a relevant physiological trait is at least partially due to the accumulation of a protein, then the gene(s) causing the genetic variation of the protein accumulation should cause the variation of the phenotypic trait. Thus, a locus that affects protein quantity should also affect the phenotypic trait. In other words, PQL(s) and QTL(s) detected on the same segregating population should be found at the same location on a genetic map. In this context, we used an F2 pedigree of maritime pine (200 two-year-old plants) to map QTLs for a series of quantitative traits (stomatal conductance, maximum photosynthesis, carbon isotope discrimination, water consumption, predawn water potential, relative water content, osmotic potential, transpiration, height growth, and biomasses) and PQLs of drought-inducible proteins. In respect to the PQL analysis, out of 163 needle proteins studied, one to four PQLs were detected for 138 proteins. They were distributed all over the genome. Determination coefficients of each PQL were rather high. Interestingly, for a given protein, PQLs detected in the non-stressed condition did not generally correspond to those detected in the drought-stressed condition. In respect to the QTL analysis, most QTLs were mapped in three distinct

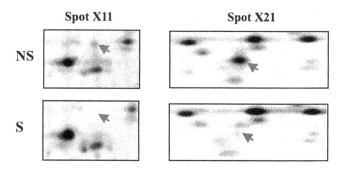

FIGURE 3.5. Detailed portion of an area of silver-stained 2DE gels from the stem of nonstressed (NS) and stressed (S) seedlings growing in hydroponic solution. Spots X11 and X21 (indicated by the arrows) were sequenced and were found to belong to the SAM-synthase gene family. The accumulation of the protein decreases for both members in the stress condition.

chromosomal regions. Colocations between QTLs and PQLs were identified in these three QTL "hot spots." Sequencing of the corresponding proteins should contribute to our understanding of these coincidences and might be helpful in making the difference between tight linkage and pleiotropy if the biochemical function is consistent with the phenotypic effect.

CONCLUSION

As can be deduced from the examples given in this chapter, proteomics has become an essential field of research in maritime pine biology, in which the combined approaches of genetics, physiology, and molecular biology will in the coming years provide critical tools to understand the mechanisms underlying plant growth and development. It can be expected that forest tree proteomics will become a very active field for many species and constitutes the next challenge for the coming years in forest tree genomics. This will be the consequence of the development of (1) the complete sequences of the *Populus* genome, (2) extensive EST sequence data in conifers and eucalyptus, and (3) the increasing use of user-friendly 2DE systems and mass spectrometers. In particular, the recent progress in MS-based quantitative proteomics (Gygi et al., 2000; Han et al., 2001) allows for the rapid and accurate quantification of differences in the level of protein expression. This technique makes it possible to quantify proteins that are excluded and are either underrepresented or not represented on 2DE gel patterns (i.e., very acidic and basic proteins, excessively large and small proteins, low-abundance proteins). By alleviating some of the limitations of 2DE (inability to detect rare proteins, difficulty of comparing gels), this technology should contribute to the deployment of proteome analysis.

However, proteomics alone may not provide complete insight into the mechanisms that establish protein expression patterns. This is because important regulatory proteins such as transcription factors and signalling proteins are usually not visualized on 2DE gels due to their low abundance. Therefore, proteomics will be most useful when combined with other functional genomics tools and approaches such as transcriptomics. Parallel studies of proteomes and transcriptomes should make it possible to understand the relationship between mRNA and protein levels and to answer the questions posed in turn by large-scale proteomic studies about the genes and proteins involved in the regulation of genome expression, from transcription to posttranslational processes. Finally, if the information obtained from DNA, RNA, and protein-based genomics studies will clearly provide valuable information for the understanding of the functioning of any organism, the measurement of RNA and protein levels will not be sufficient because of the intricacies of metabolic networks. The technology to directly measure the chemical constituents of plants is now being developed (Fiehn, 2002) and, in combination with the precise quantification of gene/protein expression, is likely to provide intensely accurate and meaningful views of the functioning of cells, tissues, and organisms.

REFERENCES

Adessi, C., Miege, C., Albrieux, C., and Rabilloud, T. (1997). Two-dimensional electrophoresis of membrane proteins: A current challenge for immobilized pH gradients. *Electrophoresis* 18: 127-135.

Ahuja, M.R. (2000). Genetic engineering of forest trees. In S.M. Jain and S.C. Minocha (Eds.), *Molecular Biology of Woody Plants* (pp. 31-50). Dordrecht, The Netherlands: Kluwer Academic Publishers.

Allona, I., Quinn, M., Shoop, E., Swope, K., Cyr, S.S., Carlis, J., Riedl, J., Retzel, E., Campbell, M.M., Sederoff, R., and Whetten, R.W. (1998). Analysis of xylem formation in pine by cDNA sequencing. *Proc. Natl. Acad. Sci. USA* 95: 9693-9698.

Anderson, L. and Seilhammer, J. (1997). A comparison of selected mRNA and protein abundances in human liver. *Electrophoresis* 18: 533-537.

Arabidopsis Genome Initiative (2000). Analysis of the genome sequence of the flowering plant *Arabidopsis thaliana*. *Nature* 408: 796-815.

Avila, C., Muñoz-Chapuli, R., Plomion, C., Frigerio, J-M., and Cánovas, F.M. (2000). Two genes encoding distinct cytosolic glutamine synthetases are closely located in the pine genome. *Febs Lett.* 477: 237-243.

Bahrman, N. and Damerval, C. (1989). Linkage relationships of loci controlling protein amounts in maritime pine (*Pinus pinaster* Ait). *Heredity* 63: 267-274.

Bahrman, N. and Petit, R.J. (1995). Genetic polymorphism in maritime pine (*Pinus pinaster* Ait.) assessed by two-dimensional gel electrophoresis of needle, bud and pollen proteins. *J. Mol. Evol.* 41: 231-237.

Bahrman, N., Plomion, C., Petit, R.J., and Kremer, A. (1997). Contribution to maritime pine genetics using protein revealed by two-dimensional electrophoresis. *Ann. Sci. For.* 54: 225-236.

Bahrman, N., Zivy, M., Damerval, C., and Baradat, P. (1994). Organization of the variability of abundant proteins in seven geographical origins of maritime pine (*Pinus pinaster* Ait). *Theor. Appl. Genet.* 88: 407-411.

Barreneche, T., Bahrman, N., and Kremer, A. (1996). Two dimensional gel electrophoresis confirms the low level of genetic differentiation between *Quercus robur* L. and *Quercus petraea* (Matt) Liebl. *For. Genet.* 3: 89-92.

Cervera, M.T., Plomion, C., and Malpica, C. (2000). Molecular markers and genome mapping in woody plants. In S.M. Jain and S.C. Minocha (Eds.), *Molecular Biology of Woody Plants* (pp. 375-394). Dordrecht, The Netherlands: Kluwer Academic Publishers.

Chang, W.W., Huang, L., Shen, M., Webster, C., Burlingame, A.L., and Roberts, J.K. (2000). Patterns of protein synthesis and tolerance of anoxia in root tips of maize seedlings acclimated to low-oxygen environment, and identification of proteins by mass spectrometry. *Plant Physiology* 122: 295-318.

Chevallet, M., Santoni, V., Poinas, A., Rouquié, D., Kieffer, S., Rossignol, M., Lunardi, J., Garin, J., and Rabilloud, T. (1998). New Zwitterionic detergents improve the analysis of membrane proteins by two-dimensional electrophoresis. *Electrophoresis* 19: 1901-1909.

Costa., P. (1999). Réponse moléculaire, physiologique et génétique du pin maritime à une contrainte hydrique. [Molecular, physiological, and genetic response of maritime pine to drought.] PhD thesis. University of Nancy, Nancy, France.

Costa, P., Bahrman, N., Frigerio, J-M., Kremer, A., and Plomion, C. (1998). Water-deficit-responsive proteins in maritime pine. *Plant. Mol. Biol.* 38: 587-596.

Costa, P., Pionneau, C., Bauw, G., Dubos, C., Bahrman, N., Kremer, A., Frigerio, J-M., and Plomion, C. (1999). Separation and characterization of needle and xylem maritime pine proteins. *Electrophoresis* 20: 1098-1108.

Costa, P. and Plomion, C. (1999). Genetic analysis of needle protein in maritime pine. 2. Quantitative variation of protein accumulation. *Silvae Genetica* 48: 146-150.

Costa, P., Pot, D., Dubos, C., Frigerio, J-M., Pionneau, C., Bodénès, C., Bertocchi, E., Cervera, M-T., Remington, D.L., and Plomion, C. (2000). A genetic map of maritime pine based on AFLP, RAPD and protein markers. *Theor. Appl. Genet.* 100: 39-48.

Cushman, J.C. and Bohnert, H. (2000). Genomic approaches to plant stress tolerance. *Cur. Opin. Plant Biol.* 3: 117-124.

Damerval, C., de Vienne, D., Zivy, M., and Thiellement, H. (1986). Technical improvements in two-dimensional electrophoresis increase the level of genetic variation detected in wheat-seedling proteins. *Electrophoresis* 7: 52-54.

Damerval, C., Maurice, A., Josse, J.M., and de Vienne, D. (1994). Quantitative trait loci underlying gene product variation: A novel perspective for analyzing regulation of genome expression. *Genetics* 137: 289-301.

de Vienne, D., Burstin, J., Gerber, S., Leonardi, A., Le Guilloux, M., Murigneux, A., Beckert, M., Bahrman, N., Damerval, C., and Zivy, M. (1996). Two-dimensional electrophoresis of proteins as a source of monogenic and codominant markers for population genetics and mapping the expressed genome. *Heredity* 76: 166-177.

de Vienne, D., Leonardi, A., and Damerval, C. (1988). Genetic aspects of variation of protein amounts in maize and pea. *Electrophoresis* 9: 742-750.

de Vienne, D., Leonardi, A., Damerval, C., and Zivy, M. (1999). Genetics of proteome variation for QTL characterization: Application to drought-stress responses in maize. *J. Exp. Bot.* 50: 303-309.

Doebley, J., Stec, A., and Gustus, C. (1995). Teosinte *branched1* and the origin of maize: Evidence for epistasis and the evolution of dominance. *Genetics* 141: 559-570.

Dubos, C. (2000). Réponse moléculaire de jeunes plants de pin maritime soumis à un stress hydrique en condition hydroponique. [Molecular response of young maritime pine seedlings subjected to water deficit in hydroponic conditions.] PhD thesis. University of Nancy, Nancy, France.

Dubos, C. and Plomion, C. (2001). Drought differentially affects expression of a PR-10 protein, in needles of maritime pine (*Pinus pinaster* Ait) seedlings. *J. Exp. Bot.* 358: 1143-1144.

Dubos, C. and Plomion, C. (2003). Identification of water-deficit responsive genes in maritime pine (*Pinus pinaster* Ait.) roots. *Plant. Mol. Biol.* 51: 249-262.

Etienne, C., Rothan, C., Moing, A., Plomion, C., Bodénès, C., Svanella-Dumas, L., Cosson, P., Pronier, V., Monet, R., and Dirlewanger, E. (2003). Candidate genes and QTLs for sugar and organic acid content in peach [*Prunus persica* (L.) Batsch]. *Theor. Appl. Genet.* 105: 145-159.

Fenyö, D. (2000). Identifying the proteome: Software tools. *Cur. Op. Biotech.* 11: 391-395.

Ferro, M., Seigneurin-Berny, D., Rolland, N., Chapel, A., Salvi, D., Garin, J., and Joyard, J. (2000). Organic solvent extraction as versatile procedure to identify hydrophobic chloroplast membrane proteins. *Electrophoresis* 21: 3517-3526.

Fiehn, O. (2002). Metabolomics – The link between genotypes and phenotypes. *Plant. Mol. Biol.* 48: 155-171.

Frary, A., Nesbitt, T.C., Grandillo, S., Knaap, E., Cong, B., Liu, J., Meller, J., Elber, R., Alpert, K.B., and Tanksley, S.D. (2000). *fw2.2*: A quantitative trait locus key to the evolution of tomato fruit size. *Science* 289: 85-88.

Fridman, E., Pleban, T., and Zamir, D. (2000). A recombinant hotspot delimits a wild QTL for tomato sugar content to 484 bp within an invertase gene. *Proc. Natl. Acad. Sci. USA* 97: 4718-4723.

Gallardo, F., Jianming F., Canton, F.R., Garcia-Gutiérrez, A., Canovas, F.M., and Kirby, E.G. (1999). Expression of a conifer glutamine synthase gene in transgenic poplar. *Planta* 210: 19-26.

Gerber, S., Lascoux, M., and Kremer, A. (1997). Relation between protein markers and quantitative traits in maritime pine (*Pinus pinaster* Ait.). *Silvae Genetica* 46: 286-290.

Gerber, S., Rodolphe, F., Bahrman, N., and Baradat, P. (1993). Seed-protein variation in maritime pine (*Pinus pinaster* Ait) revealed by two-dimensional electrophoresis: Genetic determinism and construction of a linkage map. *Theor. Appl. Genet.* 85: 521-528.

Gion, J-M. (2001). Etude de l'architecture génétique de caractères quantitatifs chez l'eucalyptus: Des marqueurs anonymes aux gènes candidats. [Genetic architecture of qualitative traits in eucalyptus: From anonymous markers to candidate genes.] PhD thesis. University of Rennes, Rennes, France.

Goldstein, D.B. and Weale, M.E. (2001). Population genomics: Linkage disequilibrium holds the key. *Cur. Biol.* 11: 576-579.

Görg, A., Obermaier, C., Boguth, G., Harder, A., Scheibe, B., Wildgruber, R., and Weiss, W. (2000). The current state of two-dimensional electrophoresis with immobilized pH gradient. *Electrophoresis* 21: 1037-1053.

Gygi, S.P., Rist, B., and Aebersold, R. (2000). Measuring gene expression by quantitative proteome analysis. *Cur. Op. Biotech.* 11: 396-401.

Gygi, S.P., Rochon, Y., Franza, B.R., and Aebersold, R. (1999). Correlation between protein and mRNA abundance in yeast. *Mol. Cell. Biol.* 19: 1720-1730.

Han, D.K., Eng, J., Zhou, H., and Aebersold, R. (2001). Quantitative profiling of differentiation-induced microsomal proteins isotope-coded affinity tags and mass spectrometry. *Nat. Biotech.* 19: 946-951.

Harvey Millar, A., Sweetlove, L.J., Giegé, P., and Leaver, C.J. (2001). Analysis of the arabidopsis mitochondrial proteome. *Plant Physiology* 127: 1711-1727.

Hughes, L. (2000). Biological consequences of global warming: Is the signal already apparent? *Trends. Ecol. Evol.* 15: 56-61.

Kersten, B., Bürkle, L., Kuhn, E., Giavalisco, P., Konthur, Z., Lueking, A., Walter, G., Eickhoff, H., and Schneider, U. (2002). Large-scale plant proteomics. *Plant. Mol. Biol.* 48: 133-141.

Klose, J. (1975). Protein mapping by combined isoelectric focusing and electrophoresis of mouse tissues. A novel approach to testing for induced point mutations in mammals. *Humangenetik* 26: 231-243.

Kruft, V., Eubel, H., Jänsch, L., Werhahn, W., and Braun H-P. (2001). Proteomic approach to identify novel mitochondrial proteins in arabidopsis. *Plant Physiology* 127: 1694-1710.

Küster, B., Mortensen, P., Andersen, J.S., and Mann, M. (2001). Mass spectrometry allows direct identification of proteins in large genomes. *Proteomics* 1: 641-650.

Lander, E.S. and Botstein, D. (1989). Mapping mendelian factors underlying quantitative traits using RFLP linkage maps. *Genetics* 121: 185-199.

Leonardi, A., Damerval, C., Hebert, Y., Gallais, A., and de Vienne, D. (1991). Association of protein amount polymorphism (PAP) among maize lines with performances of their hybrids. *Theor. Appl. Genet.* 82: 552-560.

Li, G. and Assmann, S.M. (2000). Mass spectrometry. An essential tool in proteome analysis. *Plant Physiology* 123: 807-809.

Li, G., Waltham, M., Anderson, N.L., Unsworth, E., Treston, A., and Weinstein, J.N. (1997). Rapid mass spectrometric identification of proteins from two-dimensional polyacrylamide gels after in gel proteolytic digestion. *Electrophoresis* 18: 382-390.

Mangin, B., Goffinet, B., and Rebaï, A. (1994). Constructing confidence intervals for QTL location. *Genetics* 138: 1301-1308.

O'Farrell, P.H. (1975). High resolution two-dimensional electrophoresis of proteins. *J. Biol. Chem.* 250: 4007-4021.

Panter, S., Thomson, R., de Bruxelles, G., Laver, D., Trevaskis, B., and Udvardi, M. (2000). Identification with proteomics of novel proteins associated with peribacteroid membrane of soybean root nodules. *Plant Mol. Microbe Interact.* 13: 325-333.

Paterson, A.H. (1995). Molecular dissection of quantitative traits: Progress and prospects. *Genome Res.* 5: 321-333.

Peltier, J.B., Friso, G., Kalume, D.E., Roepstorff, P., Nilsson, F., Adamska, I., and van Wijk, K.J. (2000). Proteomics of the chloroplast: Systematic identification and targeting analysis of lumenal and peripheral thylakoid proteins. *Plant Cell* 12: 319-342.

Petit, R.J., Bahrman, N., and Baradat, P. (1995). Comparison of genetic differentiation in maritime pine (*Pinus pinaster* Ait.) estimated using isozyme, total protein and terpenic loci. *Heredity* 75: 382-389.

Pflieger, S., Lefebvre, V., and Causse, M. (2001). The candidate gene approach in plant genetics: A review. *Mol. Breed.* 7: 275-291.

Pleissner, K-P., Oswald, H., and Wegner, S. (2000). Image analysis of two-dimensional gels. In M. Dunn and S. Pennington (Eds.), *Proteomics* (pp. 131-149). Oxford, UK: Bios.

Plomion, C., Bahrman, N., Durel, C-E., and O'Malley, D.M. (1995). Genomic analysis in *Pinus pinaster* (maritime pine) using RAPD and protein markers. *Heredity* 74: 661-668.

Plomion, C., Costa, P., and Bahrman, N. (1997). Genetic analysis of needle protein in maritime pine. 1. Mapping dominant and codominant protein markers assayed on diploid tissue, in a haploid-based genetic map. *Silvae Genetica* 46:161-165.

Plomion, C., Durel, C-E., and O'Malley, D. (1996). Genetic dissection of height in maritime pine seedlings raised under accelerated growth condition. *Theor. Appl. Genet.* 93: 849-858.

Plomion, C., Hurme, P., Frigerio, J-M., Ridolphi, M., Pot, D., Pionneau, C., Avila, C., Gallardo, F., David, H., Neutlings, G., et al. (1999). Developing SSCP markers in two *Pinus* species. *Mol. Breed.* 5: 21-31.

Plomion, C., Pionneau, C., Brach, J., Costa, P., and Baillères, H. (2000). Compression wood-responsive proteins in developing xylem of maritime pine (*Pinus pinaster* Ait.). *Plant Physiology* 123: 959-969.

Prime, T.A., Sherrier, D.J., Mahon, P., Packman, L.C., and Dupree, P. (2000). A proteomic analysis of organelles from *Arabidopsis thaliana*. *Electrophoresis* 21: 3488-3499.

Roberts, J.K.M. (2002). Proteomics and a future generation of plant molecular biologists. *Plant. Mol. Biol.* 48: 143-154.

Rommens, J.M., Iannuzzi, M.C., Kerem, B.S., Drumm, M.L., Melmer, G., Dean, M., Rozmahel, R., Cole, J.L., Kennedy, D., Hidaka, N., et al. (1989). Identification of the cystic fibrosis gene: Chromosome walking and jumping. *Science* 245: 1059-1065.

Rossignol, M. (2001). Analysis of the plant proteome. *Cur. Op. Biotech.* 12: 131-134.

Rothschild, M.F. and Soller, M. (1997). Candidate gene analysis to detect genes controlling traits of economic importance in domestic livestock. *Probe* 8: 13-19.

Rouquié, D., Peltier, J-B., Marquis-Mansion, M., Tournaire, C., Doumas, P., and Rossignol, M. (1997). Construction of a directory of tobacco plasma membrane proteins by combined two-dimensional gel electrophoresis and protein sequencing. *Electrophoresis* 18: 654-660.

Rowley, A., Choudharry, J.S., Marzioch, M., Ward, M.A., Weir, M., Solari, R.C., and Blackstock, W.P. (2000). Applications of protein mass spectrometry in cell biology. *Methods* 20: 383-397.

Santoni, V., Rabilloud, T., Doumas, P., Rouquié, D., Mansion, M., Kieffer, S., Garin, J., and Rossignol, M. (1999). Towards the recovery of hydrophobic proteins on two-dimensional electrophoresis gels. *Electrophoresis* 20: 705-711.

Santoni, V., Rouquié, D., Doumas, P., Mansion, M., Boutry, M., Déhais, P., Sahnoun, I., and Rossignol, M. (1998). Use of proteome strategy for tagging proteins present at the plasma membrane. *Plant J.* 16: 633-641.

Sewell, M.M. and Neale, D.B. (2000). Mapping quantitative traits in forest trees. In S.M. Jain and S.C. Minocha (Eds.), *Molecular Biology of Woody Plants* (pp. 407-424). Dordrecht, The Netherlands: Kluwer Academic Publishers.

Sterky, F., Regan, S., Karlsson, J., Hertzberg, M., Rodhde, A., Holmberg, A., Amini, B., Bhaleraos, R., Larsson, M., Villarroel, R., et al. (1998). Gene discovery in the wood-forming tissues of poplar: Analysis of 5,692 expressed sequence tags. *Proc. Natl. Acad. Sci. USA* 95: 13330-13335.

Thiellement, H., Bahrman, N., Damerval, C., Plomion, C., Rossignol, M., Santoni, V., de Vienne, D., and Zivy, M. (1999). Proteomics for genetical and physiological studies in plants. *Electrophoresis* 20: 2013-2026.

Thiellement, H., Plomion, C., and Zivy, M. (2001). Proteomics as a tool for plant genetics and breeding. In M. Dunn and S. Pennington (Eds.), *Proteomics* (pp. 289-309). Oxford, UK: Bios.

Urquhart, B.L., Atsalos, T.E., Roach, D., Basseal, D.J., Bjellqvist, B.J., Britton, W.L., and Humphery-Smith, I. (1997). Proteomic contigs' of *Mycobacterium tuberculosis* and *Mycobacterium bovis* (BCG) using novel immobilised pH gradients. *Electrophoresis* 18: 1384-1392.

van Wijk, K.J. (2000). Proteomics of the chloroplast: Experimentation and prediction. *Trends Plant Sci.* 5: 420-425.

Voiblet, C., Duplessis, S., Encelot, N., and Martin, F. (2001). Identification of symbiosis-regulated genes in *Eucalyptus globulus-Pisolithus tinctorius* ectomycorrhiza by differential hybridization of arrayed cDNAs. *Plant J.* 25: 181-192.

Wang, X.R. and Szmidt, A.E. (2001). Molecular markers in population genetics of forest trees. *Scand. J. For. Res.* 16: 199-220.

Washburn, M.P., Wolters, D., and Yates, J.R. (2001). Large-scale analysis of the yeast proteome by multidimensional protein identification technology. *Nat. Biotech.* 19: 242-246.

Wasinger, V.C., Bjellqvist, B.J., and Humphery-Smith, I. (1997). Proteomic "contigs" of *Ochrobactrum anthropi*, application of extensive pH gradients. *Electrophoresis* 18: 1373-1383.

Wilkins, M.R., Sanchez, J.C., Gooley, A.A., Apel, R.D., Humphery-Smith, I., Hochstrasser, D.F., and Williams, K.L. (1995). Progress with proteome projects: Why all proteins expressed by a genome should be identified and how to do it. *Biotechnol. Genet. Eng. Rev.* 13: 19-50.

Yates, J.R. III (1998). Mass spectrometry and the age of the proteome. *J. Mass Spectrom.* 33: 1-19.

Zhang, W. and Smith, C. (1992). Computer simulation of marker-assisted selection utilizing linkage disequilibrium. *Theor. Appl. Genet.* 83: 813-820.

Zivy, M. and de Vienne, D. (2000). Proteomics: A link between genomics, genetics and physiology. *Plant. Mol. Biol.* 44: 575-580.

Chapter 4

Exploring the Transcriptome
of the Ectomycorrhizal Symbiosis

Francis Martin
Sébastien Duplessis
Annegret Kohler
Denis Tagu

INTRODUCTION

A large and diverse community of microorganisms proliferates in the rhizosphere where they compete and interact with one another and with plant roots. Within this microbial community, mycorrhizal fungi are almost ubiquitous. Their vegetative mycelium and root tips form a mutualistic symbiosis that is the site of intense nitrogen, phosphorus, and carbon transfers. The various mycorrhizal associations allow terrestrial plants to colonize and grow efficiently in suboptimal environments. Among the various types of mycorrhizal symbioses, arbuscular endomycorrhiza, ectomycorrhiza, or ericoid associations are found on most annual and perennial plants (Smith and Read, 1997). About two-thirds of these plants are symbiotic with endomycorrhizal glomalean fungi. Ericoid mycorrhiza are ecologically important but mainly restricted to heathlands (Read, 1991). Although a relatively small number of plant species (e.g., oaks, pines) develop ectomycorrhiza, they dominate forest ecosystems in boreal, temperate, and montane regions (Read, 1991). In the different mycorrhizal associations, the hyphal networks prospecting the soil or growing in host roots are active metabolic structures that provide essential nutrient resources (e.g., phosphate and

---

We would like to thank M. Chalot, P. E. Courty, C. Delaruelle, Y. Deville, N. Encelot, M. Peter, and C. Voiblet for their collaboration and Dr. Anders Tunlid for providing us with preprints before publication. Investigations carried out in our laboratory were supported by grants from the Institut National de la Recherche Agronomique (INRA) Collaborative Research Programs in Microbiology, the INRA Lignome Genome Initiative, and the Région Lorraine.

amino acids) to the host plant (Martin and Botton, 1993; Smith and Read, 1997; Koide and Kabir, 2000). These nutrient contributions are reciprocated by the provision of a stable carbohydrate-rich niche in the roots for the fungal partner, making the relationship a mutualistic symbiosis (Nehls, Mikolajewski, et al., 2001). Anatomical features, such as the extension of the soil hyphal web or the symbiotic interface, resulting from the development of the symbiosis are of paramount importance to the ecophysiological fitness of the mature mycorrhiza (Fitter et al., 1998). The ecological performance of mycorrhizal fungi is a complex phenotype affected by many different genetic traits and by biotic and abiotic environmental factors (Robinson and Fitter, 1999; Daniell et al., 1999; Leake, 2001).

THE ECTOMYCORRHIZAL SYMBIOSIS

The first mycorrhizal associations must have been derived from earlier types of plant-fungus interactions, such as endophytic fungi in the bryophyte-like precursors of vascular plants (Redecker et al., 2000). Structures similar to arbuscular mycorrhiza have been observed in plant fossils from the Early Devonian (Selosse and Le Tacon, 1998), whereas fossil ectomycorrhiza have been found in the middle Eocene (LePage et al., 1997). The ectomycorrhizal symbiosis has evolved repeatedly over the last 130 to 180 million years and has had major consequences for the diversification of both the mycobionts and their hosts (Hibbett et al., 2000). Coniferous trees and some deciduous trees form ectomycorrhizal symbioses with a large number of basidiomycetes (e.g., agarics, boletes) and a few ascomycetes (e.g., truffles). The ectomycorrhizal fungi are not a phylogenetically distinct group, but an assemblage of very different fungal species that have independently developed a symbiotic lifestyle. The switch between saprobic and mycorrhizal lifestyles probably happened convergently, and perhaps many times, during evolution of these fungal lineages (Hibbett et al., 2000). This may have facilitated evolution of ectomycorrhizal lineages with a broad range of physiological and ecological functions reflecting partly the activities of their disparate saprotrophic ancestors. These symbioses have had major consequences for the ecological diversification of both the mycobionts and their hosts (Hibbett et al., 2000).

Ectomycorrhizal symbioses have a distinct host range, allowing formation of ectomycorrhiza on a limited set of trees and shrubs. However, a given species of ectomycorrhizal fungus is usually able to establish a mutualistic symbiosis on a broad range of species, although highly specific interactions may occur (e.g., *Rhizopogon vinicolor-Pseudotsuga mensiesii*). In temperate and boreal forests, up to 95 percent of the short roots form ectomycorrhizas

(Taylor et al., 2000). Ectomycorrhizas have a beneficial impact on plant growth in natural and agroforestry ecosystems (Grove and Le Tacon, 1993). Central to the success of these symbioses is the exchange of nutrients between the symbionts (Smith and Read, 1997). The fungus gains carbon from the plant while the plant's nutrient uptake is mediated via the fungus. In addition, the establishment of the symbiosis is required for the completion of the fungal life cycle (i.e., formation of fruiting bodies). The prospecting and absorbing extraradical hyphal web captures soil minerals (P, N, water, micronutrients) (Smith and Read, 1997) and organic nitrogen (Chalot and Brun, 1998; Näsholm et al., 1998) and assimilates and translocates a large portion of them to the growing plant (Rygiewicz and Andersen, 1994; Simard et al., 1997). Ectomycorrhizal fungi affect not only mineral and water uptake, but also adaptation to adverse soil chemical conditions (Meharg and Cairney, 2000) and susceptibility to diseases (Smith and Read, 1997) and contribute substantially to plant productivity (Grove and Le Tacon, 1993). On the other hand, the fungus within the root is protected from competition with other soil microbes and, therefore, is a preferential user of the plant carbon (ca 10 to 20 percent of the host photoassimilates). The dependence of ectomycorrhizal fungi on the current photoassimilates of their hosts has been demonstrated by a field experiment in which the host trees were girdled to arrest the flow of assimilates between shoots and mycorrhizal root systems (Högberg et al., 2001). In plots with girdled trees, the fruiting of ectomycorrhizal fungi was negligible compared to adjacent control areas. Mycorrhizae functioning alter quality and quantity of carbon allocated below ground by the host plant (Rygiewicz and Andersen, 1994). Mycorrhizal fungi represent an interface in the soil-plant system and have the ability to regulate plant metabolism. In addition, they constitute links in the chain of transfers by which carbon and nitrogen may move between plant and soil compartments (Simard et al., 1997; Fitter et al., 1999) and can thus influence carbon and nitrogen cycling rates in host plants and forest ecosystems (Leake, 2001).

THE ANATOMY AND DEVELOPMENT OF ECTOMYCORRHIZA

Development of a mature ectomycorrhiza proceeds through a programmed series of events. Fungal hyphae emerging from a soil propagule (spores, sclerotia) or an older mycorrhiza penetrate into the root cap cells and grow through them. Below the tip, the invasion of root cap cells proceeds inward until the hyphae reach the epidermal cells (Horan et al., 1988). The development of the mantle and the Hartig net (Figure 4.1), characterized by labyrin-

FIGURE 4.1. The ectomycorrhizal symbiosis. (A) A basidiomycetous bolet able to form a mutualistic symbiosis with conifers and hardwood trees (photograph courtesy of F. Martin). (B) A seedling of Douglas fir *(Pseudotsuga menziesii)* colonized by the ectomycorrhizal basidiomycete *Laccaria bicolor.* The fungal mycelium has developed ectomycorrhiza on the root system and has produced a basidiocarp aboveground (photograph courtesy of P. Klett-Frey). (C) Short roots ensheathed by an ectomycorrhizal tomentelloid fungus. The orange mantle covers the root tip of *Eucalyptus globulus* (photograph courtesy of A. Jambois). (D) Transverse section of a *Eucalyptus/Pisolithus* ectomycorrhiza showing the external (EM) and internal (IM) mantles: the fungal hyphae have begun to penetrate between the epidermal cells of the root cortex (RC) to form the Hartig net (HN). Epidermal cells are radially enlarged. Extramatrical hyphae (EH) are exploring the medium (photograph courtesy of B. Dell). (See also color photo section.)

thine branching of hyphae growing in the root apoplast, is linked to pivotal events at the hyphal tip. Hyphae located on the surface of the primary root are uniform in diameter and have frequent clamp connections, while those forming the early stages of the mantle on first-order mycorrhizal laterals were often greatly enlarged and multibranched (Kottke and Oberwinkler, 1987). Morphogenetic changes take place upon contact of the hyphae with living cortical cells, which is pivotal for initiation of mantle formation and Hartig net construction. Progression from the strongly rhizomorphic outgrowth of the free-living mycelium to the plectenchymatous structure of the ectomycorrhizal sheath and the coenocytic Hartig net hyphae is associated with a lack of septation, a loss of apical coherence, and intimate juxtaposition of hyphae (Kottke and Oberwinkler, 1987). After attachment onto epidermal cells, hyphae multiply to form a series of layers several hundred micrometers thick which differentiates to form the mature mantle. The hyphae in these structures are encased in an extracellular polysaccharide and proteinacous matrix (Bonfante, 2001). Air and water channels that allow the flow of nutrients into the symbiosis innervate these structures, although most of the nutrient transfer probably takes place via the symplastic way. An outward network of hyphae prospecting the soil and gathering nutrients radiate from the outer layers of the mantle.

Ultrastructural analysis of ectomycorrhizal ontogeny conducted on various ectomycorrhizas (Kottke and Oberwinkler, 1987; Bonfante, 2001) have indicated that the cells involved in the interface between the symbionts undergo extensive changes during ontogeny. The hyphae penetrate between epidermal and sometimes cortical root cells, inducing changes in the cell wall architecture, which presumably involve pectin hydrolysis in the middle lamella. Changes in the host cell wall may involve (according to ultrastructural observations) degradative events or the synthesis of new cell wall polymers to produce wall ingrowths (Bonfante, 2001). In addition, regular, highly organized wall branchings may be induced in the Hartig net hyphae. Thin sections of the interface have revealed that numerous mitochondria, lipid bodies, dictyosomes with proliferating cisternae, and extensive endoplasmic reticulum are contained within the coenocytic hyphae of the Hartig net. This feature suggests a highly active anabolic activity and reflects a drastic biosynthesis of new materials (i.e., biopolymers) (Kottke and Oberwinkler, 1987). In this region of the Hartig net, the hyphae contain several nuclei that may be in the process of dividing. Abundant invaginations of fungal plasmalemma have been observed. These membrane structures allow ions and metabolites to pass at a high rate between adjacent cells, providing the anatomical basis for intercellular communication and the local coordination between the symbionts. Endoplasmic reticulum with numerous ribosomes is indicative of high biosynthesis of secreted proteins. Acid phosphatases and enzymes involved in cell wall degradation (i.e., pectate lyase, peroxidase, and polygalacturonate lyase) may correspond to such secreted enzymes. The clustering of organelles within the Hartig net, presumably through modulation in the organization of the cytoskeleton, may facilitate the more efficient transport of substrates among these organelles and the symbiotic interface. There is clearly a structural and physiological heterogeneity within the ectomycorrhizal mantle, and between the mantle and the inward and outward fungal networks (Cairney and Burke, 1996).

THE MORPHOGENESIS OF ECTOMYCORRHIZA

The morphogenesis of the ectomycorrhizal symbioses encompasses a series of complex and overlapping ontogenic processes in the hyphae and the roots. This includes a general growth stimulus of the hyphae (Lagrange et al., 2001), a trophic response directing hyphal growth inward toward the plant tissues (Horan and Chilvers, 1990), and a morphogenetic effect leading to compact hyphal mantle development (Kottke and Oberwinkler, 1987). In addition to putative morphogens, the supply of nutrients, the presence of a physical support, and the supply of oxygen likely play roles in the alteration of hyphal shape and mantle aggregation (Martin et al., 2001). On the other hand, rhizospheric compounds, such as auxins and hypaphorine, released by fungal hyphae stimulate the formation of lateral roots, dichotomy of the apical meristem in conifer species, and cytodifferentiation (radial elongation, root hair decay) of root cells (Grange et al., 1997; Ditengou and Lapeyrie, 2000; Ditengou et al., 2000). Symbiosis development also leads to novel metabolic patterns in hyphae and plant cells (Martin and Botton, 1993; Buscot et al., 2000; Nehls, Mikolajewski, et al., 2001). The entire purpose of the ontogenic program in the ectomycorrhizal symbiosis is to extend the uptake capacity of the root system (up to 1,000 m of hyphae/m of root) (Rousseau et al., 1994) and to provide a carbohydrate-rich niche for the mycobiont. This mutuality of host and microbe and its implicit coevolutionary implications (Hibbett et al., 2000) are key concepts, although the underlying mechanisms are largely unknown. It appears that the distinction between saprobic and mycorrhizal behavior is not a sharp division but a continuum that must take into account ecological factors, variations in host surveillance mechanisms, and evolutionary factors that determine the cost (i.e., fitness) of the symbiosis (Hibbett et al., 2000). The ecological performance of ectomycorrhizal fungi is therefore a complex phenotype affected by many different traits and by environmental factors (Read, 1991; Leake, 2001; Tagu et al., 2001). Characterization of the primary genetic traits controlling the symbiosis development and its metabolic activity, such as nutrient scavenging and assimilation, will open the door to understanding the ecological fitness of the ectomycorrhizal symbiosis.

ALTERATION IN GENE EXPRESSION
IN ECTOMYCORRHIZA

In addition to the morphological and physiological changes described, the interaction between the ectomycorrhizal fungus and its host induces an avalanche of changes in gene expression in both partners. Detailed information on these processes is essential for the understanding of symbiotic tissue development. The current goal is to gain a greater understanding of the organization of protein networks and how they control cell function. Specific objectives of the ongoing studies encompass (1) the determination of the baseline RNA expression pattern for selected ectomycorrhizal associations at specific points during symbiosis development, (2) determination of transcript/protein expression patterns from genes that have homologues in different symbioses (e.g., hydrophobins), (3) development of protein expression patterns for specific tissues during development, and (4) determination of cellular protein interactions using the transcript and protein expression data in combination.

Plants and fungi excrete a wide range of more- or less-attractive compounds (e.g., flavanoids, alkaloids) (Paiva, 2000). Both ectomycorrhizal partners probably possess one to many types of signal receptors/sensors that may bind with several (or a number of) rhizospheric excreted signals (Horan and Chilvers, 1990; Béguiristain and Lapeyrie, 1997; Ditengou and Lapeyrie, 2000; Ditengou et al., 2000; Martin et al., 2001). In turn, signal/sensor complexes probably activate/repress expression of downstream genes including components of the signaling patways. Fungal genes, such as *PF6.2* and *ras* from *Laccaria bicolor* (Kim, Hiremath, and Podila, 1999; Sundaram et al., 2001) and *ras* from *Pisolithus microcarpus* (Duplessis et al., unpublished results), are induced before any physical contact confirming that diffusible elicitors are involved in the early steps of the ectomycorrhizal interaction. Similarly, an auxin-regulated glutathione-*S*-transferase expressed in *Eucalyptus globulus* roots is up-regulated (Nehls, Bock, et al., 1998) by hypaphorine, a tryptophan betaine, secreted by *P. microcarpus* (Béguiristain and Lapeyrie, 1997). Once the fungal hyphae are growing in the root apoplast, other trophic and developmental inputs from both symbionts are likely necessary for successful symbiosis. Alteration of the expression of genes and proteins coding for cytoskeleton and cell wall components accompany morphogenesis of hyphae (Ninii et al., 1996; Martin et al., 1999; Gorfer et al., 2001; Laurent et al., 1999) and root cells (Ninii et al., 1996; Carnero-Diaz et al.,

1996) (Table 4.1). In entering its new niche, the colonizing hyphae need to adjust their gene and protein content to the new environment. Detecting the new environment and coordinating fungal and plant developments might involve the use of signaling networks (Gorfer et al., 2001; Martin et al., 2001).

As might be anticipated, symbiosis induces significant changes in metabolic activities and enzyme levels in root and fungal tissues (Smith and Read, 1997; Blaudez et al., 1998; Martin et al., 1998). These metabolic alterations are accompanied by mineral- and nutrient-related changes in gene expression. The up-regulation of genes coding for the glyoxylate pathway enzyme, malate synthase (Balasubramanian et al., 2002), and the autophagy-involved *LB-AUT7* (Kim, Bernreuther, et al., 1999) suggests that there is a rapid catabolism and recycling of existing fungal material, and channeling toward biosynthesis of new cell components during the early phase of the interaction between *Laccaria bicolor* and *Pinus resinosa*. This metabolic shift may reflect the transition from a saprobic to symbiotic mode of growth. When the symbiotic interface is differentiated, it has been observed that hexose transporters of the symbionts in the *Amanita muscaria* associated to

TABLE 4.1. Most prevalent mRNAs in *Eucalyptus globulus-Pisolithus microcarpus* ectomycorrhiza as measured by EST redundancy. The columns represent (1) the ID of a representative clone, (2) the genomic origin (plant versus fungus), (3) the best database match and the corresponding species obtained with a WU-BLASTX search, and (4) the number of ESTs among 850 total ESTs that assembled into a contiguous sequence.

Contig ID[a]	Organism	Best database match (species)	Redundancy[b]
EgPtdB57	Fungus	No match (referred to as SRAP17)	29
5A8	Fungus	HydPt-2 *(P. microcarpus)*	19
EgPtdA12	Fungus	SRAP32-1, Type I *(P. microcarpus)*	18
st54	Fungus	Transmembrane FUN34 protein *(Saccharomyces cerevisiae)*	9
5C6	Fungus	No match	9
7A7	Fungus	Elongation factor 1-γ *(Artemia* sp.)	7
7A6	Fungus	No match	6
11A2	Fungus	Metallothionein *(Agaricus bisporus)*	7

[a] Representative EST ID from assembly contig
[b] Number of ESTs that assembled into a contiguous sequence

Picea abies (Nehls, Wiese, et al., 1998) or *Populus* (Nehls, Bock, et al., 2001) and *Paxillus involutus-Betula pendula* (Wright et al., 2000) symbioses are differentially expressed. The fungal gene coding for an hexose transporter, *AmMst1,* is up-regulated (Nehls, Wiese, et al., 1998), whereas the *Picea* hexose transporter is slightly down-regulated (Nehls et al., 2000). Similarly, the expression of *B. pendula* hexose and sucrose transporters, BpHEX1, BpHEX2, and BpSUC1 is down-regulated in mycorrhizal roots (Wright et al., 2000). As emphasized by the authors, the down-regulation of expression of these transporters is not compatible with the increased carbon fluxes taking place in the roots as a result of the carbon drain imposed by the mycobiont (Rygiewicz and Andersen, 1994). Other transporters, which remain to be identified, are likely involved in the translocation of carbohydrates between the partners. Expression of several genes controlling assimilation of carbon and nitrogen compounds is up-regulated in ectomycorrhiza (Nehls, Mikolajewski, et al., 2001). The transcript and protein levels of the root NADP-isocitrate dehydrogenase are increased in the *E. globulus/ P. microcarpus* ectomycorrhiza (Boiffin et al., 1998). *Amanita muscaria* phenylalanine ammonium lyase gene, *AmPAL,* is likely regulated in ectomycorrhiza through changes in nitrogen and sugar levels (Nehls, Ecke, and Hampp, 1999). Whether gene expression of these metabolic genes is controlled by sugar- and amino acid-dependent regulation, as reported for free-living mycelium (Lorillou et al., 1996; Jargeat et al., 2000; Nehls, Kleber, et al., 1999; Nehls, Bock, et al., 2001), or by symbiosis-related developmental signals is not known (Wright et al., 2000). Nehls, Bock, and colleagues (2001) have suggested that the expression of the *A. muscaria* hexose-transporter gene, *AmMst1,* is regulated only by the hexose concentration in the symbiotic apoplastic space of *Amanita-Populus* mycorrhiza.

The ectomycorrhizal symbiosis has a drastic effect on the phosphate absorption and assimilation. It is clear that ectomycorrhizal fungi can translocate phosphate from the soil to the plant root (Beever and Burns, 1981), and access to additional phosphate transported through hyphal networks can have a significant effect on host growth and reproduction (Smith and Read, 1997). However, no information is currently available on the regulation of genes controlling these processes (i.e., phosphate absorption, polyphosphate synthesis). High-affinity phosphate transporters (*GvPT* and *GiPT*) that share structural and sequence similarities with the high-affinity, proton-coupled phosphate transporters from yeast and *Neurospora crassa* were cloned from the endomycorrhizal fungi *Glomus versiforme* (Harrison and Van Buuren, 1995) and *G. intraradices* (Maldonado-Mendoza et al., 2001). On the basis of their phosphate transport activity and their expression in the extra-radical mycelium, it has been suggested that *GvPT* and *GiPT* play a role in soil phosphate sensing and acquisition and are involved in phosphate movement in the symbiosis (Maldonado-Mendoza et al., 2001). A high-affinity transporter has been cloned from the ectomycorrhizal fungus *Hebeloma cylindrosporum* (C. Plassard and H. Sentenac, personal communication). Whether this phosphate transporter plays a role in phosphate acquisition in ectomycorrhiza will await further investigations.

These findings at the molecular level illustrated the drastic changes in metabolic activities experienced by the partners during ectomycorrhiza development and functioning. Further work is necessary on our extremely incomplete understanding of the nutrient-sensing mechanisms and the downstream regulation networks. Related to this is the need to better understand the cellular and tissue compartmentation of assimilative enzymes and the transport of metabolites between the physiologically heterogeneous fungal compartments and between symbionts (Cairney and Burke, 1996).

THE TRANSCRIPTOME
OF ECTOMYCORRHIZAL SYMBIOSES

A Glimpse of Genes Expressed in Ectomycorrhiza

The cellular biology, biochemistry, and metabolism of ectomycorrhizal fungi have been largely investigated due to their ability to grow vegetatively on synthetic media. They have, however, lacked the coordinated development of resources aimed at sequencing the genome or producing genetic tools that are useful to the scientific community, such as large numbers of expressed sequence tags (ESTs) and insertion and expression tagged lines as are available for several model fungi. Although EST databases of trees that are able to develop ectomycorrhiza have recently been developed (Allona et al., 1998; Sterky et al., 1998; Whetten et al., 2001), they are limited to a few species (e.g., *Pinus taeda* and *Populus* spp.).

To speed the discovery of novel genes and functions involved in ectomycorrhiza development and functioning, we have developed EST databases of *Pisolithus microcarpus* 441, *Laccaria bicolor, Populus* × *interamericana,* and the *Eucalyptus globulus-P. microcarpus* ectomycorrhiza (Tagu and Martin, 1995; Voiblet and Martin, 2000; Voiblet et al., 2001; see <http://mycor.nancy. inra.fr/>). Other groups have recently reported their commitment to develop additional EST databases for *Hebeloma cylindrosporum* (Sentenac et al., unpublished results), *Tuber borchii* (Lacourt et al., 2002; Polidori et al., 2002), and *B. pendula-P. involutus* ectomycorrhiza. As the number of ESTs increases, *in silico* comparisons of transcriptomes across genera, species, ecotypes, and strains of symbiotic fungi will become possible.

In the following section, we will summarize the data obtained on the transcriptomes of the *E. globulus-P. microcarpus* (Voiblet et al., 2001; Duplessis and Martin, unpublished results) and *B. pendula-P. microinvolutus* ectomycorrhizae. About 3,500 cDNA were randomly selected from cDNA libraries of *E. globulus- P. microcarpus* ectomycorrhizae at different developmental stages and partially sequenced (see the database EctomycorrhizaDB at <http://mycor.nancy.inra.fr/EctomycorrhizaDB/index.html>). About 50 percent of the ESTs show no similarity to DNA or protein sequences in GenBank, as of February 2003 (Peter et al., 2003). These are candidates for genes that are either specific to eucalypt or to *P. microcarpus*. The largest category (30 percent) of identified sequences corresponded to genes coding for the gene-protein expression machinery, which includes transcripts such as those coding for ribosomal proteins, initiation and elongation factors, and the ubiquitin/proteasome pathway components. Cell wall proteins, such as hydrophobins (Tagu et al., 1996) and symbiosis-regulated Arg-Gly-Asp (RGD)-containing acidic polypeptides (SRAP32) (Laurent et al., 1999) that are found at the surface of interacting fungal cell, were also abundantly detected (10 percent). As expected in a symbiotic organ in which bilateral transfers are intense and assimilative activity very high, a significant portion of genes (13 percent) expressed in the eucalypt ectomycorrhiza are coding for enzymes of primary and secondary metabolism (e.g., adenosine triphosphate [ATP] synthase, glyceraldehyde-3-P dehydrogenase [G3PDH], glutamine synthetase, alanine aminotransferase). Transcripts involved in cell signaling and cell communication (e.g., adenosine diphosphate [ADP]-ribosylation factors, calmodulin, heterotrimeric guanine triphosphatases [GTPases], ras) are also abundant (13 percent) in mycorrhizal tissues (Voiblet et al., 2001). This snapshot of the transcript population emphasizes the prominence of fungal transcripts in the colonized roots confirming previous studies of the protein patterns (Hilbert et al., 1991; Burgess et al., 1995) and rRNA populations (Carnero-Diaz et al., 1997).

To analyze the transcriptome of *B. pendula-P. involutus* ectomycorrhiza, 3,555 ESTs of symbiotic tissues, 3,964 ESTs of free-living growing mycelium, and 2,532 ESTs from nonmycorrhizal birch were analyzed (see <www.biol.lu.se/funghostdb>). By assembly of all ESTs (10,051), 2,284 unique transcripts, either of fungal or plant origin, were characterized. Of those transcripts, at least 154 displayed more than a 2.5-fold differential expression (as based on cDNA redundancy) in mycorrhiza as compared to the axenic conditions and were considered as symbiosis-related transcripts. Transcripts related to protein synthesis were down-regulated, whereas transcripts related to cell rescue, defense, cell death, and aging were up-regulated in the mycorrhizal root tissues as compared to the nonsymbiotic fungus. Expression of calmodulin and hydrophobins showed a striking up-regulation. In addition, the mycorrhizal root tissue displayed an up-regulation in transcripts related to nucleotide metabolism and carbon catabolism as compared to the saprophytically growing fungus, whereas transcripts related to amino acid metabolism and lipid, fatty acid, and isoprenoid metabolism were down-regulated.

With multiple EST programs dealing with pathogenic and mutualistic fungi, in the near future we will have an unparalleled opportunity to ask which genetic features are responsible for common divergent traits involved in pathogenesis and symbiosis. A few of the many possible breakthroughs will be in characterization of common transduction networks, identification of novel surface proteins that play critical roles in plant-fungus interactions, and new insights into unique metabolic routes critical for mycorrhiza functioning.

Genes Differentially Expressed in Ectomycorrhiza

Differential cDNA screening, subtractive cDNA hybridization, and differential mRNA display were used to identify plant and fungal genes that are induced upon symbiosis development in several ectomycorrhizal symbioses: *E. globulus-P. microcarpus* (Nehls and Martin, 1995; Voiblet and Martin, 2000), *L. bicolor-Pinus resinosa* (Kim, Hiremath, and Podila, 1999), and *A. muscaria-P. abies* (Nehls, Mikolajewski, et al., 1999). These studies confirmed that ectomycorrhiza development and functioning is accompanied by drastic changes in gene expression at the transcriptional level and allowed the identification of several symbiosis-regulated (SR) genes (Table 4.2). Sequencing of these genes indicated that the differentially cloned cDNAs are highly complex. Several differentially expressed genes were found to have homology with known cell wall proteins, such as hydrophobins (Tagu et al., 1996; Duplessis et al., 2001) and cell wall RGD-containing acidic polypeptides (Laurent et al., 1999), communication and developmental genes (e.g., *ras,* Sundaram et al., 2001; auxin-induced glutathion-*S* transferase (GST) and transcriptional factors, Nehls, Béguiristain, et al., 1998; Charvet-Candela et al., 2002), autophagocytosis-involved *Lb-AUT7* (Kim, Bernreuther, et al., 1999), and metabolic genes (malate synthase, Balasubramanian et al., 2002). Other genes (e.g., *PF6.2, SC13*) (Nehls and Martin, 1995; Kim et al., 1998; Nehls, Mikolajewski, et al., 1999) demonstrated no significant matches with entries in the databases and potentially represent cDNAs from novel ectomycorrhiza-induced genes.

Sequencing genes is only the first part of a greater challenge confronting biology of symbioses. The next step is to decipher the function of genes and to unravel the complex interactions governing genetic networks. Global ap-

TABLE 4.2. Ectomycorrhiza-regulated genes identified by targeted approaches or differential screening

Function	Name	Identity	Organism	[RNA] in myc.	Remarks	Reference	Acc. No.*
Unknown	SC13	Unknown	Amanita muscaria	↗	Down-regulation in fruit bodies	Nehls, Mikolajewski, et al., 1999	AJ101141
	PF6.2	Unknown	Laccaria bicolor	↖	Genomic sequence available	Kim et al., 1998	U93505
	PF6.1 PF24.2 PF24.3	Unknown	Laccaria bicolor	↖		Kim et al., 1998	U93507 U93509 U93510
Signal transduction	SbCdc42	GTPase	Suillus bovinus	↑		Gorfer et al., 2001	AF234180
	SbRac1	GTPase	Suillus bovinus	↑		Gorfer et al., 2001	AF235004
	Lbras	GTPase	Laccaria bicolor	↖	Functionally complements its Saccharomyces homologue	Sundaram et al., 2001	AF034098
	TbSP1	Phospholipase A2	Tuber borchii	↑	Regulated by nutrient deprivation	Soragni et al., 2001	AF162269
	Pp-iaa88	Aux/IAA protein	Pinus pinaster	↖	Regulated by auxins	Charvet-Candela et al., 2002	AJ315675
Cell structure	Sbact1 Sbact2	Actin	Suillus bovinus	↑	Genomic sequence available	Tarkka et al., 2000	AF155931 AF156259
	EgTubA1	α-Tubulin	Eucalyptus globulus	↖		Carnero-Diaz et al., 1996	U37794
	Lb-AUT7	Vesicule transport protein	Laccaria bicolor	↖	Functionally complements its Saccharomyces homologue	Kim, Bernreuther, et al., 1999	U93506
	SC25	Extensin-like protein	Amanita muscaria	↖	Up-regulation in fruit bodies	Nehls, Mikolajewski, et al., 1999	AJ101142
	HydPt-1 HydPt-2	Hydrophobin	Pisolithus tinctorius	↖	Cell wall polypeptide, HydPt-1 is a surfactant polypeptide	Tagu et al., 1996	U29605
	HydPt-3	Hydrophobin	Pisolithus tinctorius	↖		Duplessis et al., 2001	AF097516

TABLE 4.2 (continued)

	Gene	Function	Species		Notes	Reference	Accession
	Srap32-1, Srap32-2, Srap32-3, Srap32-4	Unknown	Pisolithus tinctorius	↖	Cell wall polypeptides	Laurent et al., 1999	AF035678 F124722 AF124723 F124724
	Tbf-1	Unknown	Tuber borchii	↗	Cell wall polypeptide expressed in fruit body	De Bellis et al., 1998	U83996
Metabolism	AmMst1	Monosaccharide transporter	Amanita muscaria	↖	Regulation by N sources functionally complements in Saccharomyces	Nehls, Wiese, et al., 1998	Z83828
	AmAAP1	Amino acid transporter	Amanita muscaria	↑	Regulation by N sources functionally complements its Saccharomyces homologue	Nehls, Kleber, et al., 1999	AJ223504
	AmPal	Phenylalanine ammonium lyase	Amanita muscaria	↖	Regulation by C sources tissue specific	Nehls, Ecke, and Hampp, 1999	AJ010143
	Mycf102	Acid phosphatase	Pisolithus tinctorius	↗		Tagu et al., 1993	Not available
	Lb-MS	Malate synthase	Laccaria bicolor	↖		Balasubramanian et al., 2002	AF031099
	PaMst1	Monosaccharide transporter	Picea abies	↗		Nehls et al., 2000	CAB06079
	BbHEX1, BpHEX2	Hexose transporters	Betula pendula	↗		Wright et al., 2000	AF168772, AF168773
	BpSUC1	Sucrose transporter	Betula pendula	↗		Wright et al., 2000	AF168771
	Eglcdh	Isocitrate dehydrogenase	Eucalyptus globulus	↖		Boiffin et al., 1998	U80912
Plant de-fense, stress	EgHypar	Glutathione S-transferase	Eucalyptus globulus	↖	Regulated by auxins and hypaphorine	Nehls, Béguiristain, et al.,1998	U80615
	Bet v1-SC1	Pathogenesis-related protein	Betula pendula	↖		Feugey et al., 1999	AF124837

*(GenBank accession number)

proaches aimed at describing the expression levels of sets of genes under specific developmental and physiological conditions (transcriptomic), and the protein patterns that follow (proteomics) will provide extremely valuable information on the genetic networks controlling the ectomycorrhiza establishment (Martin, 2001). The ultimate goal is not only to understand how individual proteins function in vivo, but also to determine how protein networks are organized so that they impart specific function to the ectomycorrhizal organ.

Expression Profiling As a Means of Understanding Ectomycorrhiza Development

Because properties of the mycobiont and of the host seem equally important, and because multiple factors on both sides are involved, the evolving genomic studies of the fungi and of the host tree have the greatest promise of providing rapid understanding of the symbiosis gene networks. Work on a single gene, protein, or property, no matter how detailed and careful, will not be useful as the kind of comprehensive analysis now available in model trees, such as poplars and pines, as well as in several ectomycorrhizal fungi, such as *P. microcarpus, H. cylindrosporum,* and *P. involutus.* We hypothesized that genome-wide expression profiling would provide insight into the molecular and physiological cascades controlling the symbiosis formation and metabolism (Martin, 2001).

To examine gene activity changes associated with the development of the *E. globulus-P. microcarpus* symbiosis on a wider scale, we have performed expression profiling using cDNA arrays (Voiblet et al., 2001; Duplessis and Martin, unpublished results). RNA used for cDNA array hybridizations were derived from nonmycorrhizal roots, free-living mycelium, and colonized roots collected 4, 7, 12, 15, 21, and 28 days after contact between the growing root tips and the edges of the fungal colonies. Thus, these time points for RNA collection correspond to the various stages of ectomycorrhiza development: early hyphae-root contacts, root surface colonization, mantle formation, root penetration and subsequent Hartig net formation, and mature symbiotic organ. Among the 750 arrayed cDNAs, we found 79 (11 percent) to 164 (23 percent) genes having a regulation ratio ≥ 2.0 depending on the developmental stage examined. As expected from previous studies (Tagu et al., 1996; Carnero-Diaz et al., 1996; Nehls, Wiese, et al., 1998; Nehls, Béguiristain, et al., 1998; Nehls, Ecke, and Hampp, 1999; Nehls, Kleber, et al., 1999; Nehls, Mikolajewski, et al., 1999; Kim et al., 1998; Kim, Hiremath, and Podila, 1999; Kim, Bernreuther, et al., 1999; Wright et al., 2000; Sundaram et al., 2001), many functional groups were found to be involved in symbiosis development, including genes involved in cell growth, differentiation and signaling, synthesis of cell surface and extracellular matrices, and energy metabolism. Hierarchical cluster analysis (Eisen et al., 1998) was therefore used to define groups of genes having both related regulation patterns and expression amplitudes.

Genes Regulated in the Host Plant

The five transcripts found to be more abundant in nonmycorrhizal eucalypt roots are predicted to encode G3PDH, an abscisic-induced protein, an unknown protein, ubiquitin, and a tonoplastic aquaporin. The G3PDH, abscisic-induced protein, and tonoplastic aquaporin remain the most prevalent transcripts within the mycorrhizal seedling roots (Voiblet et al., 2001; Duplessis and Martin, unpublished results). As the up-regulation of *EgTuba* and *EgHypar* coding for an α-tubulin and a hypaphorine- and auxin-regulated glutathione *S*-transferase of eucalypt roots are known ectomycorrhiza-regulated genes (Carnero-Diaz et al., 1996; Nehls, Béguiristain, et al., 1998), we analyzed the cluster clade including these genes in detail (Figure 4.2, bottom cluster). It contains genes expressed strongly during most of the mycorrhiza formation. Among the identified up-regulated genes, 58 percent of them were homologues of known genes. This group contains the transcripts coding an elicitor-induced *O*-methyltransferase identified in our previous cDNA array analysis of four-day-old *E. globulus-P. tinctorius* mycorrhiza (Voiblet et al., 2001), suggesting that colonized root cells mount defense reactions to restrict the fungal invasion. Cellular functions up-regulated by mycorrhiza formation include the mitochondrial respiration (e.g., cytochrome C reductase, ATP synthase), protein synthesis, and turnover (e.g., ubiquitin/proteasome pathway, ribosomal proteins and translation factors). Approximately 33 percent of these up-regulated genes were involved in cell growth and proliferation (i.e., transcription/translation, protein synthesis, cell wall proline-rich proteins).

Other nodes in the *Eucalyptus* clustergram (Figure 4.2) define groups of transcripts with different properties. For example, the top clade is composed of transcripts strongly repressed during mycorrhiza formation. Among these transcripts, aquaporins, water stress-regulated, and abscissic-regulated proteins are abundantly represented. Metallothionein transcripts were also found in this clade. The down-regulation of these genes suggests that mycorrhiza formation facilitates, as expected, water transfer to the plant and thus relieves stresses experienced by the growing roots.

Genes Regulated in the Mycobiont

Among the most abundantly expressed genes in differentiating ecto-mycorrhiza are many genes expected to be involved in the synthesis of the fungal cell wall and symbiosis interfacial matrix (Martin et al., 1999; Bonfante, 2001). Genes accumulating in *P. microcarpus* cell walls, including genes encoding the different hydrophobins and SRAP32s (Voiblet et al., 2001), are up-regulated in symbiotic tissues. These are candidate markers for symbiosis-related changes in the cell wall, and the encoded proteins likely function during symbiosis development. Because activation of *hydPt-2* coding for hydrophobin-2 is a robust molecular marker for early symbiosis development (Tagu et al., 1996; Duplessis et al., 2001), we analyzed the cluster including this gene in detail (Figure 4.3, top clade). It contains 32 sequences (from a maximum of 11 different genes). In addition to *hydPt-2*, other members of the hydrophobin (e.g., hydrophobin-4) and *SRAP32* multigene families (e.g., type II *SRAP32* and *SRAP17*) showed a similar expression profile. These transcripts were represented on the arrays by multiple spots, and their nearly invariant clustering in the *hydPt-2* group demonstrates the internal consistency of our cDNA array analysis. Members of the *hydPt-2* cluster show a unique overall expression profile. They are strongly activated in the early stages of mycorrhiza development (4 to 12 days after contact) when the mantle formation was taking place. Moreover, these genes are moderately activated in the median stages of development (days 12 to 15), and they are going back to their constitutive level in the mature mycorrhiza (day 21). This expression pattern is consistent with previous findings (Tagu et al., 1996; Laurent et al., 1999). Other genes of less certain relationship to cell wall formation are also highly up-regulated in symbiotic tissues, such as proline-rich proteins and a proteophosphoglycan. Interestingly, genes coding for actin and a homologue of the yeast transmembrane GPR/FUN34 protein are up-regulated at the earliest stages, suggesting a coordinated expression of structural genes involved in cytoskeleton, membrane, and cell wall synthesis.

The global gene expression analyses presented here add critical new information to existing models of ectomycorrhiza development. At the different developmental stages studied, development of the *E. globulus-P. microcarpus* and *B. pendula-P. involutus* symbioses does not induce the expression of ectomycorrhiza-*specific* genes (Voiblet et al., 2001; Duplessis and Martin, unpublished results). Expression profiling showed that developmental reprogramming takes place in roots and hyphae. This induces the repression of processes occurring in free-living partners and the activation of proteins involved in hyphae proliferation and anabolism. We hypothesize

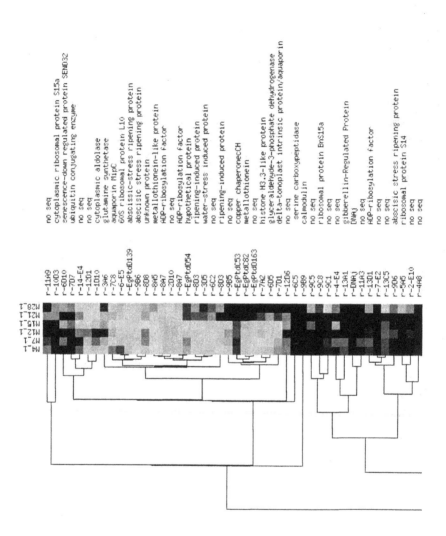

FIGURE 4.2. Cluster analysis of a set of 200 genes expressed in *Eucalyptus globulus* roots during ectomycorrhiza development. Clustering was calculated by a hierarchical method based on Euclidean distance measurements and the data represented using the method of Eisen et al. (1998). Color intensity reflects the magnitude of gene expression ratio for each gene, red indicates genes expressed at a higher level in roots colonized by *Pisolithus microcarpus* (com-

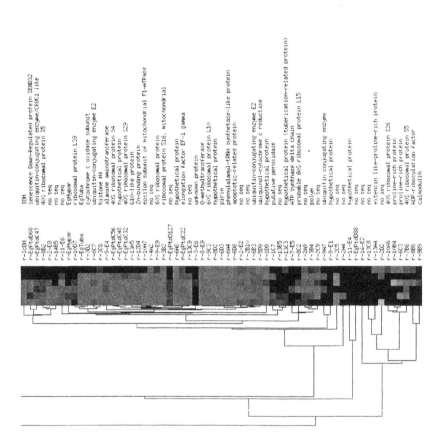

pared to nonmycorrhizal roots), green indicates genes with a down-regulated expression in symbiotic tissues, and black indicates gene expression ratio near 1. Each column represents gene expression ratios from a different developmental stage (between 0 and 28 days postcontact). No seq = sequence not available; IDH = NADP-dependent isocitrate dehydrogenase; polyA = polyA-binding protein. (See also color photo section.)

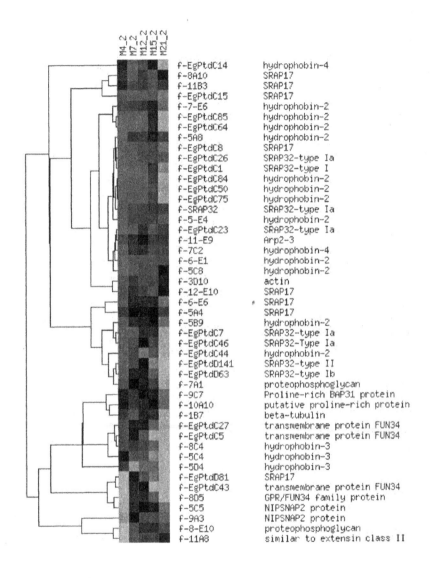

FIGURE 4.3. A group of *Pisolithus microcarpus* genes coding for structural proteins and clustering with the hydrophobines, a family of cell wall polypeptides up-regulated during *Eucalyptus globulus* ectomycorrhiza formation. For details on the hierarchical clustering, see Figure 4.2. SRAP = symbiosis-regulated acidic polypeptides; Arp = Actin-related protein. (See also color photo section.)

that these alterations in transcript profiles define the biochemical cascades underlying the progressive changes in the anatomy and metabolism of the differentiating ectomycorrhiza. A marked change in the gene expression in *E. globulus-P. microcarpus* is observed at multiple levels: (1) a general activation of the protein synthesis machinery probably supporting an intense cell division/proliferation, (2) increased accumulation of transcripts coding for cell surface and cytoskeleton proteins in fungal hyphae (hydrophobins, SRAPs, proteoglycans, tubulins) and roots (α-tubulin, extensins) probably involved in the mantle and symbiotic interface formation, and (3) the up-regulation of energy metabolism in colonized roots. Several of these cellular functions are also regulated in *B. pendula-P. involutus*. These data suggest a highly dynamic environment in which symbionts are sending and receiving signals, exposed to high levels of stress conditions and remodeling tissues. Results from these genomic analyses have underscored the inherent complexities encountered at this scale; for example, hundreds of expression differences can be attributed to stage of plant/fungus growth, variations between culture media lots, pH, and aeration variabilities (Duplessis et al., unpublished results). Attention should be focused on reproducibility in array-based studies, and statistical approaches are being developed to guide study design.

So far, changes in gene expression have been studied only on a limited number of associations, and the number of identified SR-genes is low (<100). Only hydrophobin and calmodulin genes have been found to be clearly up-regulated in different ectomycorrhizal associations (Tagu et al., 1996; Johansson et al., submitted), but not all (Mankel et al., 2000). This result suggests that many of the genes highly expressed during ectomycorrhiza formation in various ectomycorrhizal types are significantly different in sequence or relative abundance. A more complete analysis of this key question awaits the completion of larger sets of ectomycorrhiza ESTs/expression profiles, or a separate project specifically focused on addressing the question of how many unique genes may be involved in ectomycorrhiza formation involving either different host plants and/or different mycobionts. Gene profiling carried out on different ectomycorrhizal associations will allow the identification of common molecular signatures. Functional genomics in loblolly pine (Allona et al., 1998; Whetten et al., 2001) and poplar (Sterky et al., 1998; Hertzberg et al., 2001) has the potential to contribute to our understanding of ectomycorrhiza development, as these species are well-known hosts of ectomycorrhizal fungi. Yet a precise understanding of how SR genes and proteins function and interact with one another in a cellular context also requires the ability to introduce precise alterations within specific components of these genetic networks. In this respect, targeted transgenesis and knocking-out genes are not yet possible in either ectomycorrhizal fungi and their host trees. Testing the roles of candidate plant genes in ectomycorrhiza formation in pine species will be difficult, due to the formidable challenges these species present as experimental models. Functional analysis of SR genes will probably be easier in poplar, an emerging model system in plant biology (Bradshaw et al., 2000). *Populus* species, able to establish both endo- and ectomycorrhizas, are thus poised to fill a gap in the tools that are presently available for studying the function of mycorrhiza-regulated proteins in vivo.

OUTLOOK

Ectomycorrhiza development influences gene expression in the host plant and the colonizing fungus in a pleiotropic manner. A range of fungal tissue differentiates can be distinguished by a combination of anatomical and cytological features. On the other hand, root tips proliferate and root cells experience major changes in their orientation and morphology. Advances in recent years have provided insights on the molecular basis of ectomycorrhiza morphogenesis and functioning. Several studies using EST profiling (Johansson et al., submitted), mRNA display (Kim et al., 1998; Polidori et al., 2002), or a gene array approach (Voiblet et al., 2001; Duplessis and Martin, unpublished results) are currently examining the expression of genes in the whole ectomycorrhizas. It is apparent from the expression profiling studies that there is a vast complexity of genetic programs with overlapping expression patterns. This includes attachment to host surfaces and penetration into root tissues, induction of weak plant-defense reactions, the down-expression of plant protein biosynthesis, changes in hormonal balance in lateral roots by fungal auxins, morphogenetic switches of the fungal hyphae, and novel metabolic patterns. The sequence information generated from the Genome Sequence projects (Voiblet et al., 2001; Duplessis and Martin, unpublished results) provides the starting point for understanding how protein networks are organized within cells and how they control cell function. However, considering the tissue and cellular complexity of the symbiosis, one has to conclude that this is an approach with low resolution for the analysis of cellular function during development and functioning. Gene expression profiles should now be performed on purified (or at least enriched) samples from specific cell populations and tissues. The recent analysis of the differential expression of hexose-regulated fungal genes *AmPAL* and *AmMst1* within the different symbiotic tissues of the *Amanita-Populus* ectomycorrhiza (Nehls, Bock, et al., 2001) has shown that this innovative approach is within reach.

Studies have used two-dimensional (2-D) electrophoresis gels to build expression maps of ectomycorrhiza proteins (Hilbert et al., 1991; Guttenberger and Hampp, 1992; Simoneau et al., 1993; Burgess et al., 1995; Tarkka et al., 1998). These studies are clearly necessary to establish a true proteome profile for the ectomycorrhizal symbiosis. Coupling of EST studies, mass spectrometry, and preparative 2-D gel electrophoresis, the range of characterized proteins can be increased substantially.

The identification of genes and inducer molecules that integrate the actions of the multiple programs of gene expression involved in generating a mature symbiotic organ will be the source for many answers. In this respect, it is crucial to identify plant and fungal pattern mutants altered in the organization of the ectomycorrhiza structure. This would allow for the study of the specific events that take place during the different stages of symbiosis development and functioning.

A comparative study of gene expression in different types of ectomycorrhizas using expression profiling might reveal to what extent similarities and differences in the various types of ectomycorrhiza are the result of variation in the basic mechanisms underlying the respective developmental programs. The comparison of mycorrhizal fungi genomes to those in pathogenic fungi will in itself be quite illuminating.

REFERENCES

Allona, I., Quinn, M., Shoop, E., Swope, K., St. Cyr, S., Carlis, J., Riedl, J., Retzel, E., Campbell, M.M., Sederoff, R., and Whetten, R.W. (1998). Analysis of xylem formation in pine by cDNA sequencing. *P. Natl. Acad. Sci. USA* 95: 9693-9698.

Balasubramanian, S., Kim, S.J., and Podila, G.P. (2002). Differential expression of a malate synthase gene during the preinfection stage of symbiosis in the ectomycorrhizal fungus *Laccaria bicolor*. *New Phytol.* 154: 517-528.

Beever, R.E. and Burns, D.J.W. (1981). Phosphorus uptake, storage and utilization by fungi. *Adv. Bot. Res.* 8: 100-135.

Béguiristain, T. and Lapeyrie, F. (1997). Host plant stimulates hypaphorine accumulation in *Pisolithus tinctorius* hyphae during ectomycorrhizal infection while excreted fungal hypaphorine controls root hair development. *New Phytol.* 136: 525-532.

Blaudez, D., Chalot, M., Dizengremel, P., and Botton, B. (1998). Structure and function of the ectomycorrhizal association between *Paxillus involutus* and *Betula pendula*. II. Metabolic changes during mycorrhiza formation. *New Phytol.* 138: 543-552.

Boiffin, V., Hodges, M., Galvez, S., Balestrini, R., Bonfante, P., Gadal, P., and Martin, F. (1998). Eucalypt NADP-dependent isocitrate dehydrogenase: cDNA cloning and expression in ectomycorrhiza. *Plant Physiology* 117: 939-948.

Bonfante, P. (2001). At the interface between mycorrhizal fungi and plants: The structural organization of cell wall, plasma membrane and cytoskeleton. In B. Hock (Ed.), *The Mycota,* Volume IX, *Fungal Associations* (pp. 45-61). Berlin, Heidelberg: Springer-Verlag.

Bradshaw, H.D., Ceulemans, R., Davis, J., and Stettler, R. (2000). Emerging model systems in plant biology: Poplar *(Populus)* as a model forest tree. *J. Plant Growth Regul.* 19: 306-313.

Burgess, T., Laurent, P., Dell, B., Malajczuk, N., and Martin, F. (1995). Effect of fungal-isolate aggressivity on the biosynthesis of symbiosis-related polypeptides in differentiating eucalypt ectomycorrhizas. *Planta* 195: 407-417.

Buscot, F., Munch, J.C., Charcosset, J.Y., Gardes, M., Nehls, U., and Hampp, R. (2000). Recent advances in exploring physiology and biodiversity of ectomycorrhizas highlight the functioning of these symbioses in ecosystems. *FEMS Microbiol. Rev.* 24: 601-614.

Cairney, J.W.G. and Burke, R.M. (1996). Physiological heterogeneity within fungal mycelia: An important concept for a functional understanding of the ectomycorrhizal symbiosis. *New Phytol.* 134: 685-695.

Carnero-Diaz, M.E., Martin, F., and Tagu, D. (1996). Eucalypt α-tubulin: cDNA cloning and increased level of transcripts in ectomycorrhizal root system. *Plant Mol. Biol.* 31: 905-910.

Carnero-Diaz, M.E., Tagu, D., and Martin, F. (1997). Ribosomal DNA internal transcribed spacers to estimate the proportion of *Pisolithus tinctorius* and *Eucalyptus globulus* RNAs in ectomycorrhiza. *Appl. Environ. Microbiol.* 63: 840-843.

Chalot, M. and Brun, A. (1998). Physiology of organic nitrogen acquisition by ectomycorrhizal fungi and ectomycorrhizas. *FEMS Microbiol. Rev.* 22: 21-44.

Charvet-Candela, V., Hitchin, S., Ernst, D., Sandermann Jr., H., Marmeisse, R., and Gay, G. (2002). Characterization of an *Aux/IAA* cDNA up-regulated in *Pinus pinaster* roots in response to colonization by the ectomycorrhizal fungus *Hebeloma cylindrosporum. New Phytol.* 154: 769-778.

Daniell, T.J., Hodge, A., Young, J.P.W., and Fitter, A. (1999). How many fungi does it take to change a plant community? *Trends Plant. Sci.* 4: 81-82.

De Bellis, R., Agostini, D., Piccoli, G., Vallorani, L., Potenza, L., Polidori, E., Sisti, D., Amoresano, A., Pucci, P., Arpaia, G., et al. (1998). The *tbf-1* gene from the white truffle *Tuber borchii* codes for a structural cell wall protein specifically expressed in fruitbody. *Fungal Genet. Biol.* 25: 87-99.

Ditengou, F.A., Béguiristain, T., and Lapeyrie, F. (2000). Root hair elongation is inhibited by hypaphorine, the indole alkaloid from the ectomycorrhizal fungus *Pisolithus tinctorius,* and restored by indole-3-acetic acid. *Planta* 211: 722-728.

Ditengou, F.A. and Lapeyrie, F. (2000). Hypaphorine from the ectomycorrhizal fungus *Pisolithus tinctorius* counteracts activities of indole-3-acetic acid and ethylene but not synthetic auxins in eucalypt seedlings. *Mol. Plant Microbe In.* 13: 151-158.

Duplessis, S., Sorin, C., Voiblet, C., Palin, B., Martin, F., and Tagu, D. (2001). Cloning and expression analysis of a new hydrophobin cDNA from the ectomycorrhizal basidiomycete *Pisolithus. Current Genet.* 39: 335-339.

Eisen, M.B., Spellman, P.T., Brown, P.O., and Botstein, D. (1998). Cluster analysis and display of genome-wide expression patterns. *P. Natl. Acad. Sci. USA* 95: 14863-14868.

Feugey, L., Strullu, D.G., Poupard, P., and Simoneau, P. (1999). Induced defense responses limit Hartig net formation in ectomycorhizal birch roots. *New Phytol.* 144: 541-547.

Fitter, A.H., Graves, J.D., Watkins, N.K., Robinson, D., and Scrimgeour, C. (1998). Carbon transfer between plants and its control in networks of arbuscular mycorrhizas. *Funct. Ecology* 12: 406-412.

Fitter, A.H., Hodge, A., and Daniell, T.J. (1999). Resource sharing in plant-fungus communities: Did the carbon move for you? *Trends Ecol. Evol.* 14: 70.

Gorfer, M., Tarkka, M.T., Hanif, M., Pardo, A.G., Laitianen, E., and Raudaskoski, M. (2001). Characterization of small GTPases Cdc42 and Rac and the relationship between Cdc42 and actin cytoskeleton in vegetative and ectomycorrhizal hyphae of *Suillus bovinus*. *Mol Plant Microbe Int.* 14: 135-144.

Grange, O., Bärtschi, H., and Gay, G. (1997). Effect of the ectomycorrhizal fungus *Hebeloma cylindrosporum* on in vitro rooting of micropropagated cuttings of arbuscular mycorrhiza-forming *Prunus avium* and *Prunus cerasus*. *Trees* 12: 49-56.

Grove, T.S. and Le Tacon, F. (1993). Mycorrhiza in plantation forestry. *Adv. Plant Pathol.* 9: 191-227.

Guttenberger, M. and Hampp, R. (1992). Ectomycorrhizins—Symbiosis-specific or artifactual polypeptides from ectomycorrhizas? *Planta* 188: 129-136.

Harrison, M.J. and Van Buuren, M.L. (1995). A phosphate transporter from the mycorrhizal fungus *Glomus versiforme*. *Nature* 378: 626-629.

Hertzberg, M., Aspeborg, H., Schrader, J., Andersson, A., Erlandsson, R., Blomqvist, K., Bhalerao, R., Uhlen, M., Teeri, T.T., Lundeberg, J., et al. (2001). A transcriptional roadmap to wood formation. *P. Natl. Acad. Sci. USA* 98: 14732-14737.

Hibbett, D.S., Gilbert, L.B., and Donoghue, M.J. (2000). Evolutionary instability of ectomycorrhizal symbioses in basidiomycetes. *Nature* 407: 506-508.

Hilbert, J.L., Costa, G., and Martin, F. (1991). Ectomycorrhizin synthesis and polypeptide changes during the early stage of eucalypt mycorrhiza development. *Plant Physiology* 97: 977-984.

Högberg, P., Nordgren, A., Buchmann, N., Taylor, A.F.S., Ekblad, A., Högberg, M.N., Nyberg, G., Ottosson-Löfvenius, M., and Read, D.J. (2001). Large-scale forest girdling shows that current photosynthesis drives soil respiration. *Nature* 411: 789-792.

Horan, D.P. and Chilvers, G.A. (1990). Chemotropism—The key to ectomycorrhizal formation? *New Phytol.* 116: 297-301.

Horan, D.P., Chilvers, G.A., and Lapeyrie, F.F. (1988). Time sequence of the infection process in eucalypt ectomycorrhizas. *New Phytol.* 109: 451-458.

Jargeat, P., Gay, G., Debaud, J.C., and Marmeisse, R. (2000). Transcription of a nitrate reductase gene isolated from the symbiotic basidiomycete fungus *Hebeloma cylindrosporum* does not require induction by nitrate. *Mol. Gen. Genet.* 263: 948-956.

Kim, S.J., Bernreuther, D., Thumm, M., and Podila, G.K. (1999). *LB-AUT7*, a novel symbiosis-regulated gene from an ectomycorrhizal fungus, *Laccaria bicolor*, is functionally related to vesicular transport and autophagocytosis. *J. Bacteriol.* 181: 1963-1967.

Kim, S.J., Hiremath, S.T., and Podila, G.K. (1999). Cloning and characterization of symbiosis-regulated genes from the ectomycorrhizal fungus *Laccaria bicolor*. *Mycol. Res.* 103: 168-172.

Kim, S.J., Zheng, J., Hiremath, S.T., and Podila, G.K. (1998). Cloning and characterization of a symbiosis-related gene from an ectomycorrhizal fungus *Laccaria bicolor*. *Gene* 222: 203-212.

Koide, R.T. and Kabir, Z. (2000). Extraradical hyphae of the mycorrhizal fungus *Glomus intraradices* can hydrolyse organic phosphate. *New Phytol.* 148: 511-517.

Kottke, I. and Oberwinkler, F. (1987). The cellular structure of the Hartig net: Coenocytic and transfer cell-like organization. *Nord. J. Bot.* 7: 85-95.

Lacourt, I., Duplessis, S., Abbà, S., Bonfante, P., and Martin, F. (2002). Isolation and characterization of differentially expressed genes in mycelium and fruit body of *Tuber borchii*. *Appl. Environ. Microbiol.* 68: 4574-4582.

Lagrange, H., Jay-Allemand, C., and Lapeyrie, F. (2001). Rutin, the phenolglycoside from eucalyptus root exudates, stimulates *Pisolithus* hyphal growth at picomolar concentrations. *New Phytol.* 149: 349-355.

Laurent, P., Voiblet, C., Tagu, D., De Carvalho, D., Nehls, U., De Bellis, R., Balestrini, R., Bauw, G., Bonfante, P., and Martin, F. (1999). A novel class of ectomycorrhiza-regulated cell wall polypeptides in *Pisolithus tinctorius*. *Mol. Plant Microbe In.* 12: 862-871.

Leake, J.R. (2001). Is diversity of ectomycorrhizal fungi important for ecosystem function? *New Phytol.* 152: 1-8.

LePage, B.A., Currah, R.S., Stockey, R.A., and Rothwell, G.W. (1997). Fossil ectomycorrhizae from the middle Eocene. *Am. J. Bot.* 84: 410-412.

Lorillou, S., Botton, B., and Martin, F. (1996). Nitrogen source regulates the biosynthesis of NADP-glutamate dehydrogenase in the ectomycorrhizal basidiomycete *Laccaria bicolor*. *New Phytol.* 132: 289-296.

Maldonado-Mendoza, I.E., Dewbre, G.R., and Harrison, M.J. (2001). A phosphate transporter gene from the extra-radical mycelium of an arbuscular mycorrhizal fungus *Glomus intraradices* is regulated in response to phosphate in the environment. *Mol. Plant Microbe In.* 14: 1140-1148.

Mankel, A., Krause, K., Genenger, M., Kost, G., and Kothe, E. (2000). A hydrophobin accumulated in the Hartig' net of ectomycorrhiza formed between *Tricholoma terreum* and its compatible host tree is missing in an incompatible association. *J. Appl. Bot.* 74: 95-99.

Martin, F. (2001). Frontiers in molecular mycorrhizal research—Genes, loci, dots and spins. *New Phytol.* 150: 499-507.

Martin, F., Boiffin, V., and Pfeffer, P. E. (1998). Carbohydrate and amino acid metabolism in the *Eucalyptus globus-Pisolithus tinctorius* ectomycorrhiza during glucose utilization. *Plant Physiology* 118: 627-635.

Martin, F. and Botton, B. (1993). Nitrogen metabolism of ectomycorrhizal fungi and ectomycorrhiza. *Adv. Plant Path.* 9: 83-102.

Martin, F., Duplessis, S., Ditengou, F., Lagrange, H., Voiblet, C., and Lapeyrie, F. (2001). Developmental cross talking in the ectomycorrhizal symbiosis: Signals and communication genes. *New Phytol.* 151: 145-154.

Martin, F., Laurent, P., De Carvalho, D., Voiblet, C., Balestrini, R., Bonfante, P., and Tagu, D. (1999). Cell wall proteins of the ectomycorrhizal basidiomycete *Pisolithus tinctorius:* Identification, function, and expression in symbiosis. *Fungal Genet. Biol.* 27: 161-174.

Meharg, A.A. and Cairney, J.W.G. (2000). Co-evolution of mycorrhizal symbionts and their hosts to metal-contaminated environments. *Adv. Ecol. Res.* 30: 69-112.

Näsholm, T., Ekblad, A., Nordin, A., Giesler, R., Högberg, M., and Högberg, P. (1998). Boreal forest plants take up organic nitrogen. *Nature* 392: 914-916.

Nehls, U., Béguiristain, T., Ditengou, F., Lapeyrie, F., and Martin, F. (1998). The expression of a symbiosis-regulated gene in eucalypt roots is regulated by auxins and hypaphorine, the tryptophan betaine of the ectomycorrhizal basidiomycete *Pisolithus tinctorius. Planta* 207: 296-302.

Nehls, U., Bock, A., Ecke, M., and Hampp, R. (2001). Differential expression of hexose-regulated fungal genes *AmPAL* and *AmMst1* within *Amanita/Populus* ectomycorrhizas. *New Phytol.* 150: 583-589.

Nehls, U., Ecke, M., and Hampp, R. (1999). Sugar- and nitrogen-dependent regulation of an *Amanita muscaria* phenylalanine ammonium lyase gene. *J. Bacteriol.* 181: 1931-1933.

Nehls, U., Kleber, R., Wiese, J., and Hampp, R. (1999). Isolation and characterization of a general amino acid permease from the ectomycorrhizal fungus *Amanita muscaria. New Phytol.* 144: 343-349.

Nehls, U. and Martin, F. (1995). Gene expression in roots during ectomycorrhiza development. In V. Stocchi, P. Bonfante, and M. Nuti (Eds.), *Biotechnology of Ectomycorrhizae. Molecular Approaches* (pp. 125-137). New York: Plenum Press.

Nehls, U., Mikolajewski, S., Ecke, M., and Hampp, R. (1999). Identification and expression-analysis of two fungal cDNAs regulated by ectomycorrhiza and fruit body formation. *New Phytol.* 144: 195-202.

Nehls, U., Mikolajewski, S., Magel, E., and Hampp, R. (2001). Carbohydrate metabolism in ectomycorrhizas: Gene expression, monosaccharide transport and metabolic control. *New Phytol.* 150: 533-541.

Nehls, U., Wiese, J., Guttenberger, M., and Hampp, R. (1998). Carbon allocation in ectomycorrhizas: Identification and expression analysis of an *Amanita muscaria* monosaccharide transporter. *Mol. Plant Microbe In.* 11: 167-176.

Nehls, U., Wiese, J., and Hampp, R. (2000). Cloning of a *Picea abies* monosaccharide transporter gene and expression—Analysis in plant tissues and ectomycorrhizas. *Trees* 14: 334-338.

Niini, S., Tarkka, M.T., and Raudaskoski, M. (1996). Tubulin and actin patterns in Scots pine *(Pinus sylvestris)* roots and developing ectomycorrhiza with *Suillus bovinus. Physiol. Plantarum* 96: 186-192.

Paiva, N.L. (2000). An introduction to the biosynthesis of chemicals used in plant-microbe communication. *J. Plant Growth Regul.* 19: 131-143.

Peter, M., Courty, P.E., Kohler, A., Delaruelle, C., Martin, D., Tagu, D., Frey-Klett, P., Duplessis, S., Chalot, M., Podila, G., and Martin, F. (2003). Analysis of expressed sequence tags from the ectomycorrhizal basidiomycetes *Laccaria bicolor* and *Pisolithus microcarpus. New Phytologist* (in press).

Polidori, E., Agostini, D., Zeppa, S., Potenza, L., Palma, F., Sisti, D., and Stocchi, V. (2002). Identification of differentially expressed cDNA clones in *Tilia platyphyllos-Tuber borchii* ectomycorrhizae using a differential screening approach. *Mol. Gen. Genomics* 266: 858-864.

Read, D.J. (1991). Mycorrhizas in ecosystems. *Experientia* 47: 376-390.

Redecker, D., Kodner, R., and Graham, L.E. (2000). Glomalean fungi from the Ordovician. *Science* 289: 1920-1921.

Robinson, D. and Fitter, A. (1999). The magnitude and control of carbon transfer between plants linked by a common mycorrhizal network. *J Exp. Bot.* 50: 9-13.

Rousseau, J.V.D., Sylvia, D.M., and Fox, A.J. (1994). Contribution of ectomycorrhiza to the potential nutrient absorbing surface of pine. *New Phytol.* 128: 639-644.

Rygiewicz, P.T. and Andersen, C.P. (1994). Mycorrhizae alter quality and quantity of carbon allocated below ground. *Nature* 369: 58-60.

Selosse, M.-A. and Le Tacon, F. (1998). The land flora: A phototroph-fungus partnership? *Trends Ecol. Evol.* 13: 15-20.

Simard, S.W., Perry, D.A., Jones, M.D., Myrold, D.D., Durall, D.M., and Molina, R. (1997). Net transfer of carbon between ectomycorrhizal tree species in the field. *Nature* 388: 579-582.

Simoneau, P., Viemont, J.D., Moreau, J.C., and Strullu, D.G. (1993). Symbiosis-related polypeptides associated with the early stages of ectomycorrhiza organogenesis in birch (*Betula pendula* Roth). *New Phytol.* 124: 495-504.

Smith, S.E. and Read, D.J. (1997). *Mycorrhizal Symbiosis,* Second Edition. London: Academic Press.

Soragni, E., Bolchi, A., Balestrini, R., Gambaretto, C., Percudani, R., Bonfante, P., and Ottonello, S. (2001). A nutrient-regulated, dual localization phospholipase A2 in the symbiotic fungus *Tuber borchii. EMBO J.* 20: 5079-5090.

Sterky, F., Regan, S., Karlsson, J., Hertzberg, M., Rohde, A., Holmberg, A., Amini, B., Bhalerao, R., Larsson, M., Villarroel, R., et al. (1998). Gene discovery in the wood-forming tissues of poplar: Analysis of 5,692 expressed sequence tags. *P. Natl. Acad. Sci. USA* 95: 13330-13335.

Sundaram, S., Kim, S.J., Suzuki, H., Mcquattie, C.J., Hiremath, S.T., and Podila, G.K. (2001). Isolation and characterization of a symbiosis-regulated *ras* from the ectomycorrhizal fungus *Laccaria bicolor. Mol. Plant Microbe In.* 14: 618-628.

Tagu, D., Faivre Rampant, P., Lapeyrie, F., Frey-Klett, P., Vion, P., and Villar, M. (2001). Variations in the ability to form ectomycorrhizas in F1 progeny of an interspecific poplar (*Populus* spp.) cross. *Mycorrhiza* 10: 237-240.

Tagu, D. and Martin, F. (1995). Expressed sequence tags of randomly selected cDNA clones from *Eucalyptus globulus-Pisolithus tinctorius* ectomycorrhiza. *Mol. Plant Microbe In.* 8: 781-783.

Tagu, D., Nasse, B., and Martin, F. (1996). Cloning and characterization of hydrophobins-encoding cDNAs from the ectomycorrhizal basidiomycete *Pisolithus tinctorius*. *Gene* 168: 93-97.

Tagu, D., Python, M., Cretin, C., and Martin, F. (1993). Cloning symbiosis-related cDNAs from eucalypt ectomycorrhiza by PCR-assisted differential screening. *New Phytol.* 125: 339-343.

Tarkka, M., Niini, S.S., and Raudaskoski, M. (1998). Developmentally regulated proteins during differentiation of root system and ectomycorrhiza in Scots pine *(Pinus sylvestris)* with *Suillus bovinus*. *Physiol. Plantarum* 104: 449-455.

Tarkka, M.T., Vasara, R., Gorfer, M., Raudaskoski, M. (2000). Molecular characterization of actin genes from homobasidiomycetes: Two different actin genes from *Schizophyllum commune* and *Suillus bovinus*. *Gene* 25: 27-35.

Taylor, A.F.S., Martin, F., and Read, D.J. (2000). Fungal diversity in ectomycorrhizal communities of Norway spruce [*Picea abies* (L.) Karst.] and beech (*Fagus sylvatica* L.) along North-South transects in Europe. In E.D. Schulze (Ed.), *Ecological Studies: Carbon and Nitrogen Cycling in European Forest Ecosystems* (pp. 343-365), Volume 142. Berlin, Heidelberg: Springer-Verlag.

Voiblet, C., Duplessis, S., Encelot, N., and Martin, F. (2001). Identification of symbiosis-regulated genes in *Eucalyptus globulus-Pisolithus tinctorius* ectomycorrhiza by differential hybridization of arrayed cDNAs. *Plant J.* 25: 181-191.

Voiblet, C. and Martin, F. (2000). Identifying symbiosis-regulated genes in *Eucalyptus globulus-Pisolithus tinctorius* ectomycorrhiza using suppression subtractive hybridization and cDNA arrays. In P.J.G.M. de Wit, T. Bisseling, and W.J. Stiekema (Eds.), *Biology of Plant-Microbe Interactions, Volume 2* (pp. 208-213). Amsterdam: International Society for Molecular Plant-Microbe Interactions.

Whetten, R., Sun, Y.-H., Zhang, Y., and Sederoff, R. (2001). Functional genomics and cell wall biosynthesis in loblolly pine. *Plant Mol. Biol.* 47: 275-291.

Wright, D.P., Scholes, J.D., Read, D.J., and Rolfe, S.A. (2000). Changes in carbon allocation and expression of carbon transporter genes in *Betula pendula* Roth. colonized by the ectomycorrhizal fungus *Paxillus involutus* (Batsch) Fr. *Plant Cell Environ.* 23: 39-49.

PART II:
MOLECULAR BIOLOGY
OF WOOD FORMATION

Chapter 5

Genomics of Wood Formation

Jae-Heung Ko
Jaemo Yang
Sookyung Oh
Sunchung Park
Kyung-Hwan Han

INTRODUCTION

Trees undergo secondary growth and produce a woody body as a result of growth and differentiation of cells produced by the cambial meristem. This secondary growth is one of the most important biological processes on earth. For example, trees make up over 90 percent of the terrestrial biomass and play a prominent role in offsetting the gases released by the burning of fossil fuels, thus mitigating the potential effects of global warming. The total carbon uptake in the continental United States is equivalent to 20 to 40 percent of the fossil fuel emissions worldwide (Pacala et al., 2001). Trees serve as a primary feedstock for biofuel, fiber, solid wood products, and various natural compounds.

Various genomics tools, including expressed sequence tags (ESTs), cDNA microarrays, cDNA-AFLP (amplified fragment length polymorphism), enhancer/gene trap analysis, and GeneChip arrays, have been developed and used for wood formation studies. Furthermore, the discovery that *Arabidopsis* can be induced to produce secondary xylem (i.e., wood formation) allowed tree biologists to tap into the enormous genomics resource available for the model plant. In addition, the recent advances in proteomics technology promise a new tool for obtaining a more comprehensive view on wood formation. The poplar genome-sequencing project, newly launched by the U.S. government, will also provide invaluable resources to advance our understanding of the structural genomics within the tree species. Taken together, these emerging tools will undoubtedly offer a unique opportunity to overcome the problems associated with tree studies.

WOOD BIOSYNTHESIS

Wood is formed as a result of the successive addition of secondary xylem, which differentiates from the vascular cambium. This developmental continuum of secondary phloem and xylem is affected by the secondary vascular system through its organization and its rhythm of activity. Phloem and xylem differentiate radially on each side of the vascular cambium. The cells on the xylem side of the cambium first pass through a dividing zone in which the xylem mother cells continue to divide, then an expansion zone in which the derivative cells expand to their final size, next through a maturation zone in which lignification and secondary cell wall thickening occurs, and finally through a zone of programmed cell death in which all cellular processes are terminated (Chaffey, 1999). The resulting mature secondary xylem includes xylem parenchyma, fibers, vessels, and tracheary elements. The cells of tracheary elements lose their nuclei and other contents during the xylem differentiation, leaving a hollow tube that is part of a vessel.

The trunk wood of many tree species has two distinctly different regions: sapwood and heartwood. The proportion of the two varies according to species, age of trees, growth rate, and environmental conditions (Hillis, 1987). Sapwood conducts sap (water, solutes, and gases) from the root to all parts of the tree, provides structural support for the entire tree, and serves as a reservoir for water, energy, minerals, and solutes. The ray cells in sapwood are alive (Kozlowski and Pallardy, 1997) and serve as the source of raw materials for secondary substances. The ray parenchyma may also serve as radial communication channels from the cambium through the sapwood, while axial parenchyma cells function largely as storage tissue. On the other hand, the heartwood resulting from physiological cell death is defined as the "dead" central core of the woody axis and provides only passive support to the tree.

Vascular Tissue Formation

In later development of woody plants, the primary vascular tissues are replaced by secondary vascular tissues produced by a secondary meristem (i.e., the vascular cambium). The vascular cambium originates from the procambium, which is derived from the apical meristem, and occurs as a continuous ring of cells between the xylem and the phloem throughout the length of fully expanded shoots and roots (the so-called cambial zone) (Larson, 1994; Mauseth, 1998). This secondary meristem plays several crucial roles in tree growth and development (Catesson et al., 1994). It increases stem diameter by the production of functional vascular elements through tangential (or periclinal) divisions, facilitates stem enlargement and the maintenance of the meristem itself through radial (or anticlinal) and transverse divisions, serves as a bridging point for the translocation of nutrients between phloem and xylem, and acts as a communication center for the transmission of signals, such as plant growth regulators, in both the axial and radial directions. However, regulation of the cambial growth that controls wood production and diameter growth of trees is not currently understood.

In *Arabidopsis,* several mutants that interfere with various aspects of vascular development have been isolated (Dengler and Kang, 2001). Some of these mutants have been described as having auxin transport (*PIN1*; Gälweiler et al., 1998) or auxin signaling defects, and loss of tissue continuity within the vascular system (Hardtke and Berleth, 1998; Berleth and Sachs, 2001). A recessive mutation in the *WOODEN LEG (WOL)* gene results in reduced proliferation of procambial cells, altered xylem organization, and absence of phloem cells within the root vascular tissue (Scheres et al., 1995; Mahonen et al., 2000). It is notable that the *WOL* gene encodes a putative two-component *His* kinase with a receptor domain, suggesting that it functions as a signal transducer (Mahonen et al., 2000). A class of transcription factors that are involved in vascular development was reported to be the homeodomain-leucine zipper (HD-ZIP) family. Zhong and Ye (1999) isolated an *Arabidopsis* gene that affects fiber and vascular differentiation from a pendant stem phenotype. The gene *(IFL1; interfascicular fiberless 1)* was found to encode HD-ZIP protein and expressed both in the expected interfascicular regions and in the vascular bundles. Recently, it has been reported that *IFL1* and *REVOLUTA (REV)* have mutations in the same gene (Ratcliffe et al., 2000). An analysis of several alleles indicated that *IFL1/ REV* is necessary for lateral meristem initiation and normal organ development (Talbert et al., 1995; Otsuga et al., 2001), as well as for proper differentiation of vascular cells of the stem (Zhong et al., 1997). ATHB-8, -9, -14, and -15, together with *IFL1/REV,* are also members of the HD-ZIP III family (Sessa et al., 1998). Ectopic expression of ATHB-8 in *Arabidopsis* plants increased the production of xylem tissue and promoted vascular cell differentiation. These results are consistent with the hypothesis that ATHB-8 is a positive regulator of proliferation and differentiation, and participates in a positive feedback loop in which auxin signaling induces the expression of ATHB-8, which in turn positively modulates the activity of procambial and cambial cells to differentiate (Baima et al., 2001).

Secondary Cell Wall Formation

Secondary cell wall formation in xylem cells occurs after completion of radial cell expansion. The orientation and arrangement of cellulose microfibrils is random or longitudinal in the primary cell wall. This secondary cell wall deposition is obvious by the formation of a dense array of helical and almost transverse cellulose microfibrils, which limit further radial expansions.

Cellulose microfibrils, a major component of secondary cell walls, are synthesized from plasma membrane-bound enzyme complexes that are rosette-shaped in the freeze-fracture preparations of plant cells (Mueller and Brown, 1980; Brett, 2000). Recent immuno-histochemistry study has shown that a cellulose synthase of *Vigna angularis* was localized to the rosette complex (Kimura et al., 1999). In arabidopsis, there are at least 10 cellulose synthases and more than 30 cellulose synthase-like genes, all of which have the type A catalytic domain and, therefore, are called *CesA* genes (Richmond and Somerville, 2001; see Chapter 6). Among them, at least two (*RSW1* and *PRC1*) are known to have a role in primary wall biosynthesis (Arioli et al., 1998; Fagard et al., 2000). However, *IRX3* (irregular xylem 3) and *IRX1* (irregular xylem 1) are known to act on secondary wall biosynthesis (Taylor et al., 1999; Turner et al., 2001). Recently, the *CesA* genes have been cloned from the wood-forming tissues of tree species such as pine (Allona et al., 1998) and populus (Wu et al., 2000). In populus, two genes (*PtCesA1* and *PtCesA2*) are reported. Interestingly, *PtCesA2* is expressed specifically during secondary wall biosynthesis in developing xylems but not in the phloem fiber (Wu et al., 2000). It suggests that the individual *CesA* gene has a cell-specific expression pattern and role. Cellulose biosynthesis involves initiation, elongation, and termination of the sugar chain (see Chapter 6). However, biochemical mechanisms underlying each step of cellulose synthesis remain to be elucidated. Recently, Peng and colleagues (2002) identified a lipid sitosterol-β-glucoside (SG) as a primer for cellulose synthesis in plants. SG acts as a primer, initiating the polymerization of glucan chains catalyzed by CesA proteins of the cellulose synthase complex. The resulting sitosterol cellodextrins (SCDs) may be cleaved from the sitosterol primer by Korrigan cellulase. Further elongation of the cellodextrins, catalyzed by the same or different CesA proteins, produces the glucan chains of cellulose, which then coalesce into microfibrils (Peng et al., 2002).

A precursor for carbohydrate components of the cell wall is uridine diphosphate (UDP)-D-glucose (Reiter and Vanzin, 2001), which is produced by sucrose synthase (SuSy) or UDP-glucose pyrophosphorylase. While the enzyme activity of SuSy is low in the cambium, the highest activity is found in the region of secondary wall formation in developing xylem of Scots pine (Uggla et al., 2001). This supports a hypothesis that the membrane-associated isoforms of SuSy supplies UDP-D-glucose directly to cellulose synthase (Haigler et al., 2001). Matrix carbohydrates and lignin precursors for secondary cell wall deposition are probably secreted to the cell wall via exocytosis. Because numerous vesicles are usually observed near the cell wall, the process of fusion itself has been suggested to be a rate-limiting step in the cell wall biosynthesis (Mellerowicz et al., 2001). ROP1 (rho-related GTPase from Plants 1) has been known to play a regulatory role in vesicle exocytosis and pollen tube enlargement (Zheng and Yang, 2000). Interestingly, genes coding for similar proteins, called *RAC13* and *RAC9,* were specifically up-regulated in cotton fibers during the transition from the primary to secondary wall (Delmer et al., 1995). This suggests that the rho proteins may play a role in cell wall formation of the secondary xylem. Within the wood-forming tissues, cell wall composition differs between the primary and secondary wall (Mellerowicz et al., 2001).

Lignin Biosynthesis and Deposition

Lignin is a complex phenolic polymer that reinforces the walls of certain cells in the vascular tissues of higher plants. Lignin plays an important role in mechanical support, water transport, and pathogen resistance. Lignification starts in vessel elements and ray cells that have secondary cell walls, whereas fibers and isolated cells lignify later (Murakami et al., 1999).

Lignin is mainly derived from the dehydrogenative polymerization of three different hydroxycinnamyl alcohols (or monolignols), *p*-coumaryl alcohol, coniferyl alcohol, and sinapyl alcohol (H, G, and S, respectively). These monolignols differ from one another only by their degree of methoxylation and are formed by removal of water from sugars to create aromatic structures. These reactions are not reversible. The content and composition of lignin are known to vary among taxa, tissues, cell types, and cell wall layers, and to depend on the developmental stage of the plant and the environmental conditions (reviewed in Sederoff et al., 1999). Biosynthesis of lignin precursors proceeds through the common phenylpropanoid pathway, starting with the deamination of phenylalanine to cinnamic acid (for details see Chapter 7). Additional enzymatic reactions include hydroxylation of the aromatic ring, the methylation of selected phenolic hydroxyl groups, the activation of cinnamic acids to cinnamoyl-CoA esters, and the reduction of these esters to cinnamaldehydes and further to cinnamyl alcohols. The first involves the enzymes of the common phenylpropanoid pathway, phenylalanine ammonia-lyase (PAL), cinnamate 4-hydroxylase (C4H), and hydroxycinnamate CoA ligase (4CL). Second is the methylation step of monolignols, caffeate/5-hydroxyferulate-*O*-methyltransferase and ferulate-5-hydroxylase (F5H). The third level involves the last steps of monolignol biosynthesis, hydroxycinnamoyl-CoA:NADPH oxidoreductase and cinnamyl alcohol dehydrogenase (CAD).

Temporal and spatial expression of genes involved in lignin biosynthesis has been extensively studied in recent years (see Chapter 7). A strong activity of the *PAL* promoter has been detected in the developing xylem (Kawamata et al., 1997). PAL and C4H are expressed in the vascular bundles, and the expression of both of the genes is closely coregulated in parsley (Koopmann et al., 1999). Poplar genes encoding two isoforms of 4CL are differentially expressed (Hu et al., 1998). One of them is specifically expressed in developing xylem and probably involved in lignin biosynthesis, whereas the other is epidermis specific and probably associated with the biosynthesis of nonlignin phenylpropanoids. *CAD* gene expression was identified in the vascular tissues, the periderm, and the cambium by promoter analysis and in situ hybridization (Feuillet et al., 1995; Regan et al., 1999). CCoAOMT (caffeoyl-CoA *O*-methyltransferase) was immunolocalized in all cell types of the developing xylem (Zhong et al., 2000).

Little is known about the regulatory mechanisms of lignin biosynthesis genes. Recently, altered expression of the genes encoding MYB-related (Tamagnone et al., 1998) and LIM-related transcription factor (Kawaoka et al., 2000) has been shown to affect the expression of the genes involved in phenylpropanoid biosynthesis pathway, and to affect lignin content and the accumulation of phenolics in transgenic tobacco plants. Two mutants, *elp1* (ectopic lignification deposition) (Zhong et al., 2000) and *eli1* (ectopic lignification) (Cano-Delgado et al., 2000), appear to alter the normal pattern of lignin deposition and, therefore, offer an opportunity to examine the spatial control of lignin deposition. Stems of *elp1* plants had about 20 percent more lignin content than wild type. This increase in lignin content was accompanied by an increase in the activities of enzymes involved in lignin biosynthetic pathway such as PAL, CCoAOMT, and CCR. The most likely role of the ELP1 gene product is as a negative regulator of the lignin biosynthetic pathway that under normal circumstances suppresses lignin deposition in the stem (Zhong et al., 2000). In contrast to *elp1*, the primary roots of *eli1* plants demonstrate abnormal lignification patterns. This ectopic lignification pattern is associated with reduced cell elongation and shorter and thicker primary roots. Mutants with cell elongation defects (e.g., *lit, rsw1,* and *kor*) show ectopic lignification, suggesting that the ectopic lignification of *eli1* may be a consequence of a cell elongation defect (Cano-Delgado et al., 2000). Although there is increasing knowledge on the biochemical properties of the enzymes involved in the biosynthesis of monolignols, the precise order in which these reactions occur is not fully understood. Furthermore, many uncertainties surround the in vivo role of the respective enzymes in the monolignols biosynthesis pathway and polymerization of monolignols.

Programmed Cell Death

When lignification is completed, vessel elements undergo programmed cell death (PCD), which involves hydrolysis of the protoplast. Animal PCD is mediated by the family of cysteinyl aspartate-specific proteinase, known as *caspases,* which requires an aspartic-acid residue to the immediate left and at least four amino acids on the amino-terminal side of the scissile bond. Specific consensus sequences within substrate proteins are required for cleavage by caspases, limiting the number of cellular targets. Both cysteine and serine proteases have been involved in the tracheary element (TE) formation in *Zinnia* system (Beers and Freeman, 1997; Ye and Varner, 1995) and in *Pinus banksiana* (Iliev and Savidge, 1999).

Programmed cell death of TE involves the progressive degradation of organelles, along with the removal of protoplasts and parts of unlignified primary walls. Epifluorescent microscopic observations revealed that differentiating *Zinnia* TEs have all organelles such as the nucleus, vacuole, and many active mitochondria and chloroplasts, and that loss of such cell contents occurs abruptly several hours after the formation of visible secondary wall thickenings (Groover et al., 1997). The earliest sign of organelle degradation is vacuole collapse (Fukuda, 1996; Groover et al., 1997), which leads to the final degradation of cell contents. Vacuole collapse causes the release of insulated hydrolytic enzymes (such as protease, DNases, and RNases) and allows them to attack the organelles (reviewed in Fukuda, 2000). Calcium has been implicated in the regulation of both secondary cell wall formation and PCD. Groover and Jones (1999) reported that exogenously applied trypsin induced DNA degradation, and inhibition of calcium influx using chelators and channel blockers suppressed its inhibition. Also, they found a 40 kDa serine protease secreted during PCD of TE, activity of which can be prevented by a trypsin inhibitor. Based on those results, they proposed the model of cell death induction that a secreted protease digests some protein component of cell wall, the resultant protein fragments promote an influx of calcium through the plasma membrane, and then the increase in intracellular calcium causes the vacuole collapse.

Recently, Jones (2001) has reviewed the general mechanism of the plant PCD. It appears that calcium flux and collapse of the vacuole may be the universal trigger of plant cell death; however, the differences in the way death is manifested results from different mechanisms for processing the cell corpse. The different profiles of hydrolases loaded into the vacuole determine the manifestation of death. For tracheary elements, the protoplasm but not secondary cell walls are autolyzed.

Heartwood and Its Extractives Formation

Sapwood is gradually converted to inactive heartwood as the tree ages. The ray parenchyma cells at a narrow zone (called the "transition zone") adjacent to the heartwood undergo changes that result in increased synthesis of secondary products. The reserve materials in the parenchyma cells of the sapwood are used, with sucrose transported via vascular bundles from the leaves, for synthesis of heartwood extractives such as condensed tannins, terpenes, flavonoids, lignans, lipids, stilbenes, and tropolones (Burtin et al., 1998; Hillinger et al., 1996a,b; Hillis, 1987; Magel et al., 1994). The types and amounts of these secondary metabolites are dependent on species and region. Several enzymes have shown increased activity in the transition zone. For example, elevated levels of phenol-oxidizing enzymes have been observed in the transition zone of *Eucalyptus polyanthemos* (Hillis and Yazaki, 1973), *Eucalyptus elaeophora* (Wardrop and Cronshaw, 1962), and *Fagus sylvatica* (Dietrichs, 1964). Succinate dehydrogenase in *Melia azedarach* was significantly active only in the transition zone (Baqui et al., 1979). Furthermore, it has shown that two key enzymes for flavonoid biosynthesis (chalcone synthase and PAL) are highly active in the sapwood-heartwood transition zone of black locust (Magel et al., 1991). Chalcone synthase is involved in flavonoid synthesis. In addition, the stimulus for biosynthesis of robinetin, a flavonoid with strong antimicrobial activity, apparently occurs in the transition zone in *Instsia* species (Hillis, 1996). Other enzymes reported to be highly active in the transition zone include catechol oxidase, glucose-6-phosphate dehydrogenase, malic dehydrogenase, maltase, and amylase (Hillis, 1987). Previous theories suggested that the production of secondary metabolites at the transition zone was due to the elimination of excess carbohydrate. However, current thought is that these compounds provide a deterrent to pathogen attack in regions of the tree in which a biochemical response is not possible (passive resistance).

REGULATION OF WOOD FORMATION

The production of wood in trees is determined by the rate of cambial growth. Simultaneous increases in both the radial number of dividing cells and the rate of cell division in the cambial zone will increase productivity.

This cambial activity is under strict developmental control, which forms typical patterns of wood formation in time and space (Larson, 1994). Therefore, the quantity and quality of the final wood is the result of a patterned control of numbers, places, and planes of cambial cell division, and a subsequent differentiation of the cambial derivatives into tracheary elements, vessels, fibers, parenchyma, and sieve elements (Mauseth, 1998). This developmental pattern requires that every cell must express the appropriate genes in a coordinated manner upon receipt of positional information. This regulation is under strong genetic control (Zobel and Jett, 1995).

Hormonal Regulation of Secondary Growth

The plant growth regulator indole acetic acid (IAA) is a key signal for the production of xylem and phloem by the vascular cambium (Little and Sundberg, 1991). Numerous experiments have demonstrated that the application of auxin to cambial tissues stimulates xylem production (i.e., cambial cell division) (Uggla et al., 1998). A change in IAA supply to the vascular cambium results in a change in radial gradient width. Uggla and colleagues (1996) reported that there was a steep radial gradient of IAA across the cambial region of *Pinus sylvestris,* with a peak in the cambial zone. This suggests that IAA acts as a positional signal, from which cambial derivatives interpret their radial position, and regulates cambial growth rate by determining the radial population of dividing cambial-zone cells. The recent demonstration of steep concentration gradients of soluble carbohydrates across the cambium (Uggla et al., 2001), together with accumulating evidence in favor of sugar sensing in plants (Sheen et al., 1999), provides substantial evidence to the hypothesis that auxin/sucrose ratio determines the positioning of the cambium. However, no evidence supports that IAA in the cambial meristem plays an additional role in controlling rates of cell division.

Gibberellins (GAs), when applied with auxin, stimulate meristematic activity and xylem fiber elongation (Digby and Wareing, 1966) and are required for longitudinal growth. The role for GAs in wood formation and their potential in tree biotechnology has recently been demonstrated by overexpression of a GA-20 oxidase in hybrid aspen (Eriksson et al., 2000), in which the transgenic trees exhibited an increased longitudinal and radial growth as well as increased xylem fiber length.

Although cytokinins have a well-established function in cell division, their role in cambial growth is far from clear (Little and Savidge, 1987). Accurate information on endogenous cytokinin levels in the cambial region of trees is not available. Zeatin, a biologically active cytokinin, was present only in trace amounts. However, in many systems including *Zinnia* cultures, cytokinin is needed for TE induction. When cytokinin is not required, it seems likely that the experimental tissue already contains a sufficient amount of cytokinins (Milioni et al., 2001).

Exogenous ethylene stimulates cambial cell division, possibly by increasing auxin levels through interaction with auxin transporter (Eklund and Little, 1998). Previous studies showed that the number of auxin-induced TEs decreased when ethylene biosynthesis and response inhibitors were applied (Miller and Roberts, 1984), indicating that ethylene plays a role in xylogenesis. However, the need for exogenous ethylene in TE induction has not been demonstrated. Perhaps IAA induces the necessary levels of ethylene through the induction of 1-aminocyclopropane-1-carboxylic acid (ACC) synthase (Abel et al., 1995), but conceivable evidence in xylogenesis is still lacking.

Seasonal Cycle of Secondary Growth

The vascular cambium of temperate tree species follows the seasonal cycle of activity and dormancy (Baier et al., 1994). Cambial cell phenology in white ash *(Fraxinus americana)* has been studied in Canada (Zhong et al., 1995). The cambial cells were found to have no detectable mitoses until April, resumed cell division activity in the spring (between March 30 and May 1), and entered dormancy around September 10. Following springtime resumption of cambial cell-division activity, the derivatives of cambial fusiform cells expanded radially to form a zone of primary-walled cells. Then, the primary-walled derivatives underwent terminal differentiation, involving secondary wall thickening, lignification, and protoplast autolysis, to become tracheids. In poplar, the total amount of pectins per gram of cell wall was rather stable throughout the annual cycle, while its composition changed dramatically (Baier et al., 1994). Recently, Iliev and Savidge (1999) observed the highest proteolytic activity in the cambial zone and developing xylem of *Pinus banksiana* during the most active period of radial expansion of cambial derivatives over an annual cycle of growth and dormancy. Cambial dormancy in the stem of conifer species is induced by short photoperiods (Eklund et al., 1998; Mellerowicz, Coleman, et al., 1992; Mellerowicz, Riding, and Little, 1992). The molecular mechanisms regulating these events are still unknown. However, several explanations have been suggested. First, a short photoperiod may increase the concentration of the growth inhibitor abscisic acid (ABA) in the cambial region. It has been reported that exogenously provided ABA decreases tracheid radial diameter and inhibits tracheid development in conifers (Little and Eidt, 1968; Pharis et al., 1981). Second, short photoperiod could decrease the concentration of the growth regulator IAA in the cambial region. Third, resulting from high rates in both the growth and maintenance components of respiration, the O_2 concentration in the cambial region is decreased to the extent that it inhibits IAA action during either the enlargement phase of tracheid differentiation or the division of cambial cells.

ARABIDOPSIS *AS A MODEL FOR THE STUDY OF WOOD FORMATION*

In recent years, many aspects of biology common to all plants have been extensively elucidated using the model plant *Arabidopsis thaliana*. *Arabidopsis* has many advantages for genome analysis, including a short regeneration time, small size, and small nuclear genome. These advantages promoted the growth of a scientific community that has investigated the biological processes of *Arabidopsis* and has developed many genetic resources including genomics tools. In recent years, *Arabidopsis* has been shown to produce a significant quantity of secondary xylem (i.e., wood) and express all of the major components of wood development during its ontogeny, when treated properly, thereby has been used as a model for the study of wood and fiber production in trees (Lev-Yadun, 1994, 1997; Zhao et al., 2000). Zhao and colleagues (2000) demonstrated for the first time that *Arabidopsis* xylem and bark could in fact be easily separated and used for genomic research (Beers and Zhao, 2001), although Sterky and colleagues (1998) described that the xylem-forming tissue cannot be easily isolated from other tissues. Secondary xylem is derived from the vascular cambium that originates from the fascicular procambium and the interfascicular parenchyma between the strands of procambium (Lev-Yadun and Flaishman, 2001). This cambial activity is reduced by flowering. Therefore, any treatments that stop flower development promote prolonged cambial activity.

To take full advantage of *Arabidopsis* as a model plant, especially with regard to secondary growth, we established experimental conditions as described by Lev-Yadun (1994) for inducing secondary xylem development and making use of existing genomics resources, such as enhancer-trap lines, Affymetrix chips, cDNA-AFLP analysis, and cDNA microarrays. The treated *Arabidopsis* plants clearly show stem thickening through the formation and accumulation of secondary xylem as reported previously (Lev-Yadun, 1994; Zhao et al., 2000). We completed three biological replications of Affymetrix chip analysis with four different probes (xylem versus bark; xylogenesis treatment versus control) and are analyzing the data. The results from our initial analyses identified a number of genes either up- or down-regulated in the secondary xylem tissues of the treated *Arabidopsis* stems (Table 5.1; S. Oh, K.-H. Han et al., manuscript in preparation).

In addition, we are screening the *Arabidopsis* enhancer trap lines generated via T-DNA transformation in the Thomas Jack Laboratory (Campisi et al., 1999). In these lines, the enhancer or promoter proximal elements in genomic DNA adjacent to the T-DNA insertion increase the expression of the GUS (β-glucuronidase) reporter gene often in cell-, tissue-, and organ-specific expression patterns depending on the genomic location of the insertion. GUS staining is then used to detect the expression patterns. About 3,000 individual lines (approximately 26 percent of total 11,370 lines) were screened for GUS staining in developing secondary xylem. Sixty-four of the screened lines have shown stem-specific GUS expression. We are currently estimating the copy number of the 64 candidates lines by Southern blot analysis. We will then use TAIL-PCR (thermal asymmetric interlaced polymerase chain reaction; Liu and Whittier, 1995) to isolate the flanking DNA regions of the lines with single copy insertion. Our initial results confirm those reported previously (Zhao et al., 2000; Beers and Zhao, 2001) and provide further evidence that utilizing *Arabidopsis* and the wealth of its genomics resources will provide a unique opportunity to investigate the genes associated with secondary xylem development.

GENOMICS APPROACH TO WOOD FORMATION

Genomics uses a genome-wide experimental approach to assess gene expression and function. It changes and facilitates the manner in which we acquire and utilize new knowledge on fundamental biological processes in plants (Walbot, 1999; Hamadeh and Afshari, 2000). Single-pass sequence of cDNAs isolated from specialized tissues and organs offers a complementary approach to biochemical and genetic analysis (Adams et al., 1992; Allona et al., 1998; Covitz et al., 1998). Expressed sequence tags (ESTs) generated by such an effort can be used as a tool for rapid identification of new genes. DNA microarray technology provides a simple and economical way to explore the collection of genes that are expressed from genomic DNA (Brown and Botstein, 1999; Duggan et al., 1999; Lockhart and Winzeler, 2000). Furthermore, the global expression data obtained from DNA microarray hybridization experiments can be analyzed using standard statistical algorithms to arrange genes according to similarity in gene expression pattern (Eisen et al., 1998). Such cluster analysis and display systems will facilitate the identification of cambial growth-specific genes and the study of their expression patterns.

TABLE 5.1. Partial list of differentially expressed *Arabidopsis* genes identified by Affymetrix GeneChip analysis

Up-regulated in control stem (fold increase)	Up-regulated in treated stem (fold increase)	Up-regulated in bark (fold increase)	Up-regulated in xylem (fold increase)
No match (70)	Multicatalytic endopeptidase (115)	Extensin-like protein (24)	Serine carboxypeptidase isolog (33)
Pectinesterase (34)	Beta-1,3-glucanase (98)	Putative peroxidase (18)	Diphenol oxidase,
Expansin (At-EXP5) (30)	DNA-binding factor (96)	Lipoxygenase (11)	Laccase (24)
Polygalacturonase (30)	Thioredoxin H (90)	Peroxidase-like protein (10)	BURP domain-containing
Carbonic anhydrase (29)	MAP3K delta-1 protein	Hypothetical protein (10)	protein (22)
Hypothetical protein (27)	kinase (54)	Unknown protein (9)	Ferulate-5-hydroxylase
Hypothetical protein (26)	Basic chitinase gene (50)	Proline-rich cell wall	(FAH1) (18)
Lhcb4:3 protein (26)	No match (48)	protein (9)	Cellulose synthase-like
Hypothetical protein (25)	Glycine-rich protein 3	Putative protein (9)	protein (16)
Microtubule organization	(GRP3S) (47)	Vegetative storage	Germin-like protein (15)
1 (24)	Glutathione *S*-transferase	protein (8)	AT4 mRNA sequence (13)
Cell-cell signaling	(43)	Hypothetical protein (8)	Hypothetical protein (13)
protein (23)	Putative expansin (42)	Trehalose-6-phosphate	Cysteine proteinase (13)
Porphobilinogen	Fatty acid desaturase (36)	synthase (8)	Amino acid transporter (12)
deaminase (23)	Transcription factor	Aquaporin (water channel	R2R3-MYB transcription
Fimbrin 2 (22)	BBFa (36)	protein) (8)	factor (10)
Thylakoid ascorbate	Tropinone reductase (33)		Putative protein (10)
peroxidase (21)	Heat shock protein 70 (31)		
Hypothetical protein (20)	Defender against cell death		
	1 (28)		

Expressed Sequence Tag (EST) Analysis

ESTs can provide a relatively rapid means to identify genes expressed in wood. Comparison of ESTs with public databases of identified genes enables the assignment of putative identification to the cDNAs. Particularly, the ESTs derived from a specific tissue, such as wood-forming tissues, provide information on rare and tissue-specific transcripts (Allona et al., 1998; Sterky et al., 1998). A large number of ESTs have been analyzed from the wood-forming tissues of poplar (Sterky et al., 1998), black locust (Han et al., manuscript in preparation), and pine (Allona et al., 1998). Recently, Beers and Zhao (2001) produced 1,000 (500 from xylem and 500 from bark) ESTs from root-hypocotyl segments of *Arabidopsis* to identify genes with potential importance to vascular tissue differentiation and physiology. The distinct (serine, cysteine, and threonine) proteases have been implicated in the regulation of tracheary element differentiation (Zhao et al., 2000). Allona and colleagues (1998) obtained a total of 1,097 ESTs from immature xylem of loblolly pine (*Pinus taeda* L.). Among them, about 10 percent of the recognized genes encode factors involved in cell wall formation. Cell wall associated carbohydrate metabolism proteins were represented with 23 sequences, including cellulases and xyloglucan endotransglycosylases. Fifty clones corresponded to cell wall structural proteins, including extensin-like proteins along with proline-rich proteins, arabinogalactan-like proteins, and glycine-rich proteins. The lignin biosynthetic pathway genes such as PAL, C4H, OMT, 4CL, and CAD were also represented in the EST pool. Sterky and colleagues (1998) reported a total of 5,692 ESTs from the wood-forming tissues of two poplars, *Populus tremula* L. × *tremuloides* Michx. (cambial library) and *Populus trichocarpa* 'Trichobel' (developing-xylem library). Comparison of the two libraries revealed distinct differences in the relative distribution of the functional groups. The most significant difference was found in the expression of cell wall-related genes in the developing-xylem library, which was almost twice as high as in the cambial-region library. A comparison between the EST databases from wood-forming tissues of poplar and pine (Allona et al., 1998) has revealed several similarities, suggesting that the molecular control of wood formation has much in common between angiosperms and gymnosperms, despite the differences in wood structure and lignin composition (Sterky et al., 1998).We are studying the genomics of trunk wood formation and its associated extractives in a hardwood species. Such studies are made difficult by the limitation of obtaining high-quality mRNAs. We successfully isolated RNAs from the trunk wood of a ten-year-old black locust *(Robinia pseudoacacia)*. High-quality cDNA libraries (over 1.0×10^6 pfu) were constructed from sapwood,

sapwood-heartwood transition zone, and cambial region of mature black locust. Functional classification analysis of more than 3,400 ESTs indicates that primary metabolisms and protein synthesis/processing-related genes are among the most abundant, followed by those for defense, gene expression, and membrane transport in the cambial region. The genes encoding secondary metabolite biosynthetic enzymes (e.g., chalcone synthase, dihydroflavonol 4-reductase, and naringenin 3-dioxigenase) are among the most abundant in the transition zone library. The genes in all functional categories except cell wall structure and vesicular trafficking were less abundant in sapwood library, compared to cambial region and transition zone libraries (Figure 5.1). Cluster analysis of the ESTs showed that over 78 percent of the ESTs are singletons, suggesting a high level of complexity of the cDNA libraries. In other words, more than 78 percent of the randomly picked cDNA clones were unique genes (i.e., no overlapping among them). Those unique ESTs were used for construction of 2.8K unigene microarrays.

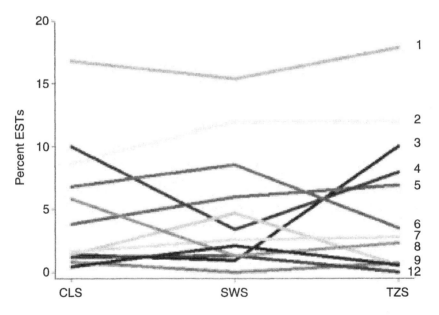

FIGURE 5.1. Percent of ESTs from 12 different functional categories represented in the three cDNA libraries: CLS (cambial region), SWS (sapwood), and TZS (transition zone). The functional categories used are (1) protein synthesis, (2) primary metabolism, (3) secondary metabolism, (4) defense, (5) signal transduction, (6) gene expression, (7) chromatin and DNA metabolism, (8) membrane transport, (9) cell division, (10) cell wall structure, (11) vesicular trafficking, and (12) cytoskeleton. (See also color photo section.)

Transcriptome Analysis by cDNA-AFLP

Many gel-based technologies have been developed for identification of differentially expressed genes (Bachem et al., 1996; Kawamoto et al., 1999; Shimkets et al., 1999; Sutcliffe et al., 2000). These methods are based on the principle that a complex starting mixture of cDNAs is fractionated into smaller subsets, in which cDNA tags are PCR amplified and separated on high-resolution gels. The observed differences in band intensity provide a good measure of the relative differences in the levels of gene expression. The most widely used method, cDNA-AFLP, has been applied with success to the systematic analysis of genes involved in particular biological processes (Durrant et al., 2000; Qin et al., 2000). In a study of the formation of tracheary elements in the *Zinnia* mesophyll cell system, over 50 partial sequences out of 600 different expressed cDNA fragments were isolated which are related to cell walls (Milioni et al., 2001). When compared to microarrays, the principal advantages of cDNA-AFLP are that it allows genome-wide expression analysis in any species without prior sequence knowledge and that both known and unknown genes can be analyzed. That is why this method is extremely valuable in tree species. However, technological improvements will most probably enhance the sensitivity and specificity of transcript detection within microarrays. Nevertheless, the high levels of redundancy in plant genomes will likely remain a major obstacle for detailed microarray studies. Fragment-based technologies, such as cDNA-AFLP, overcome the problem caused by redundancy.

We successfully isolated and extracted mRNAs from bark and xylem-enriched tissues of *Arabidopsis* after inducing secondary xylem differentiation as described previously (Lev-Yadun, 1994). Using those mRNAs, we performed cDNA-AFLP analysis for identifying differentially expressed genes between bark and xylem tissues. This experiment allowed survey of about 38,000 transcripts of *Arabidopsis*. From the initial experiments, we identified 124 gene fragments that are specific or abundant in xylem tissue and 45 gene fragments from bark tissue. For further confirmation of differential expression, reverse Northern blot analysis was adopted. The list of strongly expressed genes in secondary xylem tissue includes cellulose synthase catalytic subunit, serine carboxypeptidase, putative phytochelatin synthetase, putative glucosyltransferase, and several genes with unknown function (Ko et al., unpublished).

Microarray-Based Transcriptome Analysis

Hertzberg and colleagues (2001) have clearly demonstrated the usefulness of DNA microarray analysis by using a hybrid aspen unigene set consisting of 2,995 ESTs. This analysis revealed that the genes encoding lignin and cellulose biosynthetic enzymes, as well as a number of transcription factors and other potential regulators of xylogenesis, are under strict developmental stage-specific transcriptional regulation. Whetten and colleagues (2001) also used EST analysis and DNA microarray hybridization to identify the genes differentially regulated in the secondary xylem of loblolly pine. We constructed 2.8K unigene cDNA microarrays from the trunk wood of a mature black locust tree. These microarrays are being used to survey the gene expression profiles in different tissues such as cambial region, sapwood, and transition zone at different times of the year. Figure 5.2 shows the differential gene expression detected by microarray hybridization between cambial region and transition zone.

Gene expression profile data obtained from microarray analysis can be used to validate an existing hypothesis or generate new hypotheses. For example, heartwood formation occurs in late summer to late fall and at the beginning of dormancy in the temperate zones (Hillis, 1987). Earlier studies have indicated that heartwood formation occurs at the time of cambial dormancy in pine (Shain and Mackay, 1973), walnut, and cherry (Nelson, 1978). Studies of the cytology and coloration of the extractives suggested that heartwood formation commences in midsummer and continues into the fall and winter seasons in sugi (Nobuchi, Matsuno, and Harada, 1984) and black locust (Nobuchi, Sato, et al., 1984). Our microarray data showed that the genes encoding secondary metabolites biosynthesis enzymes were up-regulated during the summer and down-regulated in the fall or winter. This strongly suggests that the formation of heartwood and its extractives in black locust occurs during the growing season, instead of fall or beginning of dormancy.

The Affymetrix GeneChip *Arabidopsis* Genome Array contains over 8,200 sequences including putative open reading frames (ORFs) from genomic sequences and selected ESTs. Eighty percent of the sequences are predicted coding sequences from genomic bacterial artificial chromosome (BAC) entries. Twenty percent are high-quality cDNA sequences. Furthermore, the Institute for Genomic Research is currently working with the Affymetrix to develop a whole transcriptome chip based upon the most current genome annotation. This chip, representing about 23,000 of the currently predicted 25,000-plus genes in the genome, is available from Affymetrix as a standard catalog product with full Affymetrix support. The

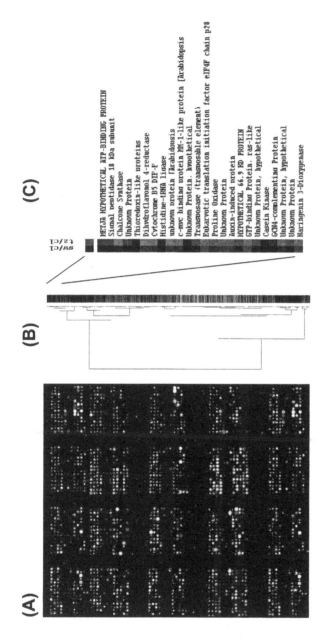

FIGURE 5.2. DNA microarray analysis of black locust. (A) Two-color overlaid scanning fluorimetric image of black locust cDNA microarray. Two hybridization data sets from the same microarray were converted into pseudo-color images and superimposed to visualize differential gene expression between the cambial region (cy3, green) and the transition zone (cy5, red). (B) Cluster analysis of microarray. (C) Exposed cluster of high-expressed clones in sapwood or transition zone compared to cambial zone. Red color indicates up-regulation in the transition zone, and green color means up-regulation in the cambial zone. (See also color photo section.)

(B)

(C)

sw/c1
t2/c1

METJA HYPOTHETICAL ATP-BINDING PROTEIN
Signal peptidase 18 KDa subunit
Chalcone Synthase
Unknown Protein
Thioredoxin-like proteins
Dihydroflavonol 4-reductase
Cytochrome B5 DIF-F
Histidine-tRNA ligase
unknown protein [Arabidopsis]
c-myc binding protein MM-1-like protein [Arabidopsis]
Unknown Protein, hypothetical
Transposase (transposable element)
Eukaryotic translation initiation factor eIF4F chain p28
Proline Oxidase
Unknown Protein
Auxin-induced protein
HYPOTHETICAL 66.9 KD PROTEIN
GTP-binding Protein, ras-like
Unknown Protein, hypothetical
Casein Kinase
GCN4-complementing Protein
Unknown Protein, hypothetical
Unknown Protein
Naringenin 3-Dioxygenase

(A)

current 8.2K Affymetrix chips were used to identify the genes differentially expressed during xylem formation with four different probes (secondary xylem versus bark; control stem versus wood formation-treated stem). Ninety-six transcripts were up-regulated more than threefold in secondary xylem and 86 transcripts more than threefold in bark tissue. Table 5.1 shows a partial list of the genes that are differentially expressed.

Proteomics of Wood Formation

The expression of genes is not necessarily proportionally related to the abundance of the corresponding proteins. Therefore, the integration of both mRNA and protein data will give a more comprehensive view on wood formation. Proteins that are preferentially produced in the developing xylem may play a substantial role in wood formation. Systematic sequencing of proteins upon excision from two-dimensional polyacrylamide gel electrophoresis (PAGE) facilitates the identification of protein functions and leads to the construction of a protein database (see Chapter 3). Proteome analysis from different tissues, developmental stages, or environmental conditions permits a fast and simultaneous comparison of variations in the abundance of a large number of proteins, while also providing useful information on posttranscriptional modifications. This proteome analysis has been used to identify the proteins produced in needles and xylem tissue of a gymnosperm tree, maritime pine (Costa et al., 1999). In addition, Vander Mijnsbrugge and colleagues (2000) performed a comparative two-dimensional PAGE on young differentiating xylem, mature xylem, and bark of poplar harvested at different times of the year. They identified xylem-preferential proteins by comparing the protein patterns from xylem and bark. All of the identified proteins were involved in the phenylpropanoid pathway, and their corresponding ESTs were present in a developing-xylem library from the same poplar clone.

Assessment of Biological Functions of Candidate Genes

Conventional genetics approaches and comparative molecular genetic studies have been effective in defining the functions of many genes involved in various aspects of biology using small herbaceous species such as the model plant *Arabidopsis thaliana*. With the *Arabidopsis* system, this approach is valuable in functional analysis of the genes identified as wood formation specific. However, *Arabidopsis* is not a tree, so it is important to conduct parallel investigations using a tree species. Unfortunately, most of the gene-disruption approaches that have been effective with herbaceous model species are not applicable to tree species, mainly due to the inherent problems of long generation time and large size, lack of a genetically pure line, and readily available mutant population. Gene "knock-in" or "knock-out" approach by ectopic expression or repression of corresponding genes in transgenic plants may provide a powerful means to assess the function of the wood formation-specific genes (see Chapter 1).

THE OUTLOOK

Wood is of primary importance to humans as timber for construction and wood pulp for paper manufacturing. During the past 25 years, the construction and fiber needs of an expanding global population have grown 36 percent (Food and Agriculture Organization, 1997), while available forestland has been rapidly diminishing. As most developed countries depend on raw material imports from developing tropical countries to meet their wood supply, overexploitation of already disappearing rain forests creates a variety of critical environment issues such as global warming and loss of biodiversity. Resolving the dilemma of achieving greater environmental protection of forest ecosystems while meeting the increasing demand for forest utilization necessitates gaining a fundamental understanding of the biochemical processes involved in tree growth and development. However, this crucial area of biology has been lagging behind that of primary growth, primarily due to the inherent difficulties of tree biology. Conventional genetics approaches and comparative molecular genetic studies have been ineffective in answering the basic questions of tree biology. The study of wood formation in a living tree is concerned with aging processes that are occurring in spatially and temporally different cells. It is extremely difficult to experimentally observe the complex metabolic changes during the processes that occur in the secondary xylem.

Use of genomics technology should facilitate our understanding of the molecular mechanisms controlling wood formation through the discovery of the novel genes that are differentially expressed and/or have biological significance in secondary growth. The recent advances in proteomics technology will also aid in gene discovery efforts. Comparative studies between gene expression and protein profiles will provide more comprehensive understanding of the biochemical and molecular events during wood formation. Characterization and functional assignment of the newly identified genes will be the most logical challenge that follows the gene discovery effort. *Arabidopsis* represents the most well-characterized model system for plant biology and has shown potential as a model for wood formation study.

However, as the relevance of *Arabidopsis* wood formation to real wood formation in trees is not certain, it is important to conduct parallel investigations using a tree species. Gene inactivation by either antisense or PTGS (posttranscriptional gene silencing) will be a powerful approach to assess the biological function of the genes. The novel genes with known function can be used to genetically manipulate the biochemical processes involved in wood production, which is responsible for producing more terrestrial biomass and stored chemical energy than any other biological process. Furthermore, biotechnological manipulation of the biochemical processes involved in wood formation can lead to significant changes in the properties of the wood produced. Along with functional genomics efforts (see Chapter 1) to unravel the molecular events underlying wood formation, the ongoing poplar genome sequencing efforts by the U.S. Department of Energy will bring invaluable resource to the tree biology community. This structural genomics effort is expected to complete the sequencing of entire genome of poplar, a widely used tree model, before the year 2004 (Tuskan et al., 2002).

REFERENCES

Abel, S., Nguyen, M.D., Chow, W., and Theologis, A. (1995). ACS4, a primary indoleacetic acid-responsive gene encoding 1-aminocyclopropane-1-carboxylate synthase in *Arabidopsis thaliana*. Structural characterization, expression in *Escherichia coli*, and expression characteristics in response to auxin. *J. Biol. Chem.* 270: 19093-19099.

Adams, M.D., Dubnick, M., Kerlavage, A.R., Moreno, R., Kelley, J.M., Utterback, T.R., Nagle, J.W., Fields, C., and Venter, J.C. (1992). Sequence identification of 2,375 human brain genes. *Nature* 355: 632-634.

Allona, I., Quinn, M., Shoop, E., Swope, K., St. Cyr, S., Carlis, J., Riedl, J., Retzel, E., Campbell, M.M., Sederoff, R., and Whetten, R.W. (1998). Analysis of xylem formation in pine by cDNA sequencing. *Proc. Natl. Acad. Sci. USA* 95: 9693-9698.

Arioli, T., Peng, L., Betzner, A.S., Burn, J., Wittke, W., Herth, W., Camilleri, C., Hofte, H., Plazinski, J., Birch, R., et al. (1998). Molecular analysis of cellulose biosynthesis in *Arabidopsis*. *Science* 279: 717-720.

Bachem, C.W., van der Hoeven, R.S., de Bruijn, S.M., Vreugdenhil, D., Zabeau, M., and Visser, R.G. (1996). Visualization of differential gene expression using a novel method of RNA fingerprinting based on AFLP: Analysis of gene expression during potato tuber development. *Plant J.* 9: 745-753.

Baier, M., Goldberg, R., Catesson, A.M., Liberman, M., Bouchemal, N., Michon, V., and Dupenhoat, C.H. (1994). Pectin changes in samples containing poplar cambium and inner bark in relation to the seasonal cycle. *Planta* 193: 446-454.

Baima, S., Possenti, M., Matteucci, A., Wisman, E., Altamura, M.M., Ruberti, I., and Morelli, G. (2001). The *Arabidopsis* ATHB-8 HD-zip protein acts as a differentiation-promoting transcription factor of the vascular meristems. *Plant Physiology* 126: 643-655.

Baqui, S., Shah, J., Pandalai, R., and Kothari, I. (1979). Histochemical changes during transition from sapwood to heartwood in Melia azedarach. *Indian J. Exp. Biol.* 17: 1032-1037.

Beers, E.P. and Freeman, T.B. (1997). Proteinase activity during tracheary element differentiation in *Zinnia* mesophyll cultures. *Plant Physiology* 113: 873-880.

Beers, E.P. and Zhao, C. (2001). Arabidopsis as a model for investigating gene activity and function in vascular tissues. In N. Morohoshi and A. Komamine (eds.), *Molecular Breeding of Woody Plants* (pp. 43-52). New York: Elsevier Science B.V.

Berleth, T. and Sachs, T. (2001). Plant morphogenesis: Long-distance coordination and local patterning. *Curr. Opin. Plant Sci.* 4: 57-62.

Brett, C.T. (2000). Cellulose microfibrils in plants: Biosynthesis, deposition, and integration into the cell wall. *Int. Rev. Cytol.* 199: 161-199.

Brown, P.O. and Botstein, D. (1999). Exploring the new world of the genome with DNA microarrays. *Nature Genet.* 21: 33-37.

Burtin, P., Jay-Allemand, C., Charpentier, J.-P., and Janin, G. (1998). Natural wood colouring process in *Juglan* sp. (*J. nigra, J. regia* and hybrid *J. nigra* 23 x *J. regia*) depends on native phenolic compounds accumulated in the transition zone between sapwood and heartwood. *Trees* 12: 258-264.

Campisi, L., Yang, Y.Z., Yi, Y., Heilig, E., Herman, B., Cassista, A.J., Allen, D.W., Xiang, H.J., and Jack, T. (1999). Generation of enhancer trap lines in *Arabidopsis* and characterization of expression patterns in the inflorescence. *Plant J.* 17: 699-707.

Cano-Delgado, A., Metzlaff, K., and Bevan, M.W. (2000). The *eli1* mutation reveals a link between cell expansion and secondary cell wall formation in *Arabidopsis thaliana*. *Development* 127: 3395-3405.

Catesson, A.M., Funada, R., Robertbaby, D., Quinetszely, M., Chuba, J., and Goldberg, R. (1994). Biochemical and cytochemical cell-wall changes across the cambial zone. *IAWA J.* 15: 91-101.

Chaffey, N. (1999). Cambium: Old challenges—New opportunities. *Trees* 13: 138-151.

Costa, P., Pionneau, C., Bauw, G., Dubos, C., Bahrmann, N., Kremer, A., Frigerio, J.M., and Plomion, C. (1999). Separation and characterization of needle and xylem maritime pine proteins. *Electrophoresis* 20: 1098-1108.

Covitz, P.A., Smith, L.S., and Long, S.R. (1998). Expressed sequence tags from a root-hair-enriched *Medicago truncatula* cDNA library. *Plant Physiology* 117: 1325-1332.

Delmer, D.P., Pear, J.R., Andrawis, A., and Stalker, D.M. (1995). Genes encoding small GTP-binding proteins analogous to mammalian *rac* are preferentially expressed in developing cotton fibers. *Mol. Gen. Genet.* 248: 43-51.

Dengler, N. and Kang, J. (2001). Vascular patterning and leaf shape. *Curr. Opin. Plant Sci.* 4: 50-56.

Dietrichs, H. (1964). Das Verhalten von Kohlenhydraten bei der. [The role of carbohydrates in wood nucleation.] *Holzforschung* 18: 14-24.

Digby, J. and Wareing, P.F. (1966). The effect of applied growth hormones on cambial division and the differentiation of the cambial derivatives. *Ann. Bot.* 30: 539-548.

Duggan, D.J., Bittner, M., Chen, Y., Meltzer, P., and Trent, J.M. (1999). Expression profiling using cDNA microarrays. *Nat. Genet.* 21: 10-14.

Durrant, W.E., Rowland, O., Piedras, P., Hammond-Kosack, K.E., and Jones, J.D. (2000). cDNA-AFLP reveals a striking overlap in race-specific resistance and wound response gene expression profiles. *Plant Cell* 12: 963-977.

Eisen, M.B., Spellman, P.T., Brown, P.O., and Botstein, D. (1998). Cluster analysis and display of genome-wide expression patterns. *Proc. Natl. Acad. Sci. USA* 95: 14863-14868.

Eklund, L. and Little, C.H.A. (1996). Laterally applied Etherel causes local increases in radial growth and indole-3-acetic acid concentration in *Abies balsamea* shoots. *Tree Physiology* 16: 509-513.

Eklund, L., Little, C.H.A., and Riding, R.T. (1998). Concentrations of oxygen and indole-3-acetic acid in the cambial region during latewood formation and dormancy development in *Picea abies* stems. *J. Exp. Bot.* 49: 205-211.

Eriksson, M.E., Israelsson, M., Olsson, O., and Moritz, T. (2000). Increased gibberellin biosynthesis in transgenic trees promotes growth, biomass production and xylem fiber length. *Nat. Biotechnol.* 18: 784-788.

Fagard, M., Desnos, T., Desprez, T., Goubet, F., Refregier, G., Mouille, G., McCann, M., Rayon, C., Vernhettes, S., and Hofte, H. (2000). PROCUSTE1 encodes a cellulose synthase required for normal cell elongation specifically in roots and dark-grown hypocotyls of *Arabidopsis. Plant Cell* 12: 2409-2424.

Feuillet, C., Lauvergeat, V., Deswarte, C., Pilate, G., Boudet, A., and Grima-Pettenati, J. (1995). Tissue- and cell-specific expression of a cinnamyl alcohol dehydrogenase promoter in transgenic poplar plants. *Plant Mol. Biol.* 27: 651-667.

Food and Agriculture Organization (FAO) of the United Nations (1997). The state of the world's forests—1997 <http://www.fao.org/docrep/W4345E/W4345E00.htm>.

Fukuda, H. (1996). Xylogenesis: Initiation, progression, and cell death. *Ann. Rev. Plant Physiol. Plant Mol. Biol.* 47: 299-325.

Fukuda, H. (2000). Programmed cell death of tracheary elements as a paradigm in plants. *Plant Mol. Biol.* 44: 245-253.

Gälweiler, L., Guan, C., Müller, A., Wisman, E., Mendgen, K., Yephremov, A., and Palme, K. (1998). Regulation of polar auxin transport by AtPIN1 in *Arabidopsis* vascular tissue. *Science* 282: 2226-2230.

Groover, A., DeWitt, N., Heidel, A., and Jones, A. (1997). Programmed cell death of plant tracheary elements differentiating *in vitro. Protoplasma* 196: 197-211.

Groover, A. and Jones, A.M. (1999). Tracheary element differentiation uses a novel mechanism coordinating programmed cell death and secondary cell wall synthesis. *Plant Physiology* 119: 375-384.

Haigler, C.H., Ivanova-Datcheva, M., Hogan, P.S., Salnikov, V.V., Hwang, S., Martin, K., and Delmer, D.P. (2001). Carbon partitioning to cellulose synthesis. *Plant Mol. Biol.* 47: 29-51.

Hamadeh, H. and Afshari, C.A. (2000). Gene chips and functional genomics. *American Scientists* 88: 508-515.

Hardtke, C.S. and Berleth, T. (1998). The *Arabidopsis* gene *MONOPTEROS* encodes a transcription factor mediating embryo axis formation and vascular development. *EMBO J.* 2: 1405-1411.

Hertzberg, M., Aspeborg, H., Schrader, J., Andersson, A., Erlandsson, R., Blomqvist, K., Bhalerao, R., Uhlen, M., Teeri, T.T., Lundeberg, J., et al. (2001). A transcriptional roadmap to wood formation. *Proc. Natl. Acad. Sci. USA* 98: 14732-14737.

Hillinger, C., Holl, W., and Ziegler, H. (1996a). Lipids and lipolytic enzymes in the trunkwood of *Robinia pseudoacacia* L. during heartwood formation. I. Radial distribution of lipid classes. *Trees* 10: 366-375.

Hillinger, C., Holl, W., and Ziegler, H. (1996b). Lipids and lipolytic enzymes in the trunkwood of *Robinia pseudoacacia* L. during heartwood formation. II. Radial distribution of lipases and phospholipases. *Trees* 10: 376-381.

Hillis, W.E. (1987). *Heartwood and Tree Exudates*. Berlin, New York: Springer-Verlag.

Hillis, W.E. (1996). Formation of robinetin crystals in vessels of *Intsia* species. *IAWA J.* 17: 405-419.

Hillis, W.E. and Yazaki, Y. (1973). Wood polyphenols of *Eucalyptus polyanthemos*. *Phytochem.* 12: 2969-2977.

Hu, W.J., Kawaoka, A., Tsai, C.J., Lung J., Osakabe, K., Ebinuma, H., and Chiang, V.L. (1998). Compartmentalized expression of two structurally and functionally distinct 4-coumarate:CoA ligase genes in aspen *(Populus tremuloides). Proc. Natl. Acad. Sci. USA* 95: 5407-5412.

Iliev, I. and Savidge, R. (1999). Proteolytic activity in relation to seasonal cambial growth and xylogenesis in *Pinus banksiana*. *Phytochem.* 50: 953-960.

Jones, A.M. (2001). Programmed cell death in development and defense. *Plant Physiology* 125: 94-97.

Kawamata, S., Shimoharai, K., Imura, Y., Ozaki, M., Ichinose, Y., Shiraishi, T., Kunoh, H., and Yamada, T. (1997). Temporal and spatial pattern of expression of the pea phenylalanine ammonia-lyase gene1 promoter in transgenic tobacco. *Plant Cell Physiol.* 38: 792-803.

Kawamoto, S., Ohnishi, T., Kita, H., Chisaka, O., and Okubo, K. (1999). Expression profiling by iAFLP: A PCR-based method for genome-wide gene expression profiling. *Genome Res.* 9: 1305-1312.

Kawaoka, A., Kaothien, P., Yoshida, K., Endo, S., Yamada, K., and Ebinuma, H. (2000). Functional analysis of tobacco LIM protein Ntlim1 involved in lignin biosynthesis. *Plant J.* 22: 289-301.

Kimura, S., Laosinchai, W., Itoh, T., Cui, X., Linder, C.R., and Brown, R.M. Jr. (1999). Immunogold labeling of rosette terminal cellulose-synthesizing complexes in the vascular plant *Vigna angularis*. *Plant Cell* 11: 2075-2086.

Koopmann, E., Logemann, E., and Hahlbrock, K. (1999). Regulation and functional expression of cinnamate 4-hydroxylase from parsley. *Plant Physiology* 119: 49-56.

Kozlowski, T. and Pallardy, S. (1997). *Physiology of Woody Plants.* San Diego, CA: Academic Press.

Larson, P.R. (1994). *The Vascular Cambium.* Berlin: Springer-Verlag.

Lev-Yadun, S. (1994). Induction of sclereid differentiation in the pith of *Arabidopsis thaliana* (L.) Heynh. *J. Exp. Bot.* 45: 1845-1849.

Lev-Yadun, S. (1997). Fibres and fibre-sclereids in wild-type *Arabidopsis thaliana. Annals Bot.* 80: 125-129.

Lev-Yadun, S. and Flaishman, M.A. (2001). The effect of submergence on ontogeny of cambium and secondary xylem and on fiber lignification in inflorescence stems of *Arabidopsis. IAWA J.* 22: 113-123.

Little, C.H.A. and Eidt, D.C. (1968). Effects of abscisic acid on budbreak and transpiration in woody species. *Nature* 220: 498-499.

Little, C.H.A. and Savidge, R.A. (1987). The role of plant growth regulators in forest tree cambial growth. *Plant Growth Regul.* 6: 137-169.

Little, C.H.A. and Sundberg, B. (1991). Tracheid production in response to indole-3-acetic-acid varies with internode age in *Pinus sylvestris* stems. *Trees* 5: 101-106.

Liu, Y.G. and Whittier, R.F. (1995). Thermal asymmetric interlaced PCR: Automatable amplification and sequencing of insert end fragments from P1 and YAC clones for chromosome walking. *Genomics* 25: 674-681.

Lockhart, D.J. and Winzeler, E.A. (2000). Genomics, gene expression and DNA arrays. *Nature* 405: 827-836.

Magel, E., Drouet, A., Claudot, A., and Ziegler, H. (1991). Formation of heartwood substances in the stemwood of *Robinia pseudoacacia* L. I. Distribution of phenylalanine ammonium-lyase and chalcone synthase across the trunk. *Trees* 5: 203-207.

Magel, E., Jay-Allemand, C., and Ziegler, H. (1994). Formation of heartwood substances in the stemwood of *Robinia pseudoacacia* L. II. Distribution of nonstructural carbohydrates and wood extractives across the trunk. *Trees* 8: 165-171.

Mahonen, A.R., Bonke, M., Kauppinen, L., Riikonen, M., Benfey, P.N., and Helariutta, Y. (2000). A novel two-component hybrid molecule regulates vascular morphogenesis of the vascular root. *Genes Dev.* 14: 2938-2943.

Mauseth, J. (1998). *Botany: An Introduction to Plant Biology.* Sudbury, MA: Jones and Bartlett Publishers.

Mellerowicz, E.J., Baucher, M., Sundberg, B., and Boerjan, W. (2001). Unraveling cell wall formation in the woody dicot stem. *Plant. Mol. Biol.* 47: 239-274.

Mellerowicz, E.J., Coleman, W.K., Riding, R.T., and Little, C.H.A. (1992). Periodicity of cambial activity in *Abies balsamea*. 1. Effects of temperature and photoperiod on cambial dormancy and frost hardiness. *Physiol. Plant.* 85: 515-525.

Mellerowicz, E.J., Riding, R.T., and Little, C.H.A. (1992). Periodicity of cambial activity in *Abies balsamea*. 2. Effects of temperature and photoperiod on the size of the nuclear genome in fusiform cambial cells. *Physiol Plant.* 85: 526-530.

Milioni, D., Sado, P-E., Stacey, N.J., Domingo, C., Roberts, K., and McCann, M.C. (2001). Differential expression of cell-wall-related genes during the formation of tracheary elements in the *Zinnia* mesophyll cell system. *Plant Mol. Biol.* 47: 221-238.

Miller, A.R. and Roberts, L.W. (1984). Ethylene biosynthesis and xylogenesis in *Lactuca* explants cultured *in vitro* in the presence of auxin and cytokinin: The effect of ethylene precursors and inhibitors. *J. Exp. Bot.* 35: 691-698.

Mueller, S.C. and Brown, R.M. Jr. (1980). Evidence for an intramembrane component associated with a cellulose microfibril-synthesizing complex in higher plants. *J. Cell Biol.* 84: 315-326.

Murakami, Y., Funada, R., Sano, Y., and Ohtani, J. (1999). The differentiation of contact cells and isolation cells in the xylem ray parenchyma of *Populus maximowiczii. Ann. Bot.* 84: 325-326.

Nelson, N. (1978). Xylem ethylene, phenol-oxidising enzymes and nitrogen and heartwood formation in walnut and cherry. *Can. J. Bot.* 56: 626-634.

Nobuchi, T., Matsuno, H., and Harada, H. (1984). Relationship between heartwood phenols and cytological structure in the transition zone from sapwood to heartwood of sugi *(Cryptomeria japonica).* In *Pacific Regional Wood Anatomy Conference (IAWA/IUFRO), Tsukuba, Japan* (pp. 132-134).

Nobuchi, T., Sato, T., Iwata, R., and Harada, H. (1984). Season of heartwood formation and the related cytological structure of ray parenchyma cells in *Robinia pseudoacacia. Mokuzai Gakkaishi* 30: 628-636.

Otsuga, D., DeGuzman, B., Prigge, M.J., Drews, G., and Clark, S.E. (2001). *REVOLUTA* regulates meristem initiation at lateral positions. *Plant J.* 25: 223-236.

Pacala, S.W., Hurtt, G.C., Baker, D., Peylin, P., Houghton, R.A., Birdsey, R.A., Heath, L., Sundquist, E.T., Stallard, R.F., and Ciais, P. (2001). Consistent land- and atmosphere-based U.S. carbon sink estimates. *Science* 292: 2316-2320.

Peng, L., Kawagoe Y., Hogan, P., and Delmer, D. (2002). Sitosterol-beta-glucoside as primer for cellulose synthesis in plants. *Science* 295: 147-150.

Pharis, R.P., Jenkins, P.A., Aoki, H., and Sassa, T. (1981). Hormornal physiology of wood growth in *Pinus radiata* D. Don: Effects of gibberrellin A_4 and the influence of abscisic acid upon $[^3H]$-gibberellin A_4 metabolism. *Aust. J. Plant Physiol.* 8: 559-570.

Qin, L., Overmars, H., Helder, J., Popeijus, H., van der Voort, J.R., Groenink, W., van Koert, P., Schots, A., Bakker, J., and Smant, G. (2000). An efficient cDNA-AFLP-based strategy for the identification of putative pathogenicity factors from the potato cyst nematode *Globodera rostochiensis. Mol. Plant Microbe Interact.* 13: 830-836.

Ratcliffe, O.J., Riechmann, J.L., and Zhang, J.Z. (2000). *INTERFASCICULAR FIBERLESS 1* is the same gene as *REVOLUTA. Plant Cell* 12: 315-317.

Regan, S., Bourquin, V., Tuominen, H., and Sundberg, B. (1999). Accurate and high resolution *in situ* hybridization analysis of gene expression in secondary stem tissues. *Plant J.* 19: 363-369.

Reiter, W.D. and Vanzin, G.F. (2001). Molecular genetics of nucleotide sugar interconversion pathways in plants. *Plant Mol. Biol.* 47: 95-113.

Richmond, T.A. and Somerville, C.R. (2001). Integrative approaches to determining Csl function. *Plant Mol. Biol.* 47: 131-143.

Scheres, B., Di Laurenzio, L., Willemsen, V., Hauser, M-T., Janmaat, K., Weisbeek, P., and Benfey, P.N. (1995). Mutations affecting the radial organisation of the *Arabidopsis* root display specific defects throughout the embryonic axis. *Development* 121: 53-62.

Sederoff, R.R., MacKay, J.J., Ralph, J., and Hatfield, R.D. (1999). Unexpected variation in lignin. *Curr. Opin. Plant Biol.* 2: 145-152.

Sessa, G., Steindler, C., Morelli, G., and Ruberti, I. (1998). The *Arabidopsis AtHB-8, 9* and *14* genes are members of a small gene family coding highly related HD-Zip proteins. *Plant Mol. Biol.* 38: 609-622.

Shain, J. and Mackay, J. (1973). Seasonal fluctuations in respiration of aging xylem in relation to heartwood formation in *Pinus radiata. Can. J. Bot.* 51: 737-741.

Sheen, J., Zhou, L., and Jang, J.-C. (1999). Sugars as signaling molecules. *Curr. Opin. Plant Biol.* 2: 410-418.

Shimkets, R.A., Lowe, D.G., Tai, J.T., Sehl, P., Jin, H., Yang, R., Predki, P.F., Rothberg, B.E., Murtha, M.T., and Roth, M.E. (1999). Gene expression analysis by transcript profiling coupled to a gene database query. *Nat. Biotechnol.* 17: 798-803.

Sterky, F., Regan, S., Karlsson, J., Hertzberg, M., Rohde, A., Holmberg, A., Amini, B., Bhalerao, R., Larsson, M., Villarroel, R., et al. (1998). Gene discovery in the wood-forming tissues of poplar: Analysis of 5,692 expressed sequence tags. *Proc. Natl. Acad. Sci. USA* 95: 13330-13335.

Sutcliffe, J.G., Foye, P.E., Erlander, M.G., Hilbush, B.S., Bodzin, L.J., Durham, J.T., and Hasel, K.W. (2000). TOGA: An automated parsing technology for analyzing expression of nearly all genes. *Proc. Natl. Acad. Sci. USA* 97: 1976-1981.

Talbert, P.B., Adler, H.T., Parks, D.W., and Comai, L. (1995). The *REVOLUTA* gene is necessary for apical meristem development and for limiting cell divisions in the leaves and stems of *Arabidopsis thaliana. Development* 121: 2723-2735.

Tamagnone, L., Merida, A., Stacey, N., Plaskitt, K., Parr, A., Chang, C.F., Lynn, D., Dow, J.M., Roberts, K., and Martin, C. (1998). Inhibition of phenolic acid metabolism results in precocious cell death and altered cell morphology in leaves of transgenic tobacco plants. *Plant Cell* 10: 1801-1816.

Taylor, N.G., Scheible, W.R., Cutler, S., Somerville, C.R., and Turner, S.R. (1999). The irregular xylem3 locus of *Arabidopsis* encodes a cellulose synthase required for secondary cell wall synthesis. *Plant Cell* 11: 769-780.

Turner, S.R., Taylor, N., and Jones, L. (2001). Mutations of the secondary cell wall. *Plant Mol. Biol.* 47: 209-219.

Tuskan, G.A., Wullschleger, S.D., Bradshaw, H.D., and Dalhman, R.C. (2002). Sequencing the *Populus* genome: Applications to the energy-related missions of DOE. *Plant, Animal and Microbe Genomes X Conference,* January 12-16, San Diego, CA.

Uggla, C., Magel, E., Moritz, T., and Sundberg, B. (2001). Function and dynamics of auxin and carbohydrates during earlywood/latewood transition in Scots pine. *Plant Physiology* 25: 2029-2039.

Uggla, C., Mellerowicz, E.J., and Sundberg, B. (1998). Indole-3-acetic acid controls cambial growth in Scots pine by positional signaling. *Plant Physiology* 117: 113-121.

Uggla, C., Moritz, T., Sandberg, G., and Sundberg, B. (1996). Auxin as a positional signal in pattern formation in plants. *Proc. Natl. Acad. Sci. USA* 93: 9282-9286.

Vander Mijnsbrugge, K., Meyermans, H., Van Montagu, M., Bauw, G., and Boerjan, W. (2000). Wood formation in poplar: Identification, characterization, and seasonal variation of xylem proteins. *Planta* 210: 589-598.

Walbot, V. (1999). Genes, genomes, genomics. What can plant biologists expect from the 1998 national science foundation plant genome research program? *Plant Physiology* 119: 1151-1156.

Wardrop, A. and Cronshaw, J. (1962). Formation of phenolic substances in ray paranchyma of angiosperms. *Nature* 193: 90-92.

Whetten, R., Sun, Y-H., Zhang, Y., and Sederoff, R. (2001). Functional genomics and cell wall biosynthesis in loblolly pine. *Plant Mol. Biol.* 47: 275-291.

Wu, L., Joshi, C.P., and Chiang, V.L. (2000). A xylem-specific cellulose synthase gene from aspen *(Populus tremuloides)* is responsive to mechanical stress. *Plant J.* 22: 495-502.

Ye, Z.H. and Varner, J.E. (1995). Induction of cysteine and serine proteases during xylogenesis in *Zinnia elegans. Plant Mol. Biol.* 30: 1233-1246.

Zhao, C., Johnson, B.J., Kositsup, B., and Beers, E.P. (2000). Exploiting secondary growth in *Arabidopsis.* Construction of xylem and bark cDNA libraries and cloning of three xylem endopeptidases. *Plant Physiology* 123: 1185-1196.

Zheng, Z.L. and Yang, Z. (2000). The Rop GTPase switch turns on polar growth in pollen. *Trends Plant Sci.* 5: 298-303.

Zhong, R., Ripperger, A., and Ye, Z.H. (2000). Ectopic deposition of lignin in the pith of stems of two *Arabidopsis* mutants. *Plant Physiology* 123: 59-70.

Zhong, R., Taylor, J.J., and Ye, Z.H. (1997). Disruption of interfascicular fiber differentiation in an *Arabidopsis* mutant. *Plant Cell* 9: 2159-2170.

Zhong, R. and Ye, Z.H. (1999). *IFL1,* a gene regulating interfascicular fiber differentiation in *Arabidopsis,* encodes a homeodomain-leucine zipper protein. *Plant Cell* 11: 2139-2152.

Zhong, Y., Mellerowicz, E.J., Lloyd, A.D., Leinhos, V., Riding, R.T., and Little, C.H.A. (1995). Seasonal-variation in the nuclear genome size of ray cells in the vascular cambium of *Fraxinus americana. Physiol. Plant.* 93: 305-311.

Zobel, B.J. and Jett, J.B. (1995). *Genetics of Wood Production.* Berlin: Springer.

Chapter 6

Molecular Genetics of Cellulose Biosynthesis in Trees

Chandrashekhar P. Joshi

INTRODUCTION

In the modern genomics era, in which each and every branch of plant research is dominated by *Arabidopsis thaliana,* working with trees is an unpopular choice. Inherent characteristics of trees, such as long generation time, slower growth, less well-known genetics, and recalcitrance to transformation, make them a difficult experimental system to work with. Irrespective of these limitations, some remarkable achievements have been made in the last few years addressing some special biological problems that can be tackled only in trees. If one wishes to understand wood development, tree is naturally a better choice than *A. thaliana.* However, it also must be admitted that the rapid progress witnessed in molecular genetics of *A. thaliana* has not yet been matched in trees that are the ultimate targets for genetic manipulations of economically important wood traits. This picture becomes much more bleak when one considers the process of cellulose biosynthesis. Identification of the first plant cellulose synthase gene from cotton was reported only seven years ago (Pear et al., 1996), and it is not surprising that little published information is available on this topic from trees. The recent revelation of the presence of a large multigene family encoding cellulose synthase genes in the *A. thaliana* genome, which appears to be a common characteristic of all higher plants, further dampens the enthusiasm of tree re-

I wish to thank Dr. Vincent Chiang and Dr. Glenn Mroz for their constant support and encouragement. I am specifically grateful to Dr. Luguang Wu who initially nurtured the dream of working in this area and performed some breakthrough research. I also wish to thank Dr. Priit Pechter and my graduate students Rajesh Chavli, Anita Samuga, and Udaya Kalluri for their assistance in preparation of this manuscript and sharing their unpublished data. This work was partially supported by the USDA-NRI Grant Program (99-35103-7986), USDA-McIntire Stennis Forestry Research Program, and Research Excellence Fund from the State of Michigan.

141

searchers working with this problem. As a result, more questions than answers are available. Many previous tree cellulose studies have devoted much attention to the description of the end product of cellulose biosynthesis than to understanding either the cause of natural genetic variations in cellulose content and quality or the mechanism of cellulose biosynthesis. The purpose of this chapter is to present the current status of our knowledge regarding the cellulose biosynthesis in trees with special emphasis on cellulose synthase genes from trees. Where it is necessary, parallels are frequently drawn from other plant species.

Wood is a highly heterogeneous biological material consisting of a variety of cell types (tracheids, vessels, parenchyma, and sclerenchyma), wall layers (P, S1, S2, and S3), and wall materials (cellulose, lignin, hemicelluloses, pectins, and extractives) (Sjostrom, 1993). About 40 to 50 percent of dry wood is cellulose, 15 to 35 percent is lignin, and 25 to 40 percent is hemicelluloses (Nevell and Zeronian, 1985). Moreover, cambial cells from hardwood and softwood species manufacture wood with varying quality every year, a process that could be amenable to genetic manipulation if we understand the role of various genes in this process and identify the key genes for such manipulations. It may appear simplistic that manipulation of a single or a few genes alone can improve wood production, but there is precedence. We have recently witnessed at least one such exception in which a lowering of lignin percentage resulted in increase in cellulose percentage and increased growth of transgenic trees with antisense-RNA-mediated inhibition of a single *4CL* gene expression (Hu et al., 1999). This kind of success could not be achieved if basic biological and biochemical information about lignin biosynthesis in trees was lacking. Thus, information about each and every economically important wood trait from forest trees must be compiled, analyzed, and utilized. Because the topic of this chapter is cellulose biosynthesis, I will focus on that aspect for further discussion, although ultimately a holistic approach of using all biological, biochemical, and economical information must be considered for effective wood improvement.

CELLULOSE HETEROGENEITY IN TREES

Cellulose is the most abundant biopolymer on earth, with over 180 billion tons of organic matter produced in nature every year (Delmer, 1999). It is a homopolymer of glucose that forms unbranched β-1,4-D-glucan chains in which every alternate glucose molecule is flipped by 180 degrees (Brown et al., 1996). It has been suggested that subsequent to synthesis, many single cellulose chains instantly associate lengthwise through hydrogen bonds to form microfibrils, and many such microfibrils come together to form macrofibrils in the secondary walls (Delmer and Amor, 1995). The number of glucan chains in each such bundle varies from 36 in the elementary fibrils to more than 1,200 in some algal species (Brown et al., 1996). The number of glucose residues per cellulose chain, or degree of polymerization (DP), also shows variation, from about 2,000 in cellulose from cotton primary walls to about 14,000 in cellulose from secondary walls of wood (Haigler, 1985). The content of cellulose also varies between these two cell-wall types. The cellulose content is 1 to 10 percent in elastic and expanding primary wall and more than 50 percent in rigid and lignin-rich secondary walls. In nature, cellulose gets aggregated in highly ordered crystalline form alternating with less-ordered amorphous form depending on the number of hydrogen bonds formed between the glucose chains. The crystalline form has more tensile strength, and wood cellulose is about 60 to 70 percent crystalline (Timell, 1986). The content, thickness, and microfibril arrangement also differ significantly in different types of cell-wall layers. The primary wall consists mainly of cellulose, hemicellulose, and pectin. It is also 0.1 to 0.2 μm in thickness and has irregularly oriented cellulose microfibrils. The secondary wall is rich in cellulose, hemicellulose, and lignin and consists of three layers, namely, S1, S2, and S3. The outer and inner, S1 and S3, layers are 0.1 to 0.3 μm thick and have orderly oriented cellulose microfibrils with a microfibril angle of 50 to 90 degrees with respect to fiber axis. The thick middle S2 layer in softwood tracheids is 1 to 5 μm thick and has a microfibril angle of 5 to 10 degrees in earlywood and 20 to 30 degrees in latewood (Sjostorm, 1993). Microfibril angle is a major determinant of stiffness in wood, and smaller angles are preferred for better wood quality. Thus, cellulose quantity and quality are highly variable features. Better understanding of molecular mechanisms controlling these variations is essential for further progress in this field.

CELLULOSE PRODUCTION
UNDER MECHANICAL STRESS

Environment plays a major role in wood formation in trees, and one of the extensively studied phenomena is reaction wood formation (Timell, 1986). When a tree or its branches are under mechanical stress due to gravity, wind, or snow, some degree of leaning or bending of the tree is observed. To compensate for such downward force, reaction wood is formed in trees. Softwood and hardwood species respond to the same stress differently. Softwoods develop compression wood in response to compression stress on the lower side of the leaning trunk, pushing the tree upward, whereas hardwoods develop tension wood in response to tension stress on the upper side of the stem, pulling the tree away from the ground. Interestingly, the cellulose and lignin properties of compression and tension wood are also contrasting. Compression wood in loblolly pine, for example, has approximately 10 percent lower cellulose content and approximately 6 percent higher lignin content than normal wood (Timell, 1986). In addition, cellulose in such compression wood has lower DP of approximately 3,000 and lower crystallinity of about only 30 percent. Tension wood, on the other hand, has 10 percent higher cellulose content and 10 percent lower lignin content. This difference in cellulose content of tension wood can be primarily attributed to a thick gelatinous layer (G) that develops inside the S1 layer, replacing the S2 and S3 layers. The G layer is almost devoid of any lignin and is made up of approximately 95 percent pure cellulose that is highly crystalline and has a 0 degree microfibril angle. Molecular events occurring in either of these reaction wood formations are still poorly understood. All these details of cellulose characteristics in wood, however, can be regarded as a snapshot of the end product of the process that makes cellulose. Our understanding of the process that leads to the finished product is still incomplete.

THE SITE OF CELLULOSE BIOSYNTHESIS

Cellulose synthesis occurs on the plasma membrane of plant cells (Haigler, 1985). In freeze-fracture replicas from angiosperms, one can easily discern particle rosettes or terminal complexes (TC) coinciding with cellulose synthesis. The history of TC visualization is also interesting. As described by Brown (1996), Roelofsen (1958) first proposed the concept of enzyme complexes present at the tips of growing cellulose microfibrils precisely regulating cellulose synthesis. Preston (1964) intuitively proposed an "ordered granule hypothesis" which stated that the intrinsic order of subunits within such enzyme complexes should dictate assembly and orientation of cellulose microfibrils. Brown and Montezinos (1976) demonstrated the presence of such highly ordered structures using the freeze-fracture technique and named these structures TCs. TC morphologies are mainly of two types, linear and rosette. Linear TCs are seen in many algal species, and rosette configuration showing sixfold symmetry is commonly observed in land plants from bryophytes to angiosperms, including trees such as pines (Mueller and Brown, 1980). Malcolm Brown and colleagues have recently used immunogold labeling of rosette TCs to confirm that cellulose synthases are present within these structures (Kimura et al., 1999). These cellulose synthase complexes are notoriously labile, and it has proven extremely difficult to isolate intact TCs and demonstrate cellulose synthase enzyme activity in vitro.

Plant TCs are part of the plasma membrane, and any disturbance in their integrity leads to an instant loss of activity. It has been proposed that TCs move in the plane of the membrane as the cellulose fiber is being extruded out of the TC pore. This movement determines the pattern of cellulose deposition. The cytoskeleton plays a major role in such TC movement because TCs act as cellulose synthesizing structures that move on the track guided by microtubules (Brown, 1996). Excellent reviews have been published in the past on these topics, and interested readers may refer to them for further information (Delmer, 1987; Delmer and Amor, 1995; Brown et al., 1996; Delmer, 1999; Richmond and Somerville, 2000; Haigler et al., 2001).

CURRENT KNOWLEDGE OF THE GENERAL PROCESS OF CELLULOSE SYNTHESIS IN PLANTS

The natural process of cellulose biosynthesis in plants is a major contributor to the earth's biomass. One would therefore ask whether it is necessary to improve cellulose production in trees. However, augmentation of cellulose content and improvement of cellulose quality in specific plant tissues such as wood or fibers holds a great challenge and promise. The process of cellulose biosynthesis can be viewed as a simple polymerization of glucose residues using the uridine diphosphate (UDP)-glucose as the substrate. The central enzyme in this process is cellulose synthase or CesA (EC 2.1.4.12), which together with several accessory enzymes including sucrose synthase (SuSy) organizes the individual glucan chains into a cellulose microfibril (Delmer and Amor, 1995; Delmer, 1999). It has been postulated that individual microfibrils are synthesized via multisubunit enzyme complexes that may be composed of different proteins. At least one of these proteins is responsible for polymerization and secretion of the glucan chains, and the other proteins may be involved in alignment of the glucan chains leading to crystallization of cellulose. In higher plants, the quantity and quality of cellulose synthesized in primary and secondary cell walls are significantly different (Haigler, 1985; Brown et al., 1996). Compared to primary walls, secondary wall cellulose shows a higher degree of polymerization, crystallinity, and organization of microfibrils (Timell, 1986; Haigler and Blanton, 1996; Taylor et al., 1999); therefore, improvements in secondary wall cellulose are economically more desirable. Such variations in primary and secondary wall cellulose are postulated to be the result of different types of CesAs (Haigler and Blanton, 1996). However, plant scientists have struggled for over thirty years to isolate candidate *CesA* genes from higher plants. One of the difficulties has been in purifying active cellulose synthases from plant membranes (Delmer, 1999), and this still appears to be a challenge.

IDENTIFICATION OF CESA GENES IN PLANTS

First Plant CesA *Gene from Cotton* Was Cloned in 1996

The first breakthrough in *CesA* gene cloning did not come from plants that make abundant amounts of cellulose but from bacteria, such as *Acetobacter,* that produce extracellular cellulose. Active cellulose synthase could be isolated from these bacteria, and the first *Acetobacter CesA* gene was cloned in the early 1990s (Saxena et al., 1990). Through an intensive search for the conserved D,D,D...QXXRW signature that was proposed to be present in many processive glycosyltransferases such as bacterial *CesAs* (Saxena et al., 1995), Delmer and colleagues isolated the first cotton *CesA* cDNAs *(GhCesAs)* from a cDNA library prepared from cotton fibers that are highly active in secondary wall cellulose synthesis (Pear et al., 1996). The full-length *GhCesA1* cDNA was 3.2 kb long and showed 70 to 80 percent identity to other partial *GhCesA2* cDNA. Specifically, temporal expression of *GhCesA1* indicated its complete absence in tissues such as root, leaf, flower, and seed but showed increased accumulation of transcripts from 17 days after anthesis to 35 days postanthesis in developing cotton fibers. Also, a small *CesA* gene family was detected by Southern analysis. The encoded 110 kDa CesA protein had eight predicted transmembrane domains (TMDs), two toward the N-terminus and six toward the C-terminus with an intervening putative catalytic domain that was proposed to be cytoplasmic. In comparison to various known processive glycosyltransferases from bacteria, a number of regions of high similarities (H-1, H-2, and H-3) and differences were recognized in cotton CesA proteins as well as some partial rice CesA proteins identified through a search for CesA homologues in rice expressed sequence tag (EST) databanks. Of special mention was the region exclusively present in plant CesAs named P-CR (plant-specific conserved region) that is absent in bacterial CesAs. Another region, HVR (hypervariable region), is also uniquely present in plant CesAs, but it showed several differences among various plant CesA proteins. Furthermore, they also speculated that the HVR region may confer some specificity of function during cellulose biosynthesis. The central catalytic domain containing the D,D,D...QXXRW motif of *GhCesAs* was shown to bind UDP-glucose, the predicted substrate for cellulose biogenesis, in a Mg^{2+}-dependent manner. Although interesting, the genetic proof of in vivo CesA functionality was lacking in these studies.

Mutant Studies Further Suggested Molecular Function of CesAs

The molecular genetic proof of CesA involvement in cellulose synthesis came from two cellulose synthesis impaired *A. thaliana* mutants, *rsw1* (radial swelling) (Baskin et al., 1992) and *irx3* (irregular xylem) (Turner and Somerville, 1997). These mutants were complemented by wild type *A. thaliana CesA* genes, *Rsw1 (AtCesA-1)* and *Irx3 (AtCesA-7)*, respectively, and these genes showed significant similarity to *GhCesAs* (Arioli et al., 1998; Taylor et al., 1999). Most interestingly, mutation in even a single *CesA* gene, *Rsw1* or *Irx3*, was proven to be detrimental for the normal synthesis of crystalline cellulose, suggesting their nonredundancy with other CesA proteins (Delmer, 1998). A possibility, therefore, exists that functional rosette formation may require involvement of more than one type of CesA subunit. Recent evidence suggests that a pair or even a group of CesAs may be involved in cellulose synthesis, and disturbance in even one subunit may lead to defective cellulose synthase machinery (Fagard et al., 2000; Taylor et al., 2000). Although localization of a pair or group of antibodies specific for each CesA subunit on rosettes as shown by Kimura and colleagues (1999) is highly desirable, other methods such as gene expression-pattern studies, mutant complementation, immunoprecipitation, or use of a yeast two-hybrid system may also provide some indirect clues about the type and number of CesA subunits or other proteins involved in this process.

Extent of Arabidopsis CesA Gene Family

With completion of the *A. thaliana* genome sequencing, at least ten distinct *CesA* genes (*AtCesA-1* to *AtCesA-10*) with similarity to cotton *CesA* genes have been identified in this model species (Richmond, 2000). Why so many *CesAs* are present in this tiny short-lived plant is still an open question, but at least some members, such as *AtCesA-4*, *AtCesA-7*, and *AtCesA-8* have been associated with secondary-wall development (Taylor et al., 1999; Holland et al., 2000; Taylor et al., 2000), indicating that different *CesAs* may be involved in cell- or tissue-specific cellulose biosynthesis. A complete molecular description of spatial and temporal regulation of various *CesA* genes is currently unavailable but highly desirable. These ten *CesA* genes are scattered on at least four out of five *A. thaliana* chromosomes, range from approximately 4.0 to 5.5 kb in size, have 10 to 14 exons, and produce transcripts of about 3 to 3.5 kb that encode about 1,000 amino acids (http://cellwall.stanford.edu). Relatively less information is available for

CesA genes from many other economically important crop and tree species (Richmond, 2000).

First CesA *Gene from Trees Was Cloned in 1998*

Wood mainly consists of cellulose, hemicellulose, and lignin that collectively impart strength and rigidity to timber products, but lignin needs to be removed from paper pulp to produce high-quality paper products. Recently, our laboratory reported that significant reduction in lignin content (up to 45 percent) by down-regulation of 4CL, a key lignin pathway enzyme, was compensated by an apparent increase in cellulose content (up to 15 percent) in transgenic aspen trees (Hu et al., 1999). These results offer another possibility to be explored, the effect of up-regulation of *CesA* genes on the lignin content of trees.

With this aim, we screened an *A. thaliana* cDNA library with *A. thaliana* ESTs showing similarity to cotton *CesA* genes and cloned many *A. thaliana CesA* cDNAs that were not available at that time (Wu et al., 1998). Using these *A. thaliana CesA* cDNAs as probes, we screened a developing xylem cDNA library from aspen *(Populus tremuloides)* available in our laboratory and cloned the first xylem-specific aspen *CesA (PtCesA)* gene, which was also shown to be tension-stress responsive (Joshi et al., 1999; Wu et al., 2000). We submitted the first *CesA* cDNA sequence from any tree species (Accession No. AF072171) to GenBank in June 1998 and subsequently characterized the full-length *CesA* cDNA from aspen. This cDNA is 3,232 base pairs (bp) long and has an open reading frame of 2,934 bp encoding a 978 amino acid-long protein of 110,278 KDa. The 5' leader and 3' trailer regions showed the presence of conserved regulatory sequence motifs typical of plant genes (Joshi, 1987a,b; Joshi et al., 1997). The predicted protein also had two TMDs toward the N-terminal and six TMDs toward the C-terminal, similar to other known CesA proteins at that time (Figure 6.1). Comparison of the PtCesA protein with other known CesA proteins indicated that the first hypervariable region (HVR I) resided in the N-terminal region immediately following the putative zinc-binding domain that was speculated to be involved in protein-protein interaction (Kawagoe and Delmer, 1997). The putative catalytic domain was present between TMD 2 and 3, and all signature motifs of processive glycosyltransferases (Saxena et al., 1995) were also present in this globular region. A second hypervariable region (HVR II), which may have some specific function in regulating the quantity and quality of cellulose produced, further interrupted the catalytic domain.

Most interestingly, PtCesA protein showed 90 percent similarity to cotton GhCesA1 that was proposed to be the secondary cell wall-specific CesA

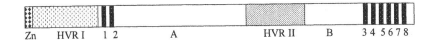

FIGURE 6.1. Diagrammatic representation of PtCesA protein. Various domains are indicated by a brief description below the diagram. Zn = zinc-binding domain; HVR I = N-terminal hypervariable region; 1-8 = transmembrane domains; subdomains A and B = highly conserved (70 to 90 percent) part of catalytic domains in relation to other CesA proteins; HVR II = central hypervariable region.

(Pear et al., 1996). This was the first indication in literature that two species, such as cotton and aspen that are widely divergent in an evolutionary sense have similar CesA proteins that may perform similar cellular functions. Moreover, it could be speculated from this similarity relationship that aspen *CesA* mRNA isolated from the xylem cDNA library would be specifically localized to developing xylem tissues. This was proved by in situ hybridization experiments in which gene-specific probes from a *PtCesA* gene were exclusively localized to developing xylem tissues (Wu et al., 2000). Transformation of tobacco plants with a *PtCesA* promoter β-glucuronidase (GUS) fusion construct further confirmed that a *PtCesA* promoter regulated the expression of *PtCesA* mRNA in developing secondary wall-rich xylem tissues. Although phloem fibers are known to have secondary cell walls, no GUS expression was observed in these tissues.

Tension wood tissue is known to produce cellulose-rich xylem and phloem fiber cells. We postulated that bending of transgenic plants expressing *PtCesA* promoter-GUS fusion would show tension-stress-responsive expression of the *gus* gene. To examine this possibility, we bent the transgenic stems from four to 40 hours, and GUS expression at the point of bending was microscopically examined. Within four hours of tension stress, phloem fibers started expressing *gus* gene when xylem fibers continued to show presence of GUS protein. Thus the *PtCesA* promoter appears to induce GUS expression in developing xylem and phloem fibers in response to even brief tension stress. After an extended 20 hours of tension stress, GUS expression was visible only in the upper half of the vascular tissue when the lower half of the vascular ring showed no GUS expression. After 40 hours of tension stress, GUS expression was visible only at a single point on the upper side of stem where maximum tension stress was applied. These observations clearly indicated that the *PtCesA* gene promoter is tension-stress responsive and may play an active role in cellulose-rich tension wood fiber production. To test the hypothesis that CesA overexpression may lead to cellulose overproduction, we already have produced a large number of transgenic aspen trees overexpressing the *PtCesA* gene, and detailed molecular genetic and chemical characterization of transgenic tissues is currently underway.

In all the plant genomes examined so far, *CesA* appears to be a large multigene family. Southern blot analysis of aspen genomic DNA with a gene-specific *PtCesA* probe indicated that at least three to four homologous members are present in the aspen genome. However, use of similar probes from an *Arabidopsis AtCesA1* gene indicated that a large *CesA* gene family is also present in aspen. In addition to our aspen *PtCesA* gene, only one more full-length *CesA* cDNA from hybrid poplar *(PcCesA)* is currently available (Wang and Loopstra, 1998). This cDNA was also isolated from the xylem cDNA library of hybrid poplar trees and showed about 70 percent similarity to *PtCesA,* indicating that they both are members of two separate gene families in poplars. No commercially important trees with improved cellulose content and quality can be produced unless more information about the structure, organization, expression patterns, and function of tree genes involved in cellulose biosynthesis becomes available. We have started collecting this information from aspen trees.

Are Distant Relatives of CesA Genes Present in the Plant Genomes?

The previous discussion suggests *CesA* genes may belong to a large multigene family of ten or more members, in which different members are likely to perform distinct functions. However, the actual size of the *CesA* gene family becomes much larger if one considers several other genes that show limited similarity to *CesA* genes. There are about 40 genes in this category in the *A. thaliana* genome, necessitating the upgrading of the status of the *CesA* multigene family to a superfamily. Cutler and Somerville (1997) proposed a new gene family, cellulose synthase-like *(CSL)* genes encoding proteins that show only limited similarity to CesA proteins (about 30 to 40 percent), mainly in the central putative UDP-glucose binding catalytic domain. Six *CesA*-like *(CSL)* gene families, *CSLA, CSLB, CSLC, CSLD, CSLE,* and *CSLG,* have been predicted to encode CSL proteins with only limited identity (30 to 45 percent) with the "true" CesA proteins (Richmond and Somerville, 2000). Among these six CSL families, predicted protein structure and topology of CSLD proteins is most similar to true CesA proteins, although both these proteins share only about 45 percent overall sequence identity. Full-length genomic sequences are available for these genes in *A. thaliana*. However, many ESTs from many other plant species show high similarity to predicted coding regions of *A. thaliana CSL* genes. The recent survey of plant *CesA* and *CSL* sequences suggests there are over 1,200 entries from about 40 plant species (Richmond and Somerville, 2000). One can get a crude estimate of abundance of *CesAs* and *CSLs* from the EST database, which indicates that over 50 percent of ESTs can be classified as *CesAs*, 17 percent as *CSLA*, 1 percent as *CSLB*, 10 percent as *CSLC*, *6 percent as CSLD*, 4 percent as *CSLE*, and 7 percent as *CSLG* (Richmond and Somerville, 2000). Although we do not clearly know what functions any of these *CSL* genes may perform, some suggestions at least for some members of the *CSLA* gene family exist, such as production of other noncellulosic polysaccharides in the cell wall as described in an international patent (WO98/55596). However, each *CSL* gene family may perform completely different functions based on its extensive diversity. Alternatively, all CesAs and CSLs may form a functional enzyme complex, but experimental evidence is currently lacking for this hypothesis. A systematic molecular genetic approach is, therefore, required to decipher functions of each of the *CesA* and *CSL* gene families by isolation of various members of these gene families. Currently we have focused our attention on *CesA* and *CSLD* gene families from aspen.

ISOLATION OF NEW CESA *AND* CSLD *GENES*

The *CesA* gene family shares some very interesting features with *CSLD* members that are not shared with any other *CSL* family. Moreover, *CSLD* genes and proteins have distinct features of their own. I will first describe the relationship between *A. thaliana CesA* and *CSLD* genes and proteins and later summarize the information available from other plant species.

Six distinct *CSLD* genes (*AtCSLD-1* to *AtCSLD-6*) have been identified in *A. thaliana* that are scattered all over the genome (on chromosomes 1, 2, 3, 4, and 5) similar to the *CesA* genes (on chromosomes 2, 3, 4, 5), suggesting that genes on different chromosomes may perform some nonredundant functions and may not have been a result of recent gene duplications (Richmond, 2000). More intriguing is the clustering of some *CSLD* genes with some *CesA* genes in the *A. thaliana* genome map, leading to the suggestion that during evolution either *CesAs* might have given rise to *CSLDs* or vice versa. For example, *AtCSLD2* and *AtCesA7* are located on chromosome 5 at 33 cM; *AtCSLD4, AtCesA1,* and *AtCesA2* on chromosome 4 at 100, 82, and 105 cM, and *AtCSLD1, AtCesA9,* and *AtCeSA10* are all on chromosome 2 at 63, 40, and 47 cM, respectively. Although some of these map positions may change as more genome information becomes available, the distribution and location of these two gene family members does suggest they might be evolutionarily more related to each other than any other *CSL* families, and *CSLD* genes might actually be a progenitor of modern *CesA* genes (Richmond and Somerville, 2000). Moreover, most of the *A. thaliana CSLD* genes have three exons, with the exception of *AtCSLD1* which has six exons. Most of the true *CesA* genes have 10 to 14 exons, and as recently suggested by Richmond and Somerville (2000), *CSLD*s may predate *CesA*s and other *CSL* genes based on their smaller number of phase zero introns (De Souza et al., 1998). Apart from this genomic information, there are many more interesting structural similarities in the primary amino acid sequences of the predicted proteins.

Most of the *A. thaliana* CesA proteins are predicted to have 1,026 to 1,088 amino acids, while the CSLD proteins are predicted to be slightly larger with 1,036 to 1,181 amino acids. Other *CSL* genes encode much smaller proteins (Richmond and Somerville, 2000). The predicted proteins show about 45 percent identity between CesAs and CSLDs. Thus CesA and CSLD proteins are more similar to each other than other CSL members, which show only about 30 percent identity with CesAs.

Multiple sequence alignment of ten complete CesA proteins from *A. thaliana* and six CSLD proteins clearly indicate that both of these proteins are made up of well-interspersed, highly conserved and diverged domains as shown in Figure 6.2, and the arrangement of domains is also highly conserved between CesA and CSLD proteins. In both these proteins, there is a N-terminal extension of about 250 amino acids, followed by two predicted TMDs. Between TMD 2 and 3 is the globular region of about 500 to 600 amino acids that is believed to contain the catalytic sites with the highly conserved D,D,D...QXXRW signature, indicating that both of these proteins are likely to be processive glycosyltransferases, as suggested by Saxena and colleagues (1995), and belong to family 2 of glycosyltransferases (Campbell et al., 1997). Following the globular region, there are six TMDs toward the C-terminal of the predicted CesA and CSLD proteins. TMD regions are believed to anchor the cytoplasmic globular domain to the plasma membrane during cellulose biosynthesis (Delmer, 1999). Although CSLA, CSLB, CSLG, and CSLE appear to localize to the Golgi apparatus in the cytoplasm (Richmond and Somerville, 2000), clear information about where CSLD is localized is still lacking. Favery and colleagues (2001) have shown that a majority of the AtCSLD3 proteins are likely to be localized on the endoplasmic reticulum and predicted that these proteins may be involved in synthesis of noncellulosic wall polysaccharides. However, no experimental data on noncellulosic polysaccharide alteration in the AtCSLD3 mutant is still available. On the basis of its structural conservation with CesAs, it is tempting to speculate that at least some CSLDs may also localize at the plasma membrane and even in the functional rosettes, similar to what was observed recently for CesA proteins (Kimura et al., 1999). In case CSLDs do not colocalize with CesAs in plasma membrane fractions, we would at least like to know what these so-called "ancestors of modern day CesAs" did before becoming involved in cellulose biosynthesis. This situation would certainly pose many intriguing questions about the origin and evolution of the *CesA* superfamily.

The N-terminal HVR I that is said to be absent in many CSL proteins is actually present in CSLD proteins. It is highly diverged, showing only about 25 to 50 percent identity among various CesAs except a small, highly conserved putative zinc-binding domain consisting of four pairs of CX_2C (Kawagoe and Delmer, 1997; Arioli et al., 1998; Wu et al., 2000), where C stands for cysteine residue and X stands for any nonconserved amino acid. It is also suggested that this conserved region may be involved in protein-protein interaction during cellulose biosynthesis (Delmer, 1999). Similar to CesAs, the N-terminal region of the CSLD proteins is also highly diverged, showing only about 25 to 50 percent identity among various CSLDs. Comparison of N-terminal HVR I regions from CesAs and CSLDs shows only about 11 to 30 percent identity between these proteins. It is true that CSLDs do not have a putative zinc-binding domain but interestingly CSLD proteins also have a highly conserved motif with four pairs of C residues separated by a variable number of nonconserved amino acids, $CX_{1-4}C$. Could it be the ancestor of modern zinc-binding domains in CesA proteins?

Multiple sequence alignment of the about 550 to 590 amino acid-long globular region or so-called catalytic domain shows that three distinct regions can be clearly identified in both CesA and CSLD proteins (Figure 6.2). The subdomain A is composed of 323 to 380 amino acids and shows 60 to 70 percent identity between CesA and CSLD proteins, which is followed by a hypervariable region (HVR II) of 45 to 90 amino acids sharing only 15 to 25 percent identity, which is then followed by a short subdomain B of 96 to 150 amino acids sharing about 60 to 70 percent identity. The first two of the three invariant D residues of the conserved processive glycosyltransferase signature reside in subdomain A, and the third conserved D residue and QVLRW sequence reside in subdomain B.

In addition to the 10 *CesA* and 6 *CSLD* gene sequences from *A. thaliana,* 15 more complete *CesA* genes are available from cotton, rice, poplar, tobacco, and corn, and three more *CSLD* genes are available from rice and tobacco. Interestingly, all these additional 18 proteins share features similar to the *A. thaliana* proteins, indicating that this modular structure was established before monocot and dicot diversion and is a common plant CesA/CSLD character. Thus one could easily make gene-specific probes from the HVR I and HVR II regions due to their high level of divergence and use them for in situ hybridization experiments to detect their expression patterns or raise

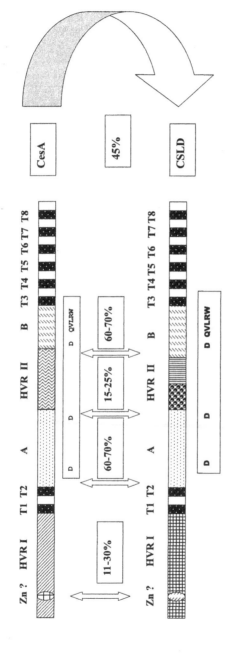

FIGURE 6.2. Comparison of the modular structure of CesA (upper panel) and CSLD (lower panel) proteins with percentage identity (shown in open boxes) between different modules (as indicated by arrows). T1 to T8 blocks indicate the predicted transmembrane domains. Highly diverged N-terminal extension shows the presence of a putative zinc-binding motif in CesAs, but presence of a similar motif in CSLDs is questionable. The HVR II region in CSLD is highly different from CesA, but among CSLDs it shows some conservation toward the C-terminal as indicated by a different box with vertical stripes. The subdomains A and B are 60 to 70 percent identical between CesAs and CSLDs. The processive glycosyltransferase signature proposed by Saxena and colleagues (1995) is indicated in a box below subdomains A, HVR II, and B.

antibodies for gene-specific protein localization studies. For the isolation of HVR II regions from any unknown members of a new species, one could design polymerase chain reaction (PCR) primers from highly conserved subdomains A and B and amplify the intervening hypervariable region. We used this strategy for isolation of at least one new CesA-specific and two distinct CSLD-specific HVR II sequences from aspen as described in the next section.

Cloning of New CesA *and* CSLD *Gene Sequences in Aspen*

We decided to build upon the information collected from CesA and CSLD HVR II regions. Because HVR II sequences have shown distinct gene-specific variations in *A. thaliana* CesA and CSLD sequences, we first performed a multiple sequence alignment of the globular domain from 25 predicted CesA proteins and nine CSLDs from *A. thaliana,* cotton, poplar, tobacco, corn, and rice. Toward the end of the subdomain A, a short peptide sequence, CYVQFPQ, is highly conserved in all 34 CesA and CSLD proteins. Moreover, toward the beginning of subdomain B, GWIYGS is 100 percent conserved. Based on the DNA sequence information encoding these two short peptides from all CesA/CSLD proteins, we synthesized two degenerate primers and used these primers (HVR2F and HVR2R) and aspen genomic DNA for PCR-mediated amplification of the genomic sequences homologous to CesA and CSLD members. Interestingly, two distinct bands of 1.2 kb and 0.6 kb were reproducibly amplified. This perfectly fits with what one would expect if *CesA* and *CSLD* gene structures were conserved between *A. thaliana* and aspen. In *A. thaliana,* HVR II regions from *CesA* genes flanked by HVR2F and HVR2R primers are expected to consist of four exons and three short introns spanning about 0.8 to 1.2 kb, and HVR II regions from *CSLD* genes flanked by HVR2F and HVR2R primers are expected to consist of only one exon spanning about 0.6 kb. These amplified PCR products were cloned in pCR2.1 plasmid from a TOPOTA Cloning Kit (Invitrogen), and over 25 PCR clones from both the size classes (0.6 kb and 1.3 kb) have so far been sequenced. We have identified two distinct groups of 0.6 kb putative CSLD HVR II clones, *PtCSLD1* and *PtCSLD2* (GenBank accession No. AF417486 and AF417487), and one distinct group of 1.3 kb putative CesA HVR II clones, *PtCesA2* (GenBank accession No. AF417485). Sequence analysis further confirmed that these new CesA and CSLD-HVR II clones indeed belong to the *CesA* and *CSLD* gene families, respectively. PtCSLD1 and PtCSLD2 HVR II proteins share only about 66 percent identity with each other and may be regarded as two distinct CSLD family members. More clones are currently being sequenced to get some rough idea about the extent of the *CSLD* gene family in aspen. One of the 1.3 kb-long PtCesA2 sequences was also fully sequenced, and it does indeed belong to *CesA* gene family in aspen and has three introns (Chavli, 2001). It is also a new member of the aspen *CesA* family based on the sequence analysis showing 53 percent and 62 percent identity with known poplar CesA proteins, PtCesA1 and PcCesA1 (Wu et al., 2000; Wang and Loopstra, 1998). Furthermore, we performed Southern analysis of aspen genomic DNA using PtCesA2, PtCSLD1, and PtCSLD2 HVR II clones to see if their copy number and hybridization patterns are different. They all showed distinctly different patterns indicative of their probe specificity. There may be at least two to three members of each of these gene families in the aspen genome.

To confirm that these two classes of CSLD and one new class of CesA HVR II sequences are actually expressed in aspen tissues, we performed reverse transcription (RT)-PCR reactions using total RNA from aspen leaf and xylem using HVR2F and HVR2R primers. The RT-PCR reaction would theoretically amplify all CesA and CSLD HVR II regions from cDNAs expressed in the respective tissues. Therefore, the amplified products were first transferred to nitrocellulose membranes and hybridized separately with PtCesA2, PtCSLD1, and PtCSLD2 HVR II clones. Both the *CSLD1* and *CSLD2* genes are differentially expressed in leaf and xylem tissues. The aspen CSLD1 clone is expressed more in xylem than leaf, while the aspen CSLD2 clone is expressed more in leaf than xylem. PtCesA2 is highly expressed in both leaf and xylem as compared to PtCSLD sequences, but it is expressed more in xylem than leaf, indicating its strong expression in woody tissues. Although this strategy has given information about three new members of the *CesA* gene family in aspen, new PCR primer design or PCR conditions are currently attempted to amplify all CesA and CSLD HVR II regions from the aspen genome. To this end, change in annealing temperature of PCR primers HVR2F and HVR2R from 44 to 60°C during PCR using aspen genomic DNA yielded different sizes of amplified PCR products, and these are currently being sequenced and analyzed (Figure 6.3).

In a parallel approach, we also used the strategy described earlier (Wu et al., 1998) of using the heterologous EST probes from *A. thaliana* for identification of new aspen *CesA* and *CSLD* cDNA clones for screening available cDNA libraries in our laboratory. At least one new CesA clone (PtCesA3) has so far been identified (Chavli, 2001), from which the HVR II region was amplified and sequenced (GenBank accession No. AY0055724). We have employed PtCesA2, PtCesA3, PtCSLD1, and PtCSLD2 HVR II regions for cDNA library screening and have so far obtained four new full-length *CesA* and two new *CSLD* cDNAs from the aspen genome. Further work with sequencing and other molecular characterization is currently in progress.

M   44° C   48° C   52° C   56° C   60° C   M

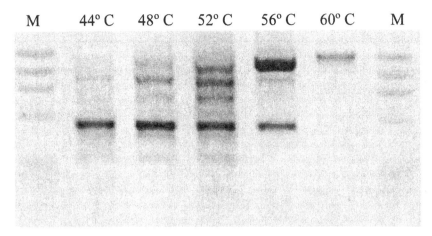

FIGURE 6.3. PCR amplification of HVR II regions from aspen genomic DNA using HVR2F and HVR2R primers at different annealing temperatures from 44 to 60°C. Lanes M indicate molecular weight markers (1 kb ladder). Note that different-sized PCR products are amplified by change in the annealing temperature.

Expressed Sequence Tag Sequences Encoding CesA and CSL Proteins from Trees

Although only two full-length *CesA* cDNAs from trees are currently available, a significantly large number of ESTs homologous to these genes are publicly available for further research. For an up-to-date listing of such sequences, please visit <http://cellwall.stanford.edu/php/species.php>. Currently a total of 119 ESTs from *Cryptomeria,* loblolly pine, poplar species, and black locust trees are currently available in the GenBank as shown in Table 6.1.

With the two ongoing public EST projects in loblolly pine (http://www. cbc.umn.edu/ResearchProjects/Pine/NCSU.pine/NCSU_top.html) and poplars (http://mlsc.ffr.mtu.edu), more CesA/CSL ESTs are expected to become available in the future (see Chapter 2). Most of these ESTs are derived from a xylem cDNA library originated from normal wood, compression wood, or sapwood. This is not surprising considering the importance of wood for the forest product industries. The majority of ESTs (98) represent one of the subfamilies of *CesAs*, while eight, nine, three, and one ESTs represent *CSLA, CSLC, CSLD,* and *CSLE* families, respectively. No representatives of *CSLB* and *CSLG* families have so far been identified, but that does not mean they are absent in these trees. It must be emphasized that these ESTs were identified through random sequencing of clones rather than directed *CesA/*

TABLE 6.1. Number of available CesA and CSL EST sequences from forest trees

Tree species	CesA/CSL subfamily	Number of EST sequences
Cryptomeria japonica	CesA	1
Pinus taeda	CesA	84
Pinus taeda	CSLA	7
Pinus taeda	CSLC	6
Pinus taeda	CSLD	1
Populus balsamifera	CesA	1
Populus tremula × *P. tremuloides*	CesA	11
Populus tremula × *P. tremuloides*	CSLA	1
Populus tremula × *P. tremuloides*	CSLC	3
Populus tremula × *P. tremuloides*	CSLD	2
Populus tremula × *P. tremuloides*	CSLE	1
Robina pseudoacacia	CesA	1

Source: Data compiled from <http://cellwall.stanford.edu>.

CSL gene cloning as we are attempting in our laboratory. Currently more than 50,000 EST sequences from the same tree species are available in the GenBank. These *CesA/CSL* clones represent only a very small portion of the total available DNA sequences from trees, and in the absence of tree genome sequencing projects, the real size of *CesA/CSL* gene families from any tree species currently remains unknown. More research should be directed in the future for the isolation and characterization of full-length *CesA/CSL* genes from economically important trees.

Phylogenetic Analysis of Available Plant CesA Genes

Several full-length *CesA* cDNAs/genes are currently available from a number of plant species from which information about encoded proteins can be easily obtained. We have compiled a total of 24 CesA proteins (ten *A. thaliana,* eight corn, two cotton, two aspen/poplar, one rice, and one tobacco). The unrooted phylogenetic tree developed from these 24 sequences using 1,000 bootstrap values is shown in Figure 6.4. The most important observation is separation of a secondary cell wall-specific clade on the right side as indicated by a delimiting line. As observed earlier by Holland and colleagues (2000), AtCesA7, which is shown to be associated with secondary wall development in *A. thaliana* (Taylor et al., 1999), does not pair with any other secondary CesAs but remains close to the other members of the group. Hybrid poplar CesA (PcCesA1) from xylem tissues groups with *Arabidopsis* AtCesA4 and rice OsCesA6. Similar to PcCesA1, expression of AtCesA4 has been recently shown to be associated with vascular tissues (Holland et al., 2000), but nothing is known about OsCesA6. The other branch of the secondary clade consists of three CesAs, GhCesA1 from developing cotton fibers (Pear et al., 1996), PtCesA1 from developing xylem (Wu et al., 2000), and AtCesA8 associated with secondary wall development (Taylor et al., 2000). Thus, all the known CesAs in this clade seem to be associated with secondary wall development. Absence of any monocot CesA members showing no secondary growth in the secondary clade is also remarkable.

Another interesting conclusion can be derived from this analysis. Assuming that this phylogenetic tree reflects structural (and possibly functional) differences among various *CesA* gene family members, one can trace the possible minimum number of functional classes of *CesA* genes in plants. Each of the six rosette subunits of the hexagonal structure is made up of six cellulose-synthesizing sites, in total 36 sites per rosette. Each of these sites could be made up of similar CesA subunits or a combination of different CesA subunits. Several authors have suggested similar organization of CesA isoforms (Saxena and Brown, 2000; Perrin, 2001; Scheible et al., 2001). CesA members from individual branches may form homo-rosettes (all six sites of the same kind of CesA isoforms) that are responsible for cellulose biosynthesis in different tissues and cell types, for example, in shoot and root (Figure 6.5a and 6.5b). Apart from the anomalous positioning of AtCesA7, the remaining nine *A. thaliana* CesAs form only six groups (Group 1: AtCesA1 and AtCesA10; Group 2: AtCesA3; Group 3: AtCesA5 and AtCesA6; Group 4: AtCesA2 and AtCesA9; Group 5: AtCesA8; and Group 6: AtCesA4). All other known *CesA* genes also roughly cluster around these six groups, although not all *A. thaliana* CesA homologues are identified from each of the non-*Arabidopsis* species. Considering the hypothesis that presence of more than one CesA is involved in cellulose synthesis in plants (Taylor et al., 2000; Fagard et al., 2000) and six rosette subunits make up the hexagonal rosette complex (Brown et al., 1996), it is tempting to speculate that one member of each of these six groups may form a functional hetero-rosette as shown in Figure 6.5c.

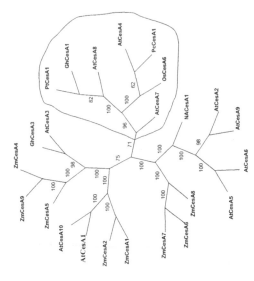

FIGURE 6.4. Phylogenetic analysis of 24 full-length CesA proteins. DNA and protein sequence analysis was performed using various program routines from Genetics Computer Group (GCG) package version 10.0. Multiple sequence alignment of various CesA proteins was done with "Pileup" and "Pretty" programs from the GCG package. Cladograms were developed using PAUP 4.0b8 version (Phylogenetic Analysis Using Parsimony, Sinaur Associates, Sunderland, MD) with parsimony analysis and heuristic search algorithm. Bootstrap analysis was done with 1,000 replicates, and bootstrap values above 50 were considered for the development of unrooted tree. GenBank accession numbers (underlined) for CesA genes included in this figure are AtCesA1, AF027172; AtCesA2, AF027173; AtCesA3, AF027174; AtCesA4, AB006703; AtCesA5, AB016893; AtCesA6, AF062485; AtCesA7, AF088917; AtCesA8, AL035526; AtCesA9, AC007019; AtCesA10, AC006300; ZmCesA1, AF20025; ZmCesA2, AF20026; ZmCesA4, AF20028; ZmCesA5, AF20029; ZmCesA6, AF20030; ZmCesA7, AF20031; ZmCesA8, AF20032; ZmCesA9, AF20033; GhCesA1, U58283; GhCesA3, AF150630; OsCesA6, AC022457; PtCesA1, AF072131; PcCesA1, AF081534; NaCesA1, AF30474. (The nomenclature used was proposed by Delmer [1999] where At = *Arabidopsis thaliana*; Gh = *Gossypium hirsutum*; Na = *Nicotiana alata*; Os = *Oryza sativa*; Pc = *Populus × canescens*; Pt = *Populus tremuloides*; and Zm = *Zea mays*.)

a) Shoot rosette subunit   b) Root rosette subunit   c) Six CesAs/rosette subunit

d) AtCesA1 and AtCesA6 forming primary rosette subunit   e) AtCesA7 and AtCesA8 forming secondary rosette subunit   f) CesA and CSLs forming rosette subunit

FIGURE 6.5. Various theoretical scenarios in CesA and CSL involvement in rosettes or TC formation

This may also explain why even slight alteration in even one *CesA* gene could seriously impact the process of cellulose biosynthesis in *A. thaliana* (Arioli et al., 1998). It is possible that these six groups perform specific nonredundant functions in cellulose synthesis. Alternatively, as proposed by Fagard and colleagues (2000), one or more nonredundant members of some of the phylogenetic branches discerned here (such as AtCesA1 and AtCesA6) form the rosettes responsible for cellulose biosynthesis in the primary wall (Figure 6.5d) and rosettes made up of AtCesA7 and AtCesA8 would synthesize cellulose in a secondary cell wall-specific manner as proposed by Taylor and colleagues (2000) (Figure 6.5e). The number and order of these CesA subunits in a functional TC could be another variable. Involvement of CSLDs and other CSLs in such rosette terminal complexes cannot yet be ruled out (Figure 6.5e). Which of these theoretical scenarios is correct needs to be investigated in the immediate future, and this could influence the possible strategies for genetic manipulation in trees. As compared to the tiny and short-lived model *Arabidopsis* plants, perennial trees produce large amounts of wood year after year. Are similar genes involved in cellulose biosynthesis in tree species and *Arabidopsis,* or are different genes specific to woody species? Perhaps the differential regulation of aspen *CesA* genes that are structurally and functionally similar to *Arabidopsis CesAs* is more important for wood formation. Alternatively, genes other than *CesAs* and *CSLs* such as *SuSy* may play a major role in this process.

FUTURE PERSPECTIVES

Understanding the process of cellulose biosynthesis in trees has numerous basic and applied implications, and isolation of novel cellulose biosynthesis-related genes from an economically important tree species is a first step toward achieving that goal. This is a very young branch of science that is poised for many breakthroughs. How do trees make their cellulose, and how is this process related to other metabolic processes that impart rigidity and strength to wood? How temporal and spatial regulation of various members of the *CesA* superfamily contribute to high-quality wood production is still unknown. What controls the microfibril angle/orientation, cellulose DP, or cellulose crystallinity? Currently available routine molecular techniques may be inadequate for tracking these quantitative traits, and better tools are needed and must be developed. Do proteins other than CesAs and CSLs regulate any of these economically important wood traits? It certainly is a strong possibility. Although proven in *A. thaliana,* molecular genetic proof of *CesA* gene involvement in tree development is still lacking, and overexpression and underexpression genetic experiments are necessary. Biochemical proof of *CesA* as the key gene in cellulose production by in vitro cellulose production is currently lacking in any plant species and should be provided in the future. All of these developments in molecular genetics of cellulose biosynthesis in trees could be applied for production of genetically improved trees producing large quantities of high-quality cellulose that has tremendous economical importance.

REFERENCES

Arioli, T., Peng, L., Betzner, A. S., Burn, J., Wittke, W., Herth, W., Camilleri, C., Hofte, H., Plazinski, J., Birch, R., et al. (1998). Molecular analysis of cellulose biosynthesis in *Arabidopsis. Science* 279: 717-720.

Baskin, T. I., Betzner, A. S., Hoggart, R., Cork, A., and Williamson, R. E. (1992). Root morphology mutants in *Arabidopsis thaliana. Aust. J. Plant. Physiol.* 19: 427-437.

Brown, R. M. Jr. (1996). The biosynthesis of cellulose. *J. Macromol. Sci.* 33: 1345-1373.

Brown, R. M. Jr. and Montezinos, D. (1976). Cellulose microfibrils: Visualization of biosynthetic and orienting complexes in association with the plasma membrane. *Proc. Natl. Acad. Sci. USA* 73: 143-147.

Brown, R. M. Jr., Saxena, I. M., and Kudlicka, K. (1996). Cellulose biosynthesis in higher plants. *Trends Plant Sci.* 1: 149-156.

Campbell, J. A., Davies, G. J., Bulone, V., and Henrissat, B. (1997). A classification of nucleotide-diphospho-sugar glycosyltransferases based on amino acid sequence similarities. *Biochem. J.* 326: 929-942.

Chavli, R. V. S. S. (2001). Molecular cloning of hypervariable II regions of cellulose synthase and cellulose synthase-like genes in *Populus tremuloides* (aspen). MS thesis. Michigan Technological University, Houghton, MI.

Cutler, S. and Somerville, C. (1997). Cellulose synthesis: Cloning *in silico. Current Biol.* 7: R108-R111.

De Souza, S. J., Long, M., Klein, R. J., Roy, S., Lin, S., and Gilbert, W. (1998). Toward a resolution of the introns early/late debate: Only phase zero introns are correlated with the structure of ancient proteins. *Proc. Natl. Acad. Sci. USA* 95: 5094-5099.

Delmer, D. P. (1987). Cellulose biosynthesis. *Ann. Rev. Plant. Physiol.* 38: 259-290.

Delmer, D. P. (1998). A hot mutant for cellulose biosynthesis. *Trends Plant Sci.* 3: 164-165.

Delmer, D. P. (1999). Cellulose biosynthesis: Exciting times for a difficult field of study. *Ann. Rev. Plant Physiol. Plant Mol. Biol.* 50: 245-276.

Delmer, D. P. and Amor, Y. (1995). Cellulose biosynthesis. *Plant Cell* 7: 987-1000.

Fagard, M., Desnos, T., Desprez, T., Goubet, F., Refregier, G., Mouille, G., McCann, M., Rayon, C., Vernhettes, S., and Hofte, H. (2000). PROCUSTE1 encodes a cellulose synthase required for normal cell elongation specifically in roots and dark-grown hypocotyls of *Arabidopsis. Plant Cell* 12: 2409-2424.

Favery, B., Ryan, E., Foreman, J., Linstead, P., Boudonck, K., Steer, M., Shaw, P., and Dolan, L. (2001). KOJAK encodes a cellulose synthase-like protein required for root hair cell morphogenesis in *Arabidopsis. Genes Dev.* 15: 79-89.

Haigler, C. (1985). The functions and biogenesis of native cellulose. In T. P. Nevell and S. H. Zoronian (Eds.), *Cellulose Chemistry and Applications* (pp. 30-83). Chichester, UK: Ellis Horwood Ltd.

Haigler, C. and Blanton, R. L. (1996). New hopes for old dreams: Evidence that plant cellulose synthase genes have finally been cloned. *Proc. Natl. Acad. Sci. USA* 93: 12082-12085.

Haigler, C., Ivanova-Datcheva, M., Hogan, P. S., Salnikov, V. V., Hwang, S., Martin, K., and Delmer, D. P. (2001). Carbon partitioning to cellulose synthesis. *Plant Mol. Biol.* 47: 29-51.

Holland, N., Holland, D., Helentjaris, T., Dhugga, K., Xoconostle-Cazares, B., and Delmer, D. P. (2000). A comparative analysis of the cellulose synthase *(CesA)* gene family in plants. *Plant Physiology* 123: 1313-1323.

Hu, W. J., Harding, S. A., Lung, J., Popko, J. L., Ralph, J., Stokke, D. D., Tsai, C. J., and Chiang, V. L. (1999). Repression of lignin biosynthesis promotes cellulose accumulation and growth in transgenic trees. *Nat. Biotechnol.* 17: 808-812.

Joshi, C. P. (1987a). An inspection of the domain between putative TATA box and translation start site in seventy-nine plant genes. *Nucleic Acids Res.* 15: 6643-6653.

Joshi, C. P. (1987b). Putative polyadenylation signals in nuclear genes of higher plants: A compilation and analysis. *Nucleic Acids Res.* 15: 9627-9640.

Joshi, C. P., Wu, L., and Chiang, V. L. (1999). A novel xylem-specific and tension stress responsive cellulose synthase gene from quaking aspen. In M. Campbell (Ed.), *Forest Biotechnology 99* (Abstract p. 19). Oxford, UK: Oxford University Press.

Joshi, C. P., Zhou, H., Huang, X., and Chiang, V. L. (1997). Context sequences of translation initiation codon in plants. *Plant Mol. Biol.* 35: 993-1001.

Kawagoe, Y. and Delmer, D. P. (1997). Pathways and genes involved in cellulose biosynthesis. In J. K. Setlow (Ed.), *Genetic Engineering,* Volume 19 (pp. 63-87). New York: Plenum Press.

Kimura, S., Laosinchai, W., Itoh, T., Cui, X., Linder, C. R., and Brown, R. M. (1999). Immunogold labeling of rosette terminal cellulose-synthesizing complexes in the vascular plant *Vigna angularis. Plant Cell* 11: 2075-2085.

Mueller, S. C. and Brown, R. M. Jr. (1980). Evidence for an intramembrane component associated with a cellulose microfibril-synthesizing complex in higher plants. *J. Cell Biol.* 84: 315-326.

Nevell, T. P. and Zeronian, S. H. (1985). Cellulose chemistry fundamentals. In T. P. Nevell and S. H. Zeronian (Eds.), *Cellulose Chemistry and Its Applications* (pp. 15-29). Chichester, UK: Ellis Horwood Ltd.

Pear, J. R., Kawagoe, Y., Schreckengost, W. E., Delmer, D. P., and Stalker, D. M. (1996). Higher plants contain homologs of the bacterial CelA genes encoding the catalytic subunit of cellulose synthase. *Proc. Natl. Acad. Sci. USA* 93: 12637-12642.

Perrin, R. M. (2001). Cellulose: How many cellulose synthases to make a plant? *Curr. Biol.* 11: R213-R216.

Preston, R. D. (1964). Structural and mechanical aspects of plant cell walls with particular evidence to synthesis and growth. In M. H. Zimmermann (Ed.), *Formation of Wood in Trees* (pp. 169-201). London: Academic Press.

Richmond, T. A. (2000). Higher plant cellulose synthases. *Genome Biol. 1: Reviews* 3001.1-3001.6.

Richmond, T. A. and Somerville, C. R. (2000). The cellulose synthase superfamily. *Plant Physiology* 124: 495-498.

Roelofsen, P. A. (1958). Cell wall structure as related to surface growth. *Acat. Bot. Neerl.* 7: 77-89.

Saxena, I. M. and Brown, R. M. (2000). Cellulose synthases and related enzymes. *Curr. Opin. Plant Biol.* 3: 523-531.

Saxena, I. M., Brown, R. M., Fevre, M., Geremia, R. A., and Henrissat, B. (1995). Multidomain architecture of β-glycosyltransferases: Implications for mechanism of action. *J. Bacteriol.* 177: 1419-1424.

Saxena, I. M., Lin, F. C., and Brown, R. M. (1990). Cloning and sequencing of the cellulose synthase catalytic subunit gene of *Acetobacter xylinum*. *Plant Mol. Biol.* 15: 673-683.

Scheible, W. R., Eshed, R., Richmond, T., Delmer, D., and Somerville, C. R. (2001). Modifications of cellulose synthase confer resistance to isoxaben and thiazolidinone herbicides in Arabidopsis Ixr1 mutants. *Proc. Natl. Acad. Sci. USA* 98: 10079-10084.

Sjostrom, E. (1993). *Wood Chemistry: Fundamentals and Applications,* Second Edition. San Diego, CA: Academic Press.

Taylor, N. G., Laurie, S., and Turner, S. R. (2000). Multiple cellulose synthase catalytic subunits are required for cellulose synthesis in *Arabidopsis. Plant Cell* 12: 2529-2540.

Taylor, N. G., Scheible, W. R., Cutler, S., Somerville, C. R., and Turner, S. R. (1999). The irregular xylem3 locus of *Arabidopsis* encodes a cellulose synthase required for secondary cell wall synthesis. *Plant Cell* 11: 769-779.

Timell, T. E. (1986). *Compression Wood in Gymnosperms.* Berlin: Springer-Verlag.

Turner, S. R. and Somerville, C. R. (1997). Collapsed xylem phenotype of *Arabidopsis* identifies mutants deficient in cellulose deposition in the secondary cell wall. *Plant Cell* 9: 689-701.

Wang, H. and Loopstra, C. (1998). Cloning and characterization of a cellulose synthase cDNA (Accession No. AF081534) from xylem of hybrid poplar (*Populus tremula* × *Populus alba*). (PGR98-179) *Plant Physiology* 118: 1101.

Wu, L., Joshi, C. P., and Chiang, V. L. (1998). AraxCelA, a new member of cellulose synthase gene family from *Arabidopsis thaliana* (Accession No. AF062485). *Plant Physiology* 117: 1125.

Wu, L., Joshi, C. P., and Chiang, V. L. (2000). A xylem-specific cellulose synthase gene from aspen *(Populus tremuloides)* is responsive to mechanical stress. *Plant J.* 22: 495-502.

Chapter 7

Tuning Lignin Metabolism Through Genetic Engineering in Trees

Lise Jouanin
Thomas Goujon

INTRODUCTION

Lignin (from the Latin *lignum:* wood) constitutes the second component of wood after cellulose. This complex cell wall biopolymer is not present in each plant cell but is representative of secondary cell walls. It encrusts specific cells such as vessels and fibers, and its appearance in plants is responsible for the development of upright plants adapted to a terrestrial habitat. The functional significance of lignins in plant development is associated with mechanical support, water conduction, and defense mechanisms. The biosynthesis of this phenolic biopolymer involves many enzymatic steps and results in a highly complex structure. Three major monomers, the monolignols, are the precursors of the lignin polymer. These three monolignols, or hydroxycinnamyl alcohols, differ in the degree of methoxylation of the aromatic ring. *p*-Coumaryl, coniferyl, and sinapyl alcohols possess, respectively, zero, one, and two methoxyl groups (Figure 7.1). These monolignols give rise respectively to the hydroxyphenyl (H), guaiacyl (G), and syringyl (S) units of the polymer. In the lignin molecule, these monomeric units are linked by different bond types (Davin and Lewis, 1992; Hatfield and Vermerris, 2001). The most frequent intermonomeric linkages are β-O-4 ether bonds (also called noncondensed) which are labile linkages and, consequently, the target of most techniques used to analyze the chemical structure of lignins and of delignification processes. The other linkages are β-5, β-1, β-β, and 5-5 (C-C or condensed linkages) and 4-O-5 (ether) linkages. The content and composition of lignins are known to vary according to taxa, tissues, cell types, and cell wall layers. In angiosperms, the syringyl/guaiacyl ratio (S/G) calculated from the labile lignin fraction is often considered to give information on the lignin structure. Gymnosperm plants (conifers) possess G-type lignins, and angiosperm trees possess G+S-type

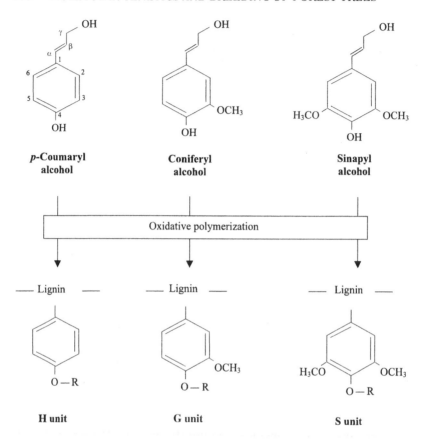

FIGURE 7.1. Structure of the three main lignin precursors or monolignols and the resulting monomers found in the lignin polymer. H = *p*-hydroxyphenyl unit; G = guaiacyl unit; S = syringyl unit.

lignins. Lignin structure also depends on the developmental plant stage and on the environmental conditions (Campbell and Sederoff, 1996). As a result of such complexity and variety, the term *lignins* rather than *lignin* is sometimes used.

The lignin pathway can be divided into different parts: the general phenylpropanoid pathway (from phenylalanine to the hydroxycinnamic acids [HCA] and their CoA-esters), and the monolignol-specific pathway (reduction of the HCA-CoA esters into monolignols and their polymerization). HCA are central to the metabolism of plant phenolics (flavonoids, coumarins, stilbenes, etc.) which are involved in defense mechanisms, in inhibition or activation of enzymes, and in morphogenesis (Baucher et al., 1998). The first enzyme of the general phenylpropanoid pathway is phenylalanine ammonia-lyase (PAL), which deamines phenylalanine, an aromatic amino acid derived from the shikimate pathway. Further enzymatic reactions include hydroxylation of the aromatic ring, methylation of selected phenolic hydroxyl groups, activation of cinnamic acids to cinnamoyl-CoA esters, and the reduction of these esters first to cinnamaldehydes and then to cinnamyl alcohols (Figure 7.2). This pathway allows the synthesis of the monomer precursors, which are then linked via oxidative coupling catalyzed by both peroxidases and laccases. Although this pathway is often presented as a "metabolic grid" (Higuchi, 1990), such a model appears to be inconsistent with recent findings (Dixon et al., 2001; Li et al., 2001; Humphreys and Chapple, 2002), and despite a number of published works, some uncertainties remain.

The aim of this chapter is not to describe lignin deposition and its variability in detail (excellent recent reviews exist in this field: Baucher et al., 1998; Grima-Pettenati and Goffner, 1999; Boudet, 2000; Dixon et al., 2001; Donaldson, 2001; Mellerowicz et al., 2001; Plomion et al., 2001; Humphreys and Chapple, 2002) but to demonstrate the potential of using genetic engineering to modify lignin content and/or composition. This knowledge is of major importance for basic research but could also be of interest in the industrial exploitation of forest trees. It could allow the identification of the most appropriate gene(s) to target to obtain trees more adapted to industrial uses. The main industrial use concerns the pulp and paper industry, in which wood is the major raw material. During the paper pulp-making process, lignins must be chemically eliminated, which is both economically expensive and environmentally unfriendly.

Modifications of specific steps of the lignin biosynthetic pathway have been tested mainly in model or annual plants such as *Arabidopsis* and tobacco, and most of the published works on lignin modifications via genetic engineering in woody plants concern poplar (*Populus* spp.). This tree is of economic importance in many countries because it can be used for many purposes (packing materials, plywood, paper pulp production, fuel, and furniture). It is also the model plant for forest tree biotechnology studies, and, indeed, poplar offers several advantages: it grows quickly, is easy to propagate, and can be readily transformed. *Agrobacterium* transformation procedures are efficient for different genotypes (reviewed in Leplé et al., 1999). Its genome is small (1.1pg/2C, 550 Mb), only five times that of *Arabidopsis thaliana*. This chapter will mainly report results obtained on this tree with some references to model plants to compare the consequences of modulation of the same enzymatic step in annual and woody plants. Modification of

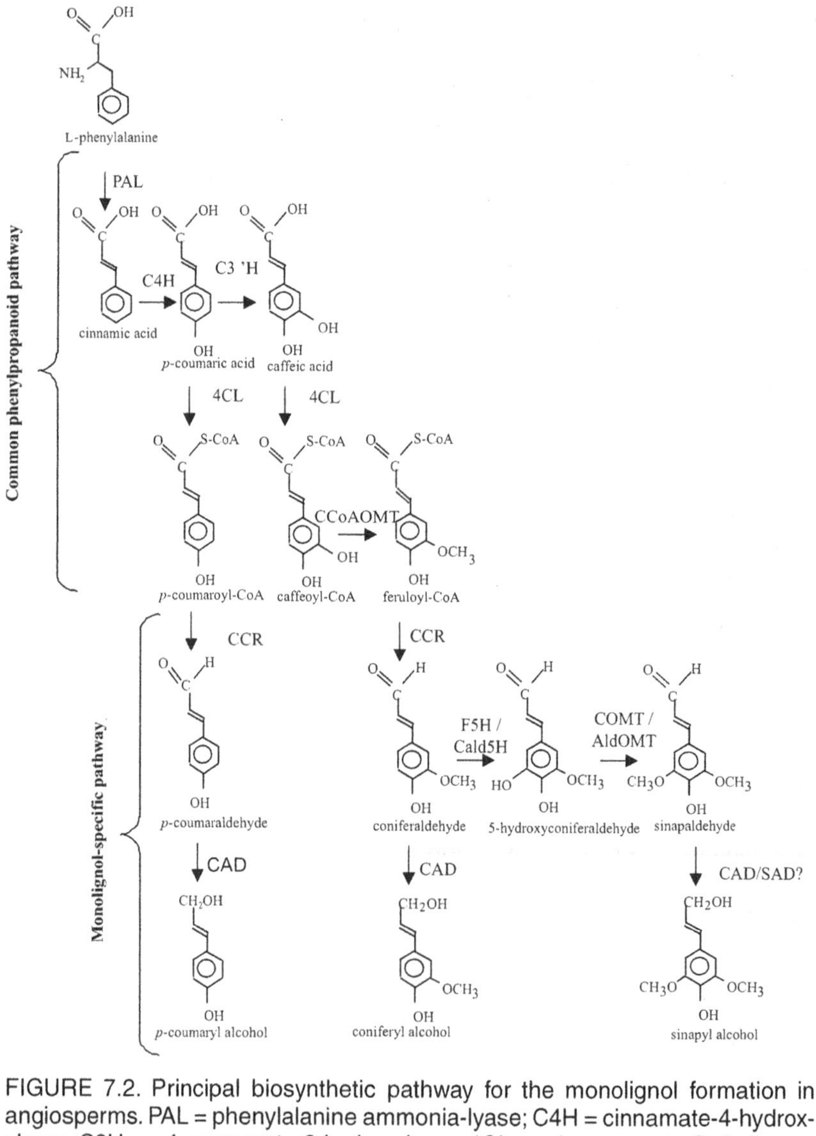

FIGURE 7.2. Principal biosynthetic pathway for the monolignol formation in angiosperms. PAL = phenylalanine ammonia-lyase; C4H = cinnamate-4-hydroxylase; C3H = 4-coumarate-3-hydroxylase; 4CL = 4-coumarate:CoA ligase; CCoAOMT = caffeoyl-CoA O-methyltransferase; CCR = cinnamoyl-CoA reductase; F5H = ferulate-5-hydroxylase or coniferyl aldehyde hydroxylase; COMT = caffeic acid O-methyltransferase or aldehyde O-methyltransferase; CAD = coniferyl alcohol dehydrogenase; SAD = sinapyl alcohol dehydrogenase. (*Source:* Adapted from Schoch et al., 2001 and Li et al., 2001.)

lignin content/structure could also represent an interest for other tree species such as *Eucalyptus,* which is the main species grown for paper pulp making (Hervé et al., 2001), and conifers, which possess a G-type lignin less adapted to the paper pulp industry than G+S lignins that need a lower amount of chemicals for their elimination (Chiang and Funaoka, 1990).

A large part of our improved understanding of lignification comes from the characterization of transgenic plants underexpressing specific genes of this metabolic pathway. Underexpression of a specific enzyme can be obtained using the antisense strategy (reviewed in Bourque, 1995). In this strategy, a cDNA (or part of it) coding for the target enzyme is introduced into a plant in an antisense orientation under the control of regulatory sequences such as the constitutive CaMV 35S promoter. The levels of mRNA repression, and consequently of enzyme activity, vary from one transgenic line to another, and lines with low residual activity can be selected. The impact on the phenotype can be more or less important depending on the level of residual activity. An alternative strategy for modifying the phenotype is overexpression mediated by the introduction of the entire cDNA in sense orientation. This strategy, however, is not always successful because high homology between the endogenous gene and the transgene can induce the silencing or cosuppression of the homologous genes (Fagard and Vaucheret, 2000).

Genetic modification of the lignin biosynthetic pathway has been primarily studied in transgenic plants obtained via sense or antisense strategies. In the case of woody plants, the main target of these studies has been poplar, but some studies concern *Eucalyptus,* another fast-growing hardwood forest tree species, and conifers. The main results of the studies in this field are summarized in Table 7.1 and detailed in the following sections.

GENETIC MODIFICATION OF PAL AND C4H

Most of the genes of the phenylpropanoid biosynthetic pathway belong to very small gene families with high homologies among the different members (reviewed in Baucher et al., 1998; Dixon et al., 2001). Large reductions in the activity of enzymes catalyzing the initial reactions of the general pathway reduce lignin biosynthesis but also have an impact on the biosynthesis of other compounds essential for plant development. Antisense strategies aimed at reducing phenylalanine ammonia-lyase (PAL; deamination of phenylalanine), or cinnamic acid 4-hydroxylase (C4H; CYP73A5; hydroxylation at the C4 position of cinnamic acid) activities have been used with tobacco or *Arabidopsis* (reviewed in Baucher et al., 1998; Dixon et al., 2001). As well as showing reduction in lignin content, these plants were

TABLE 7.1. Lignin characteristics of transgenic trees deregulated for one step of the lignin biosynthetic pathway

Tree species	Target gene	Strategy	Residual enzyme activity (%)	Lignin content reduction (%)	Lignin composition		Incorporation of abnormal units in lignins	Reference
					S/G	β-O-4 bonds		
Populus tremuloides	4CL	Antisense	<10	45	–	–	No	Hu et al. (1999)
Populus tremula × P. alba	CCoAOMT	Antisense and sense	10	12	Increase	Decrease	p-Hydroxybenzoic acid	Meyermans et al. (2000)
Populus tremula × P. alba	CCoAOMT	Antisense	30	40	Increase	Decrease	p-Hydroxybenzoic acid	Zhong et al. (2000)
Populus tremula × P. alba	F5H	Sense	ND > 100	0	Increase	Increase	No	Franke et al. (2000)
Populus tremula × P. alba	COMT	Antisense	5	0	Decrease	Increase	5-Hydroxyguaiacyl unit	Van Doorsselaere, Baucher, Chognot, et al. (1995)
Populus tremuloides	COMT	Sense	5	0	Decrease	Increase	5-Hydroxyguaiacyl unit	Tsai et al. (1998)
Populus tremula × P. alba	COMT	Sense	<3	17	Decrease	Increase	5-Hydroxyguaiacyl unit	Jouanin et al. (2000)
Populus tremula × P. alba	CCR	Antisense and sense	20-30	Reduced	Increase	Decrease	Ferulic and sinapic acids	Mellerowicz et al. (2001)

Species	Gene							Reference
Populus tremula × P. alba	CAD	Antisense and sense	20-30	0 (young tree)	Decrease	Decrease	Nonobserved	Baucher et al. (1996)
				10 (older tree)	Decrease	Decrease	Sinapaldehyde	Lapierre et al. (1999)
Eucalyptus camaldulensis	CAD	Antisense	17	0	–	–	No	Valério et al. (2003)
E. grandis × E. urophylla	CAD	Antisense	26	ND	ND	ND	ND	Tournier et al. (2003)
Populus kitakamiensis	POX	Antisense	0-44	3-26	Increase	Decrease	No	Yahong et al. (2001)
Populus tremula × P. alba	POX	Sense	800	0	–	–	No	Mellerowicz et al. (2001)
Populus tremula × P. alba	LAC3	Antisense	ND	0	–	–		Ranocha et al. (2002)
Eucalyptus camaldulensis	LIM	Antisense	20	20	ND	ND	ND	Kawaoka et al. (2001)
Pinus radiata	CAD	Antisense and sense	Reduced but variable	ND	ND	ND	ND	Walter et al. (2001)

ND = nondetermined; S/G = syringyl/guaiacyl ratio; – = similar to wildtype; 4CL = 4-coumarate:CoA ligase; CCoAOMT = caffeoyl-CoA O-methyltransferase; CCR = cinnamoyl-CoA reductase; F5H = ferulate-5-hydroxylase; COMT = caffeic acid O-methyltransferase; CAD = coniferyl alcohol dehydrogenase; PEX = peroxidase; LAC = laccase.

173

also affected in their development. Similarly, *REF3* (for reduced epidermal fluorescence), an *Arabidopsis* mutant deficient in sinapoylmalate accumulation (a HCA-derived secondary metabolite) with a low lignin content and a dwarf phenotype, has been identified (Ruegger and Chapple, 2001). *REF3* is mutated in the *C4H* gene leading to developmental consequences (Franke, Hemm, et al., 2002). Down-regulation of PAL and C4H in forest trees has not yet been reported, and in light of the undesired effects on development, it appears that genes encoding these enzymes do not represent good targets for improving wood quality.

GENETIC MODIFICATION OF 4CL

4-coumarate:Coenzyme A ligase (4CL) catalyzes the formation of CoA thioesters of hydroxycinnamic acids. *4CL* has been down-regulated by antisense inhibition in aspen *(Populus tremuloides)*. Two functionally distinct aspen *4CL (Pt4CL1* and *Pt4CL2)* genes have been isolated (Hu et al., 1998). The cDNA of *Pt4CL1* in antisense orientation with respect to a duplicated-enhancer CaMV 35S promoter was introduced into aspen. Trees with suppressed *Pt4CL1* expression exhibited up to a 45 percent reduction of lignin with no effects on the lignin composition. The lack of lignin is compensated for by a 15 percent increase in cellulose, and plant development is substantially enhanced (Hu et al., 1999). The preservation of the lignin structure may be due to the key position that *PtCL1* occupies, thereby controlling the flux of HCA substrates into the monolignol pathway. Genetic manipulation of *4CL* in trees could, therefore, be a promising strategy for reducing lignin content without modifying lignin structure. Nevertheless, the reported results have been observed only in young poplars grown in greenhouse conditions and need to be confirmed in older trees grown under natural conditions. In addition, their behavior in the field when confronted with biotic (pathogen infestation) and abiotic (resistance to wind, freezing, etc.) stresses must also be evaluated.

GENETIC MODIFICATION OF C3H

The "coumaroyl-coenzyme A 3-hydroxylase" (C3H) is a cytochrome P450 enzyme (CYP98A3) that catalyzes the hydroxylation at the C3 position. The gene encoding this enzyme has been recently identified in *Arabidopsis thaliana* (Schoch et al., 2001). C3H catalyzes the 3'-hydroxylation of coumaroyl-quinate/shikimate leading to caffeoyl CoA and then to lignin monomers. An *Arabidopsis* mutant for the C3H gene, named *REF8*, has recently been isolated (Franke, Humphreys, et al., 2002), and characterization shows that the lignins of this mutant are formed from *p*-coumaryl alcohol, a minor component of lignins in normal plants. The *REF8* mutant also displays development effects (Franke, Hemm, et al., 2002). Because this gene has been identified only very recently, no work related to deregulation of C3H in forest trees has been published.

GENETIC MODIFICATION OF CCoAOMT

Caffeoyl coenzyme A *O*-methyltransferase (CCoAOMT) converts caffeoyl CoA into feruloyl-CoA and was first characterized in parsley cell suspension culture (Pakusch et al., 1989). CCoAOMT was later hypothesized to be involved in lignification on the basis of its expression profile during lignification (Ye et al., 1994, 1995). Zhong and colleagues (2000) and Meyermans and colleagues (2000) both underexpressed CCoAOMT in poplar (*Populus tremula* × *P. alba*) using the poplar *CCoAOMT* cDNA in sense or antisense orientations under the control of the CaMV 35S promoter. Transgenic poplar plants with reduced CCoAOMT were observed to show either a 12 percent (Meyermans et al., 2000) or a 40 percent (Zhong et al., 2000) reduction in lignin content. This reduction is due to a decrease in both G and S lignins, confirming the role of this enzyme in the biosynthesis of G and S lignins. Wood from these poplars is pink-red (Meyermans et al., 2000) or light orange (Zhong et al., 2000) in color. Chemical analysis also showed an increased accumulation of *p*-hydroxybenzoic acid in the lignins together with the accumulation of soluble phenolics (O^3-β-D-glucopyranosyl-vanillic acid and O^4-β-D-glucopyranosyl-sinapic acid) (Meyermans et al., 2000). No major developmental impacts were noticed in these CCoAOMT-underexpressing poplars, although an increase in vessel fluorescence (Meyermans et al., 2000) and minor changes in vessel wall shapes in the early stage of xylem development (Zhong et al., 2000) were observed. Once again, these studies have been performed on young trees grown in greenhouse conditions, and no reports of their behavior when grown in field condition are available.

GENETIC MODIFICATION OF CCR

Cinnamoyl coenzyme A reductase (CCR) catalyzes the reduction of the hydroxycinnamoyl-CoA esters to their corresponding aldehydes and is considered to be the first committed enzyme of the monolignol-specific pathway. The *CCR* cDNA has been isolated by PCR cloning following *Eucalyptus* CCR protein sequencing (Lacombe et al., 1997). Transgenic plants underexpressing CCR have been reported in tobacco (Piquemal et al., 1998) and *Arabidopsis* (Goujon et al., 2003). Chemical analyses of these plants show that they have a lower lignin content, a more condensed lignin (less β-O-4 linkages), and an increased S/G ratio. Ferulic and sinapic acids are also incorporated into lignins, and an increase in soluble phenolic compounds is observed. Severely *CCR*-depressed tobacco and *Arabidopsis* lines exhibited marked alterations of their development (reduced size, collapsed vessels, etc). Transgenic poplars underexpressing CCR have also been obtained (Mellerowicz et al., 2001), and wood from some of these poplars possess an irregular orange-brown coloration. Lignins isolated from these colored areas show the same perturbations as those from antisense tobaccos (reduction of lignin content, increased S/G ratio, and more condensed lignins). Down-regulation of CCR does not, therefore, seem to be a promising strategy to improve wood quality because the reduction in lignin content is also linked to developmental problems.

GENETIC MODIFICATION OF F5H

Ferulate 5-hydroxylase (F5H), or coniferaldehyde 5-hydroxylase (Cald5H), is a P450 enzyme (CYP84A1) which catalyzes the hydroxylation at the C5 position. This enzyme was first supposed to act at the level of ferulic acid, but recent studies have demonstrated that F5H preferentially catalyzes hydroxylation of coniferyl-aldehyde and -alcohol (Humphreys et al., 1999; Osakabe et al., 1999). An *Arabidopsis* line mutated in the *F5H* gene (*fah* mutant) and shown to possess only G-lignins was isolated in 1992 (Chapple et al., 1992), and the *F5H* gene was isolated by Meyer and colleagues (Meyer et al., 1996). The introduction of the *Arabidopsis F5H* cDNA under the control of either the CaMV 35S or the C4H promoters in *fah-1* leads to complementation of the mutant (Meyer et al., 1998). When *F5H* expression was driven by the C4H promoter, lignins of some overexpressing lines were composed almost entirely of S-units linked predominantly in β-O-4 bonds (Marita et al., 1999). This *C4H-F5H* construct was introduced in *Populus tremula* × *P. alba*, and transgenic poplars with enhanced S-lignins were also obtained (Franke et al., 2000). Thus, F5H can be considered to be the major control point in S-lignin synthesis, and overexpressing this enzyme represents a promising strategy for increasing the S/G ratio in trees. However, as mentioned previously, it appears to be important to introduce genes possessing a low homology with the endogene in order to avoid cosuppression and ensure the stability of the overexpression. In addition, when F5H is highly overexpressed, the next methylation step can become limiting, and the precursor of the S unit (5-hydroxyconiferyl alcohol) is observed in lignins of

C4H-F5H Arabidopsis lines (Ralph, Lapierre, Lu, Marita, Pilate, et al., 2001). Coexpression of F5H and of the *O*-methyltransferase involved in S-unit synthesis could represent a better strategy to increase the S-unit content in tree lignins. This strategy appears to be promising for increasing S-lignins in order to improve delignification in pulp and paper processing (Chiang and Funaoka, 1990).

Conifers do not possess S-lignins, and this could be due to the absence of the *F5H* gene. An efficient *Agrobacterium* transformation procedure (Pilate et al., 1999) was used to introduce a poplar *F5H* cDNA (Sterky et al., 1998) under the control of an enhanced CaMV 35S promoter into larch (*Larix* × *leptoeuropaea*). However, expression of the poplar *F5H* in larch does not lead to the synthesis of S-units (Gatineau et al., 2001).

GENETIC MODIFICATION OF COMT

Caffeic acid *O*-methyltransferase (COMT) or 5-hydroxyconiferaldehyde *O*-methyltransferase (CaldOMT) was supposed to be bifunctional and therefore to methylate caffeic and 5-hydroxyferulic acids (Bugos et al., 1991). However, analysis of lignins from plants transformed with COMT antisense constructs suggested that COMT was preferentially involved in the second methylation step responsible for the formation of S-units. More recently, it has been demonstrated that COMT preferentially uses 5-hydroxyconiferaldehyde instead of caffeic acid or 5-hydroxyferulic acid (Osakabe et al., 1999; Li et al., 2000).

Antisense Strategy in Poplar

A *Populus trichocarpa* × *P. deltoides* COMT cDNA (*PtOMT1;* Dumas et al., 1992) was used to create several sense and antisense constructs with fragments covering the 5' and 3' parts of this cDNA. These constructs were introduced into the hybrid poplar (*Populus tremula* × *P. alba*), and two lines (ASOMT) containing a 900 base pair (bp) fragment corresponding to the 3' part of the *PtOMT1* gene in antisense orientation under the control of the CaMV 35S promoter were shown to possess a very low OMT activity (about 5 percent of the wild type) in the xylem (Van Doorsselaere, Baucher, Chagnot, et al., 1995). The xylem of these lines was pale rose when the bark was peeled off, and lignin analysis in six-month-old trees grown in greenhouse conditions demonstrated that although the lignin content was not changed, the lignin composition was highly modified. The noncondensed lignin has a lower S/G ratio due to a reduction in the S-lignin fraction and presence of the S-unit precursor, the 5-hydroxyguaiacyl unit (5-OH-G).

Lignins of these transgenic poplars are more condensed, and new structures, benzodioxanes, the result of the incorporation of 5-OH-G in lignins, have been identified by thioacidolysis (Lapierre et al., 1999) and NMR (Ralph, Lapierre, Lu, Marita, Pilate, et al., 2001; Ralph, Lapierre, Lu, Marita, Kim, et al., 2001). The suitability of the modified wood from these ASOMT trees for the paper industry was tested using the kraft procedure and shown to be unsuitable. This is due to the high percentage of condensed linkages and lignins of these ASOMT poplars being similar to those of conifers (Lapierre et al., 1999). These ASOMT lines were planted in the field in two locations (France and United Kingdom) for a period of four years. Their performances (growth, resistance to natural pathogen infestations, etc.) were evaluated and found to be similar to those of wild-type poplars grown in the same locations (Pilate et al., 2002).

Sense Strategies in Poplar

Two studies (Tsai et al., 1998; Jouanin et al., 2000) have reported *OMT* homologous sense suppression. In each case, the cDNA expression was driven by the CaMV 35S promoter. The xylem of these poplars has a red-brown coloration, and 5-OH-G units are incorporated in lignins. In addition, Jouanin and colleagues (2000) noticed a 17 percent reduction in lignin content associated with a 6 percent increase in cellulose for a line with COMT activity close to zero. Lignin structure was highly modified with an increase in condensed linkages and the presence of benzodioxane structures (Ralph, Lapierre, Lu, Marita, Pilate, et al., 2001; Ralph, Lapierre, Lu, Marita, Kim, et al., 2001). The absence of COMT activity had a positive effect on the pulp yield (10 percent relative increase) due to the lignin content reduction but made lignins less amenable to industrial degradation (Jouanin et al., 2000).

The original aim of overexpressing *PtOMT1* was to evaluate the interest of an increase in S-unit biosynthesis. However, even if some OMT overexpression was observed in poplars expressing the *PtOMT1* cDNA under the control of the CaMV 35S or the lignin-specific *Eucalyptus* cinnamyl alcohol dehydrogenase (CAD) promoters (Feuillet et al., 1995), no change in the S/G ratio was observed (Jouanin et al., 2000). The results suggest that OMT is not the limiting enzyme for S-unit biosynthesis. Recent work (Jouanin et al., 2001) concerning poplar *COMT* overexpression in *Arabidopsis* confirms these results. It can be concluded that overexpressing COMT does not appear to be a promising strategy for increasing S-units in lignins and, consequently, is not suitable for improving angiosperm trees for the paper industry. The expression of this poplar *COMT* alone, or associated with a poplar *F5H,* in larch does not lead to S-unit containing lignins (Gatineau et al., 2001).

GENETIC MODIFICATION OF CAD

Cinnamyl alcohol dehydrogenase catalyzes the reduction of cinnamyl aldehydes to cinnamyl alcohols, the last step in the monolignol synthesis pathway. Many studies aimed at down-regulating this enzyme in different tree species have been conducted.

CAD Underexpression in Poplar

A full-length *Populus deltoides* × *P. trichocarpa CAD* cDNA (Van Doorsselaere, Baucher, Feuillet, et al., 1995) was introduced in sense and antisense orientations into *Populus tremula* × *P. alba* (Baucher et al., 1996). As previously observed in antisense tobacco (Halpin et al., 1994), wood of CAD underexpressing poplar lines (ASCAD) exhibits a bright red color in the developing xylem. Although no reduction of lignin content was noticed in young trees, a 10 percent lower lignin content was observed in two-year-old trees (Lapierre et al., 1999) and older trees grown for a period of four years under natural conditions (field trials in France and United Kingdom; Pilate et al., 2002). Analyses showed that cinnamaldehydes (with a higher amount of sinapaldehyde than of coniferaldehyde), which are the direct precursors of the coniferyl and sinapyl alcohols, were incorporated. In addition, an increase in lignin-free phenolic end groups facilitating lignin solubilization in alkali conditions was also observed. The lower lignin content and the higher frequency of free phenolic units improve wood quality for the pulp and paper industry, as demonstrated in young trees (Baucher et al., 1996; Lapierre et al., 1999) and confirmed in older trees grown under field conditions for a period of four years (Pilate et al., 2002). It should be emphasized that all the traits related to lignin structure, including the red color of the wood observed in young ASCAD poplars, were confirmed in older plants of the same lines grown in natural conditions. No negative impact of the transformation was observed on growth rate and on resistance to natural infestations of fungi and insects (Pilate et al., 2002).

Interestingly, it was observed that high amounts of cinnamyl alcohols are still incorporated in lignins from these plants, suggesting that the CAD isoform targeted is not the only one involved in the reduction of the cinnamaldehydes. A similar result has also been obtained for a CAD knockout *Arabidopsis* line (Sibout et al., 2003). Sequencing of the entire genome of *Arabidopsis* (Arabidopsis Genome Initiative, 2000) has allowed the identification of 9 *CAD*-like genes (Tavares et al., 2000), and a multigene *CAD* family may well exist in other plants including poplar. Li and colleagues (2001) have identified an alcohol dehydrogenase in poplar with a higher affinity for sinapaldehyde than the already identified poplar CAD. They have named this novel enzyme SAD (for sinapyl alcohol dehydrogenase). They proposed that the former cinnamyl alcohol dehydrogenase can still be named CAD (for coniferyl alcohol dehydrogenase). SAD is supposed to be involved in the biosynthesis of S-units, whereas CAD would be specific for the biosynthesis of G-units. Down-regulation of SAD in transgenic plants will be highly informative as to the exact role of this enzyme *in planta*. However, the situation may be even more complex because CAD down-regulation in poplars and *Arabidopsis* has a larger impact on sinapaldehyde incorporation as compared to coniferaldehyde incorporation. Such an observation contradicts the idea that CAD is involved only in G-unit biosynthesis. More work is required in order to elucidate the exact roles of CAD and SAD in cinnamyl alcohol reduction. At the moment, CAD down-regulation seems to be the best strategy for improving wood quality for the paper industry. Attempts to overexpress the poplar CAD in poplar were unsuccessful and only resulted in cosuppression (Baucher et al., 1996).

CAD Down-Regulation in Eucalyptus

CAD underexpression has been attempted in different *Eucalyptus* species. Valério et al. (2003) and Tournier et al. (2003) have introduced a *Eucalyptus gunnii CAD* full-length cDNA in antisense orientation under the control of the CaMV 35S promoter into *E. calmaldulensis* or *E. grandis* × *E. urophylla*, respectively. In both cases, transgenic lines underexpressing CAD have been selected; however, the consequences on lignin content and structure have only been evaluated in ten-month-old *E. calmaldulensis* trees grown in greenhouse conditions. CAD repression has apparently no effect on lignin quantity or quality, and kraft pulping experiments revealed that lignin extraction was not altered (Valério et al., 2003). The young age of the tested plants or the high residual CAD activity (only determined for in vitro grown plantlets) could be the possible explanation for such an absence of the effect. Additional works on older trees are needed before conclusions can be drawn concerning the interest of a CAD-repression strategy in *Eucalyptus*.

CAD Down-Regulation in Conifers

Conifers possess G-lignins, and according to the revised lignin pathway (Li et al., 2001), this could be due to the lack of the S-unit biosynthesis branch involving the three enzymes F5H, COMT, and SAD. Very few reports on lignin modifications in conifers are available in the published literature. This is probably due to the difficulties of conifer transformation (reviewed in Tzfira et al., 1998; Pena and Seguin, 2001). Transgenic *Pinus radiata* lines containing a CAD sense or antisense construct were generated (Walter et al., 2001) and CAD activity determined at different stages of development. No information concerning the lignin content and composition in these transgenic trees has yet been reported. Nevertheless, the consequences of underexpressing CAD in conifers can perhaps be speculated since one natural CAD mutant *(cad-n1)* has been identified in loblolly pine *(Picea taeda)*. This mutant possesses severely reduced CAD activity (MacKay et al., 1997), and analyses of the lignin composition of this mutant reveal dramatic modifications, including increased incorporation of *p*-coumaryl alcohol (H-units), coniferaldehyde, vanillin, dihydroconiferyl alcohol, and free phenolic groups (Ralph et al., 1997; Lapierre et al., 2000). The overall lignin content is also reduced, and the wood color is brown. Some of these changes, in particular the increase in free phenolic groups, makes this pine wood more amenable to kraft delignification.

GENETIC MODIFICATION OF PEROXIDASES AND LACCASES

Generally, the last step in the synthesis of lignin is the polymerization of cinnamyl alcohols, although more unusual units, such as cinnamaldehydes, can also be used. Lignin polymerization proceeds through the dehydrogenation of cinnamyl alcohols followed by the coupling of phenoxy radicals. Two distinct classes of enzymes, peroxidases and laccases, have been hypothesized to be involved in this step. In plants, these enzymes are encoded by large multigenic families (about 80 peroxidase and 17 laccase genes in *Arabidopsis*). Several cDNA encoding these enzymes have been identified in lignifying tissues (Osakabe et al., 1994, 1995; Christensen et al., 1998; Christensen et al., 2001 for peroxidases; Lafayette et al., 1999; Ranocha et al., 2002 for laccases). The pattern of expression in the secondary cell wall of these genes is not well determined, although Hertzberg and colleagues (2001) have determined the exact expression pattern of several of the previously identified xylem-specific expressed sequence tags (ESTs) (Sterky et al., 1998) in different zones of the developing secondary poplar xylem.

A peroxidase gene *(prx3a)* in antisense orientation under the control of its own promoter (previously shown to direct expression in stems) has been introduced in *Populus kitakamiensis* (Yahong et al., 2001). The transformants showed lower peroxidase activity (0 to 44 percent) in the stem and had a lower lignin content (3 to 26 percent), associated with a higher S/G ratio, than the control plants (Yahong et al., 2001). In contrast, high overexpression of another peroxidase (PXP-3-4) in *Populus tremula* × *P. alba* has no impact on lignin content (Christensen et al., unpublished results cited in Mellerowicz et al., 2001).

Underexpression in poplar *(Populus tremula* × *P. alba)* of three laccases previously shown to be expressed in xylem has no effect on lignin content and composition (Ranocha et al., 2002). However, one of the transgenic populations (underexpression of LAC3) exhibited a two- to threefold increase in total soluble phenolic content and a dramatic alteration of xylem fiber cell walls (Ranocha et al., 2002). These results demonstrate the difficulty of obtaining phenotypes when using antisense and sense strategies to deregulate individual members of multigene families. In such a situation it is possible that either another isoform or a different enzyme could replace the down-regulated enzyme, therefore preventing a modification in the phenotype (Grima-Pettenati and Goffner, 1999).

GENETIC MODIFICATION
OF TRANSCRIPTION FACTORS

The manipulation of individual genes of the monolignol pathway has proven to be a promising approach for modifying lignin content and/or composition. However, another strategy could be to modify the activity of regulatory genes controlling part of or the entire lignin pathway. Several studies have shown that genes of the lignin pathway are transcriptionally regulated (reviewed in Grima-Pettenati and Goffner, 1999). Binding sites for MYB transcription factors are located in the promoter region of these genes (Martin, 1997; Tamagnone et al., 1998). Overexpression of *AmMYB308* and *AmMYB330* from *Antirrhinum majus* in tobacco represses lignin biosynthesis but is also associated with alterations of growth and development (Tamagnone et al., 1998). No report on the use of this strategy in trees has been published so far. Nevertheless, a cDNA *(Ntlim1)* encoding a PAL-box binding protein has been isolated, and its underexpression in tobacco shown to repress the expression of several genes of the phenylpropanoid pathway and to induce a reduction in lignin content with no phenotypic alterations (Kawaoka et al., 2000). This antisense *Ntlim1* construct was introduced in *E. camaldulensis,* and transgenic lines with normal growth characteristics but showing a reduction of about 20 percent lignin content were obtained. In these plants, the level of expression of *PAL, 4CL,* and *CAD* mRNA was lower than in control plants (Kawaoka et al., 2001). More studies are necessary to determine whether the modification of expression levels of transcription factors is an interesting strategy for deregulating lignin biosynthesis.

OUTLOOK

Paper production is increasing, having more than tripled over the last 35 years. The paper industry suffers from the high costs associated with removing lignins from cellulose, which also has a negative environmental impact. For this reason, many biotechnological programs have focused on the modulation of the biosynthesis of lignins in plants, with the goal of improving the delignification process. Almost all the genes of the complex lignin pathway from phenylalanine to the cinnamyl alcohols have been cloned and deregulated in different species, including trees (Grima-Pettenati and Goffner, 1999; Boudet, 2000; Dixon et al., 2001; Mellerowicz et al., 2001). Several different lignin parameters could be targeted by transgenic strategies for improving wood characteristics: reduction of lignin content, enrichment in S-units, labile linkages, and free phenolic groups. In order to make such modifications commercially interesting, growth and resistance to biotic and abiotic stresses of transgenic trees should not be negatively affected. This is extremely important in light of observations showing that a reduction in lignin content had a negative effect on the biomass and survival of four perennial herbaceous species (Casler et al., 2002). This lack of effect will need to be evaluated for every target enzyme because each modification induces specific changes in lignin characteristics. The level of residual activity also needs to be considered, and the use of threshold repression levels could be more suitable so as to avoid negative effects. This could be valid in the case of *CCR* and of *CCoAOMT* down-regulation (and possibly *4CL*). No or little impact on lignin content is noted when *COMT* is down-regulated or F5H is overexpressed. Only the latter strategy is interesting for the paper industry because it increases S-unit content, which could allow a easier delignification of paper pulp. Most of the analyses (except for some ASOMT and ASCAD poplars) have been performed on juvenile trees grown in greenhouse conditions. Long-term field trials of transgenic trees are needed to fully evaluate the effects of any given genetic modification (Strauss et al., 1998).

Many studies concern CAD, the last enzyme of the lignin monolignol pathway, and promising results have obtained following its down-regulation. This strategy has been investigated in poplar, *Eucalyptus,* and pine. In poplar, lignin structure modifications resulting from CAD repression lead to lignins more soluble in alkali and to a slight decrease in lignin content. These traits allow the use of less alkali during the kraft process. It is particularly interesting that results obtained on immature poplars grown in the greenhouse (Baucher et al., 1996; Lapierre et al., 1999) were confirmed on mature trees grown under field conditions at two independent locations where the environmental conditions (climate, soil, etc.) were different (Pilate et al., 2002). These promising results must be confirmed in other trees of industrial interest such as *Eucalyptus*. In each case, long-term field trials need to be considered to ensure the stability of the lignin structural modifications and the absence of a negative impact on growth rate and resistance to environmental stresses. In the future, one benefit of using plantations of transgenic trees more adapted to the paper industry could be to help prevent the destruction of native forests by restricting plantations to specific areas (tree farms) (Strauss et al., 1998). Public acceptance will obviously be of great importance in the development of transgenic trees in the future.

Fundamental studies are still necessary to improve the possible strategies, including identification of new target genes such as those coding for transcription factors and those involved in monolignol polymerization. A more systematic search for new genes could be of particular interest. Random sequencing of ESTs from xylem-specific cDNA libraries would be a source of novel genes for conifers (Allona et al., 1998), poplar (Sterky et al., 1998), or *Eucalyptus* (Sato et al., 2001). In addition, the use of promoters restricting the antisense mRNA synthesis to a specific target tissue could be of great interest for avoiding pleiotropic problems. Some promoters of lignin monolignol biosynthesis genes have already been used and contrasting results obtained. The *Arabidopsis C4H* promoter, when used to overexpress F5H, is better than the CaMV 35S promoter for increasing S-unit content in *Arabidopsis* (Meyer et al., 1998), tobacco, and poplar (Franke et al., 2000). In contrast, the use of the *Eucalyptus CAD* promoter did not lead to promising results with COMT overexpression in poplar (Jouanin et al., 2000) or with CAD underexpression in poplar (Pilate et al., unpublished results). To restrict lignin structural modifications to developing vessels and contact ray cells, the poplar *CCoAOMT* promoter could also prove useful (Chen et al., 2000). The determination of limiting steps of the lignin biosynthetic pathway could also be useful for molecular marker-assisted breeding programs. The genes shown, via genetic engineering, to be essential for determining lignin characteristics provide a pool of candidate genes to analyze potential collocation with quantitative trait loci (QTL) for wood and end-use properties (Plomion et al., 2001). Although this review has mainly focused on the use of trees in the paper and pulp industry, other transgenic strategies could provide useful information for other aspects of wood- and tree-related traits (Tzfira et al., 1998; Pena and Seguin, 2001).

REFERENCES

Allona, I., Quinn, M., Shoop, E., Swope, K., St. Cyr, S., Carlis, J., Riedl, J., Retzel, E., Campbell, M.M., Sederoff, R.R., and Whetten, R.W. (1998). Analysis of xylem formation in pine by cDNA sequencing. *Proc. Natl. Acad. Sci. USA* 95: 9693-9698.

Arabidopsis Genome Initiative (2000). Analysis of the genome sequence of the flowering plant *Arabidopsis thaliana*. *Nature* 208: 796-815.

Baucher, M., Chabbert, B., Pilate, G., Van Doorsseleare, J., Tollier, M.-T., Petit-Conil, M., Cornu, D., Monties, B., van Montagu, M., Inzé, D., et al. (1996). Red xylem and higher lignin extractability by down-regulating a cinnamyl alcohol dehydrogenase in poplar. *Plant Physiology* 112: 1479-1490.

Baucher, M., Monties, B., van Montagu, M., and Boerjan, W. (1998). Biosynthesis and genetic engineering of lignin. *Critical Rev. Plant Sci.* 17: 125-197.

Boudet, A.M. (2000). Lignins and lignification: Selected issues. *Plant Physiol. Biochem.* 38: 81-96.

Bourque, J.E. (1995). Antisense strategies for genetic manipulations in plants. *Plant Sci.* 105: 125-149.

Bugos, R.C., Chiang, V.L.C., and Campbell, W.H. (1991). cDNA cloning, sequence analysis and seasonal expression of lignin-bispecific caffeic acid/5-hydroxyferulic acid *O*-methyltransferase of aspen. *Plant Mol. Biol.* 31: 1203-1215.

Campbell, M.M. and Sederoff, R.R. (1996). Variation in lignin content and composition. Mechanisms of control and implications for the genetic improvement of plants. *Plant Physiology* 110: 3-13.

Casler, M.D., Buxton, D.R., and Vogel, K.P. (2002). Genetic modification of lignin concentration affects fitness of perennial herbaceous plants. *Theo. Appl. Genet.* 104: 127-131.

Chapple, C.C.S., Vogt, T., Ellis, B.E., and Sommerville, C.S. (1992). An *Arabidopsis* mutant defective in the general phenylpropanoid pathway. *Plant Cell* 4: 1413-1424.

Chen, C., Meyermans, H., Burggraeve, B., De Rycke, R.E., Onoue, K., De Vleesschauwer, V., Steenackers, M., Van Montagu, M., Engler, G.J., and Boerjan, W. (2000). Cell-specific and conditional expression of caffeoyl coenzyme A-3-*O*-methyltransferase in poplar. *Plant Physiology* 123: 853-867.

Chiang, V.L. and Funaoka, M. (1990). Comparison of softwood and hardwood kraft pulping. *Holzforschung* 44: 309-313.

Christensen, J.H., Bauw, G., Welinder, K.J., Van Montagu, M., and Boerjan, W. (1998). Purification and characterization of peroxidases correlated with lignification in poplar xylem. *Plant Physiology* 118: 125-135.

Christensen, J.H., Overney, S., Rohde, A., Diaz, W.A., Bauw, G., Simon, P., Van Montagu, M., and Boerjan, W. (2001). The syringaldazine-oxydizing peroxidase PXP 3-4 from poplar xylem: cDNA isolation, characterization and expression. *Plant Mol. Biol.* 47: 581-593.

Davin, L.B. and Lewis, N.G. (1992). Phenylpropanoid metabolism: Biosynthesis of monolignols, lignans, lignins and suberins. In H.A. Stafford and R.K. Ibrahim (Eds.), *Recent Advances in Phytochemistry* (Volume 26, pp. 325-375). New York: Plenum Press.

Dixon, R.A., Chen, F., Guo, D., and Parvathi, K. (2001). The biosynthesis of monolignols: A "metabolic grid," or independent pathway to guaiacyl and syringyl units? *Phytochem.* 57: 1069-1084.

Donaldson, L.A. (2001). Lignification and lignin topochemistry—An untrastructural view. *Phytochem.* 57: 859-873.

Dumas, B., Van Doorsselaere, J., Gielen, J., Legrand, M., Fritig, B., Van Montagu, M., and Inzé, D. (1992). Nucleotide sequence of a complementary DNA encoding *O*-methyltransferase from poplar. *Plant Physiology* 98: 796-797.

Fagard, M. and Vaucheret, H. (2000). (Trans)Gene silencing in plants: How many mechanisms? *Ann. Rev. Plant Physiol. Plant Mol. Biol.* 51: 167-194.

Feuillet, C., Lauvegeat, V., Deswarte, C., Pilate, G., Boudet, A.M., and Grima-Pettenati, J. (1995). Tissue- and cell-specific expression of a cinnamyl alcohol dehydrogenase promoter in transgenic poplar plants. *Plant Mol. Biol.* 27: 651-667.

Franke, R., Hemm, R.M., Denault, J.W., Ruegger, M.O., Humphreys, J.M., and Chapple, C. (2002). Changes in secondary metabolism and deposition of an unusual lignin in the *ref8* mutant of *Arabidopsis. Plant Journal* 30: 47-59.

Franke, R., Humphreys, J.M., Hemm, M.R., Denault, J.W., Ruegger, M.O., Cusumano, J.C., and Chapple, C. (2002). The *Arabidopsis REF8* gene encodes the 3-hydroxylase of phenylpropanoid metabolism. *Plant Journal* 30: 33-95.

Franke, R., McMichael, C., Meyer, K., Shirley, A.M., Cusumano, J.C., and Chapple, C. (2000). Modified lignin in tobacco and poplar plants over-expressing the *Arabidopsis* gene encoding ferulate 5-hydroxylase. *Plant Journal* 22: 223-234.

Gatineau, M., Sibout, R., Jouanin, L., Lelu, M.-A., and Pilate, G. (2001). Toward better understanding regarding the absence of syringyl monomers in conifers. Abstract of the 9th International Cell Wall Meeting, Toulouse, p. 149.

Goujon, T., Ferret, V., Mila, B., Pollet, B., Ruel, K., Burlat, J.P., Barrière, Y., Lapierre, C., and Jouanin, L. (2003). Down-regulation of AtCCR1 in *Arabidopsis thaliana:* Effects on phenotype, lignins and cell degradability. *Planta* (in press).

Grima-Pettenati, J. and Goffner, D. (1999). Lignin genetic engineering revisited. *Plant Sci.* 145: 51-65.

Halpin, C., Knight, M.E., Foxon, G.E., Campbell, M.M., Boudet, A.M., Boon, J., Chabbert, B., Tollier, M.-T., and Schuch, W. (1994). Manipulation of lignin quality by down-regulation of cinnamyl alcohol dehydrogenase. *Plant Journal* 6: 339-350.

Hatfield, R. and Vermerris, W. (2001). Lignin formation in plants. The dilemma of linkage specificity. *Plant Physiology* 126: 1351-1357.

Hertzberg, M., Aspeborg, H., Schrader, J., Andersson, A., Erlandsson, R., Blomqvist, K., Bhalerao, R., Uhlen, M., Teeri, T.T., Lundeberg, J., et al. (2001). A transcriptional road map to wood formation. *Proc. Natl. Acad. Sci. USA* 98: 14732-14737.

Hervé, P., Jauneau, A., Pâques, M., Marien, J.N., Boudet, A.M., and Teulières, C. (2001). A procedure for shoot organogenesis *in vitro* from leaves and nodes of an elite *Eucalyptus gunnii* clone: Comparative histology. *Plant Sci.* 161: 645-653.

Higuchi, M. (1990). Lignin biochemistry: Biosynthesis and degradation. *Wood Sci. Technol.* 24: 23-63.

Hu, W.-J., Harding, S.A., Lung, J., Popko, J.L., Ralph, J., Stokke, D.D., Tsai, C.-J., and Chiang, V.L. (1999). Repression of lignin biosynthesis promotes cellulose accumulation and growth in transgenic trees. *Nature Biotechnol.* 17: 808-812.

Hu, W.-J., Kawaoka, A., Tsai, C.-J., Lung, J., Osakabe, K., Ebinuma, H., and Chiang, V.L. (1998). Compartmentalized expression of two structurally and functionally distinct 4-coumarate:CoA ligase genes in aspen. *Proc. Nat. Acad. Sci. USA* 95: 5407-5412.

Humphreys, J.M. and Chapple, C.C. (2002). Rewriting the lignin roadmap. *Current Opinion in Plant Biology* 5: 224-229.

Humphreys, J.M., Hemm, M.R., and Chapple, C.C. (1999). New routes for lignin biosynthesis defined by biochemical characterization of recombinant ferulate 5-hydroxylase, a multifunctional cytochrome P450-dependent monooxygenase. *Proc. Natl. Acad. Sci. USA* 96: 10045-10050.

Jouanin, L., Goujon, T., de Nadaï, V., Martin, M.-T., Mila, I., Vallet, C., Pollet, B., Yoshinaga, A., Chabbert, B., Petit-Conil, M., and Lapierre, C. (2000). Lignification in transgenic poplars with extremely reduced caffeic acid O-methyltransferase activity. *Plant Physiology* 123: 1363-1373.

Jouanin, L., Goujon, T., Sibout, R., Pollet, B., Mila, I., Maba, B., Ralph, J., Petit-Conil, M., and Lapierre, C. (2001). Tuning lignin structure through genetically silencing, restoring or increasing caffeic acid O-methyltransferase activity: Evaluation with poplar and *Arabidopsis thaliana* models. *Proceedings of the 11th International Symposium on Wood and Pulping Industry,* Nice, France, pp. 25-28.

Kawaoka, A., Kaothien, P., Yoshida, K., Endo, S., Yamada, K., and Ebinuma, H. (2000). Functional analysis of tobacco LIM protein Ntlim1 involved in lignin biosynthesis. *Plant Journal* 22: 289-301.

Kawaoka, A., Nanyo, K., Sugita, K., Endo, S., Yamada-Watanabe, K., Matsunaga, E., and Ebinuma, H. (2001). Transcriptional regulation of lignin biosynthesis by tobacco LIM protein in transgenic woody plant. In N. Moroshoshi and A. Komamine (Eds.), *Molecular Breeding of Woody Plants* (pp. 205-210). Amsterdam, The Netherlands: Elsevier Science.

Lacombe, E., Hawkins, S.W., van Doorsselaere, J., Piquemal, J., Goffner, D., Poeydomenge, O., Boudet, A.M., and Grima-Pettenati, J. (1997). Cinnamoyl CoA reductase, the first committed enzyme of the lignin branch biosynthetic pathway: Cloning, expression and phylogenetic relationships. *Plant Journal* 11: 429-441.

Lafayette, P.R., Eriksson, K.-E.L., and Dean, J.F.D. (1999). Characterization and heterologous expression of laccase cDNAs from xylem tissues of yellow poplar *(Liriodendron tulipifera). Plant Mol. Biol.* 40: 23-35.

Lapierre, C., Pollet, B., MacKay, J.J., and Sederoff, R.R. (2000). Lignin structure in a mutant pine deficient in cinnamyl alcohol dehydrogenase. *J. Agri. Food Chem.* 48: 2326-2331.

Lapierre, C., Pollet, B., Petit-Conil, M., Toval, G., Romero, J., Pilate, G., Leplé, J.-C., Boerjan, W., Ferret, V., de Nadai, V., and Jouanin, L. (1999). Structural alterations of lignins in transgenic poplars with depressed cinnamyl alcohol dehydrogenase or caffeic acid *O*-methyltransferase activity have an opposite impact on the efficiency of industrial kraft pulping. *Plant Physiology* 119: 153-163.

Leplé, J.-C., Pilate, G., and Jouanin, L. (1999). Transgenic poplar trees (*Populus* species). In Y.B.S. Bajaj (Ed.), *Biochemistry in Agriculture and Forestry* (Volume 44, pp. 221-244) Berlin, Heidelberg, New York: Springer-Verlag.

Li, L., Cheng, X.F., Leshkevich, J., Umezawa, T., Harding, S.A., and Chiang, V.L. (2001). The last step of syringyl monolignol biosynthesis in angiosperms is regulated by a novel gene encoding sinapyl alcohol dehydrogenase. *Plant Cell* 13: 1567-1585.

Li, L., Popko, J.L., Umezawa, T., and Chiang, V.L. (2000). 5-hydroxyconiferyl aldehyde modulates enzymatic methylation for syringyl monolignol formation, a new view of monolignol biosynthesis in angiosperms. *J. Biol. Chem.* 275: 6537-6545.

MacKay, J.J., O'Malley, D.O., Presnell, T., Booker, F.L., Campbell, M.M., Whetten, R.W., and Sederoff, R.R. (1977). Inheritance, gene expression, and lignin characterization in a mutant pine deficient in cinnamyl alcohol dehydrogenase. *Proc. Natl. Acad. Sci. USA* 94: 8255-8260.

Marita, J.M., Ralph, J., Hatfield, R.D., and Chapple, C. (1999). NMR characterization of lignins in *Arabidopsis* altered in the activity of ferulate 5-hydroxylase. *Proc. Natl. Acad. Sci. USA* 96: 12328-12332.

Martin, C. (1997). Transcription factors and the manipulation of plant traits. *Current Opinion in Biotechnology* 7: 130-138.

Mellerowicz, E., Baucher, M., Sundberg, B., and Boerjan, W. (2001). Unravelling cell wall formation in the woody dicot stem. *Plant Mol. Biol.* 47: 239-274.

Meyer, K., Cusumano, J.C., Somerville, C., and Chapple, C.C.S. (1996). Ferulate-5-hydroxylase from *Arabidopsis thaliana* defines a new family of cytochrome P450-dependent monooxygenases. *Proc. Natl. Acad. Sci. USA* 93: 6869-6874.

Meyer, K., Shirley, A.M., Cusumano, J.C., Bell-Lelong, D.A., and Chapple, C. (1998). Lignin monomer composition is determined by the expression of a cytochrome P450-dependent monooxygenase in *Arabidopsis. Proc. Natl. Acad. Sci. USA* 95: 6619-6623.

Meyermans, H., Morreel, K., Lapierre, C., Pollet, B., De Bruyn, A., Busson, R., Herdewijn, P., Devreese, B., van Beeumen, J., Marita, J.M., et al. (2000). Modifications in lignin and accumulation of phenolic glucosides in poplar xylem upon down-regulation of caffeoyl coenzyme A *O*-methyltransferase, an enzyme involved in lignin biosynthesis. *J. Biol. Chem.* 275: 36899-36909.

Osakabe, K., Koyama, H., Kawai, S., Katayama, Y., and Morohoshi, N. (1994). Molecular cloning and the nucleotide sequences of two novel cDNAs that encode anionic peroxidases of *Populus kitakamensis. Plant Sci.* 103: 167-175.

Osakabe, K., Koyama, H., Kawai, S., Katayama, Y., and Morohoshi, N. (1995). Molecular cloning of two tandemly arrranged peroxidase genes from *Populus kitakamensis* and their differential regulation in the stem. *Plant Mol. Biol.* 28: 677-689.

Osakabe, K., Tsao, C.C., Li, L., Popko, J.L., Umezawa, T., Carraway, D.T., Smeltzer, R.H., Joshi, C.P., and Chiang, V.L. (1999). Coniferyl aldehyde 5-hydroxylation and methylation direct syringyl lignin biosynthesis in angiosperms. *Proc. Natl. Acad. Sci. USA* 96: 8955-8960.

Pakusch, A.E., Kneusel, R.E., and Matern, U. (1989). *S*-Adenosyl-L-methionine: trans-caffeoyl-coenzymeA 3-*O*-methyltransferase from elicitor-treated parsley cell suspension cultures. *Arch. Biochem. Biophys.* 271: 488-494.

Pena, L. and Seguin, A. (2001). Recent advances in the genetic transformation of trees. *Trends in Biotechnology* 19: 500-506.

Pilate, G., Guiney, E., Holt, K., Petit-Conil, M., Lapierre, C., Leplé, J.C., Pollet, B., Mila, I., Webster, E.A., Marstorp, H.G., et al. (2002). Field performances of transgenic trees with altered lignification. *Nat. Biotechnol.* 20: 607-612.

Pilate, G., Leplé, J.C., Cornu, D., and Lelu, M.A. (1999). Transgenic larch (*Larix* species). In Y.B.S. Bajaj (Ed.), *Biochemistry in Agriculture and Forestry* (Volume 44, pp. 125-141). Berlin, Heidelberg, New York: Springer-Verlag.

Piquemal, J., Lapierre, C., Myton, K., O'Conell, A., Schuch, W., Grima-Pettinati, J., and Boudet, A.M. (1998). Down-regulation of cinnamoyl CoA reductase induces significant changes of lignin profiles in transgenic tobacco plants. *Plant Journal* 13: 71-83.

Plomion, C., Leprovost, G., and Stokes, A. (2001). Wood formation in trees. *Plant Physiology* 127: 1513-1523.

Ralph, J., Lapierre, C., Lu, F., Marita, J.M., Kim, H., Lu, F., Hatfield, R.D., Ralph, S., Chapple, C., Franke, R., et al. (2001). Elucidation of new structures in lignins of CAD- and COMT-deficient plants by NMR. *Phytochem.* 57: 993-1003.

Ralph, J., Lapierre, C., Lu, F., Marita, J.M., Pilate, G., Van Doorsselaere, J., Boerjan, W., and Jouanin, L. (2001). NMR evidence of benzodioxane structures resulting from incorporation of 5-hydroxyconiferyl alcohol into lignins of *O*-methyltransferase-deficient poplars. *J. Agri. Food Chem.* 49: 86-91.

Ralph, J., MacKay, J.J., Hatfield, R.D., O'Malley, R.M., Whetten, R.W., and Sederoff, R.R. (1997). Abnormal lignin in a loblolly pine mutant. *Science* 277: 235-239.

Ranocha, P., Chabannes, M., Danoun, S., Jauneau, A., Boudet, A.M., and Goffner, D. (2002). Laccase down-regulation causes alterations in phenolic metabolism and cell structure in poplar. *Plant Physiology* 129: 145-155.

Ruegger, M. and Chapple, C. (2001). Mutations that reduce sinapoylmalate accumulation in *Arabidopsis thaliana* define loci with diverse roles in phenylpropanoid metabolism. *Genetics* 159: 1741-1749.

Sato, S., Horikiri, K., Miyashita, K., Ishige, N., Asada, T., and Hibino, T. (2001). Analysis of wood development with a genomic approach: *Eucalyptus* ESTs and TAC genomic library. In N. Moroshoshi and A. Komamine (Eds.), *Molecular Breeding of Woody Plants* (pp. 223-228). Amsterdam, The Netherlands: Elsevier Science.

Schoch, G., Goepfert, S., Morant, M., Hehn, A., Meyer, D., Ullmann, P., and Werck-Reichhart, D. (2001). CYP98A3 from *Arabidopsis thaliana* is a 3'-hydroxylase of phenolic esters, a missing link in the phenylpropanoid pathway. *J. Biol. Chem.* 276: 36566-36574.

Sibout, R., Eudes, A., Pollet, B., Goujon, T., Mila, I., Granier, F., Séguin, A., Lapierre, C., and Jouanin, L. (2003). Expression pattern of two paralogues encoding cinnamyl alcohol dehydrogenases in *Arabidopsis thaliana:* Isolation and characterization of the corresponding mutants. *Plant Physiology* (in press).

Sterky, F., Regan, S., Karlsson, J., Hertzberg, M., Rohde, A., Holmberg, A., Amini, B., Bhalerao, R., Larsson, M., Villarroel, R., et al. (1998). Gene discovery in the wood-forming tissues of poplar: Analysis of 5,692 expressed sequences tags. *Proc. Natl. Acad. Sci. USA* 95: 13300-13335.

Strauss, S., Boerjan, W., Cairney, J., Campbell, M., Dean, J., Ellis, D., Jouanin, L., and Sundberg, B. (1998). Forest biotechnology makes its position known. *Nat. Biotechnol.* 17: 1145.

Tamagnone, L., Merida, A., Parr, A., Mackay, S., Culianez-Macia, F., Roberts, K., and Martin, C. (1998). The AmMYB308 and AmMYB330 transcription factors from *Antirrhinum* regulate phenylpropanoid and lignin biosynthesis in transgenic tobacco. *Plant Cell* 10: 135-154.

Tavares, R., Aubourg, S., Lecharny, A., and Kreis, M. (2000). Organization and structural evolution of four multigene families in *Arabidopsis thaliana: AtLCAD, AtLGT, AtMYST* and *AtHD-GL2. Plant Mol. Biol.* 42: 703-717.

Tournier, V., Grat, S., Marque, C., Penchel, R., Bettolucco, F., Boudet, A.-M., and Teulières, C. (2003). Towards modulation of lignification of a "pulp tree" *E. grandis* × *E. urophylla:* Generation of transgenic *Eucalyptus* down regulated for a lignification enzyme. *Transgenic Res.* (in press).

Tsai, C.-J., Popko, J.L., Mielke, M.R., Hu, W.-J., Podila, G.K., and Chiang, V.L. (1998). Suppression of *O*-methyltransferase gene by homologous sense transgene in quaking aspen causes red-brown wood phenotypes. *Plant Physiology* 117: 101-112.

Tzfira, T., Zuker, A., and Altman, A. (1998). Forest-tree biotechnology: Genetic transformation and its application to future forests. *Trends Biotechnol.* 16: 439-446.

Valério, L., Tournier, V., Marque, C., Boudet, A.-M., Carter, D., Maunders, M., and Teulières, C. (2003). Towards modulation of lignification in the "pulp tree" *Eucalyptus.* Down regulation of cinnamyl alcohol dehydrogenase in *Eucalyptus camaldulensis. Mol. Breeding* (in press).

Van Doorsselaere, J., Baucher, M., Chognot, E., Chabbert, B., Tollier, M.-T., Petit-Conil, M., Leplé, J.C., Pilate, G., Cornu, D., Monties, B., et al. (1995). A novel lignin in poplar trees with reduced caffeic acid/5-hydroxyferulic acid *O*-methyltransferase activity. *Plant Journal* 8: 855-864.

Van Doorsselaere, J., Baucher, M., Feuillet, C., Boudet, A.M., Van Montagu, M., and Inzé, D. (1995). Isolation of cinnamyl alcohol dehydrogenase cDNAs from two important economic species: alfalfa and poplar. Demonstration of a high homology of the gene within angiosperms. *Plant Physiol. Biochem.* 33: 105-109.

Walter, C., Bishop-Hurley, S., Charity, J., Find, J., Grace, L., Höfig, K., Holland, L., Möller, R., Moody, J., Wagner, A., and Walden, A. (2001). Genetic engineering of *Pinus radiata* and *Picea abies,* production of transgenic plants and gene expression studies. In N. Moroshoshi and A. Komamine (Eds.), *Molecular Breeding of Woody Plants* (pp. 211-222). Amsterdam, The Netherlands: Elsevier Science.

Yahong, L., Tsuji, Y., Nishikubo, N., Kajita, S., and Morohoshi, N. (2001). Analysis of transgenic poplar in which the expression of peroxidase gene is suppressed. In N. Moroshoshi and A. Komamine (Eds.), *Molecular Breeding of Woody Plants* (pp. 195-204). Amsterdam, The Netherlands: Elsevier Science.

Ye, Z.-H., Kneusel, R.E., Matern, U., and Varner, J.E. (1994). An alternative methylation pathway in lignin biosynthesis in *Zinnia*. *Plant Cell* 6: 1427-1439.

Ye, Z.-H., Kneusel, R.E., Matern, U., and Varner, J.E. (1995). Differential expression of two *O*-methyltransferases in lignin biosynthesis in *Zinnia elegans*. *Plant Physiology* 108: 459-467.

Zhong, R., Morrison III, H., Himelsbach, D.S., Poole II, F.L., and Ye, Z.-H. (2000). Essential role of caffeoyl coenzyme A *O*-methyltransferase in lignin biosynthesis in woody poplar plants. *Plant Physiology* 124: 563-577.

Chapter 8

In Vitro Systems for the Study of Wood Formation

Mathew A. Leitch
Gerd Bossinger

INTRODUCTION

Wood formation (xylogenesis) is a result of cambial activity, which constitutes secondary growth in many perennial plants. The differentiation of cambial initials into woody tissue and later into heartwood cells in a mature cambium which might be many years old is not directly accessible to experimental investigation, particularly in the area of molecular biology. This is partly due to the long time frames that are involved, but also to the presence of extractives and other chemical compounds as well as physical constraints (e.g., size and internal pressures) that present obstacles preventing easy microscopic and molecular assessment. The process of wood formation has, therefore, been broken down by many researchers into its various components, which could be analyzed in an isolated manner using a number of model species and in vitro systems. Consequently, recent progress in studying the molecular basis of lignin and cellulose biosynthesis, for example, has come from weedy model plants such as *Arabidopsis thaliana* (Reiter et al. 1993; Potikha and Delmer 1995; Freshour et al. 1996; Reiter 1998) rather than from woody tree species. In this chapter we concentrate on in vitro systems and describe how these relate to various aspects of wood formation. We also propose and describe a more integrating in vitro wood formation model system and provide a staged classification of wood formation in culture.

The authors would like to thank work colleagues for discussions and the University of Melbourne, School of Resource Management, for support.

CAMBIAL ACTIVITY

The use of the term *cambium* originated prior to the fourteenth century (see Larson 1994) and has since been used to describe tissues that give rise to secondary xylem and phloem. Because we are concerned with wood formation in this chapter, we will concentrate on xylem formation and will not discuss the differentiation and development of phloem tissue and associated cell types.

In gymnosperms and woody dicotyledonous angiosperms, wood forms through activity of the vascular cambium, which provides the basis for secondary growth and wood formation. In theory, cambial differentiation begins with the fertilization of an egg cell; however, for the purpose of this chapter, we begin with the vacuolated meristematic cells of the cambium, which arose under the controlling influence of the apical meristem (shoot or root) (Phillips 1976) already present.

The cambial zone (CZ) is composed of cambial initials and all descending mother cells. Two distinct cell types make up the vascular cambium of most plants, namely the tapered fusiform initials and the almost isodiametric ray-cell initials (Zimmermann and Brown 1971).

Periclinal divisions of fusiform initials give rise to longitudinally oriented secondary xylem cells, while divisions of cambial ray-cell initials give rise to radially oriented cells that constitute the ray system. Anticlinal divisions assist in the lateral expansion of the cambium by accommodating the increase in circumference of a tree as it grows. Cell division activity occurs along the stem at various rates, depending largely on environmental conditions. Following division of a mother cell, a complete primary wall around the daughter protoplast forms (Newman 1956). The pattern of cambial reactivation varies greatly, and it has been shown in tropical and subtropical regions that cambial initiation and activity is less defined due to developmental events during the year lacking separation by intense activity and rest periods (Brown 1971).

WOOD FORMATION

Following successive divisions of xylem mother cells, elongation of cambial derivatives occurs, adding microfibril layers to the primary wall, increasing wall area and rigidity. At this stage of elongation, the cambial derivatives are uncommitted to a particular pathway or cell fate (Savidge 1993a,b). Wounding during this phase promotes cells to differentiate into new vascular elements that develop around a wounded area and reestablish continuity in the vascular system (Savidge and Farrar 1984; Lev-Yadun and Aloni 1993; Aloni et al. 1995; Cui et al. 1995; Nobuchi et al. 1995; Fukuda 1997). The ability of tissues to regenerate allowing for normal growth is of particular importance in the area of tissue culture with respect to tree species studied in vitro.

Subsequent to cell elongation, secondary wall formation and lignification occur by apposition of additional wall material to the inside of primary walls (Panshin and DeZeeuw 1980). The last phase of cell development is cell autolysis, in which death of the protoplasm occurs and cell residues are deposited as an amorphous layer or as a warty layer on the inner cell wall (Panshin and DeZeeuw 1980).

In gymnosperms, the xylem consists of tracheids, parenchyma, and resin canals axially, and parenchyma and tracheids radially. Tracheids in gymnosperms are responsible for conduction and support (Panshin and DeZeeuw 1980; Bamber 1985). Angiosperms display a larger number of cell types including fiber tracheids, libriform fibers, vasicentric tracheids, vessel elements, and parenchyma axially, and parenchyma and tracheids radially. Fibers in angiosperms are responsible mainly for support, while vessel elements are responsible for conduction (Bamber 1985). Rays in some species can make up a substantial part of the wood (0 percent to 75 percent in extremes, generally 10 percent to 40 percent) (Iqbal 1990).

Parenchyma cells (axial and radial), unlike other cells, retain their nucleus at maturity and remain alive for many years. In many species the aging of wood coincides with the death of parenchyma cells, the translocation of extractives into the surrounding cell walls, and the formation of tylosis (angiosperms) or pit aspirations (gymnosperms) leading to a distinct sapwood (conductive) and heartwood (nonconductive) transition zone (Wilson and White 1986). Such a distinction, however, cannot be made in all species, including some *Populus, Salix, Abies,* and others.

THE FORMATION OF TRACHEARY ELEMENTS

The ability to investigate the physiological effects of nutrients, plant hormones, and other chemical constituents in defined culture media under controlled conditions is one of the most important applications of tissue culture in aiding researchers to elucidate processes during the formation of plant tissues. This is particularly the case for studying tracheary element formation (see review by Aloni et al. 2000) which is a major development during wood formation.

The Zinnia *System*

One model system that has been extensively used and been instrumental in our understanding of tracheary element differentiation is the *Zinnia elegans* mesophyll cell system developed by Fukuda and Komamine (1980) based on earlier results by Kohlenbach and Schmidt (1975). These authors used media containing auxin (NAA) and cytokinin (BA) to synchronize the differentiation of mesophyll cells into tracheary elements without preceding mitotic activity or cell division. This makes the process accessible not only to developmental studies but also to biochemical analysis (Harrison et al. 1997), to physiological studies (Roberts and Haigler 1994; Roberts et al. 1997; Yamamoto et al. 1997; Domingo et al. 1998; Ohdaira et al. 2002), and to molecular investigations (Demura and Fukuda 1993, 1994; Ye and Varner 1993; Ye and Droste 1996; Yoshimura et al. 1996; McCann 1997; McCann et al. 2000; Motose et al. 2001). In essence, the process of tracheary element formation in vitro is inducible by wounding and the addition of auxin and cytokinin to the medium. The process involves the rearrangement of microtubules, the deposition of cellulose and other secondary wall components according to resulting cortical banding patterns, lignification, which follows secondary thickening initiation, and finally cell autolysis.

Fukuda (1992, 1994) and later McCann and colleagues (2000) reviewed the literature relating to the *Zinnia* system, detailing advances in our understanding of differentiation processes, both genetically and physiologically.

HEARTWOOD FORMATION
(TYLOSIS AND PIT ASPIRATION)

Heartwood formation marks the final step of stem tissue development in many perennial woody plants. In angiosperms it includes the expansion of ray parenchyma cells through pit pairs into adjacent vessels (tylosis formation), eventually closing them off and making them less accessible to microorganisms than the sapwood (Leitch et al. 1999). This is proceeded by death of all ray cells and the deposition of extractives into the surrounding tissue, providing further protection (Leitch et al. 1999). In gymnosperms it includes the closing and sealing of bordered pits (pit aspirations), eventually affecting all tracheary elements, similarly decreasing accessibility by microorganisms (Bolton and Petty 1977; Brett and Waldron 1990). Primarily the deposition of extractives results in this portion of the stem becoming much darker in color than the sapwood (Esau 1965; Kozlowski 1981).

Tylosis Formation in Stem Chips

A tylosis in hardwoods is defined as an outgrowth of the protoplasm of a living parenchyma cell through a cross-field pit into the adjacent vessel element (Panshin and DeZeeuw 1980). Formed usually in the aging inner sapwood, tylosis development has been regarded as a normal physiological process marking the transformation from sapwood to heartwood in many hardwood species (Meyer 1967; Panshin and DeZeeuw 1980; Wilson and White 1986). Tyloses in vessels of commercial species create difficulties during wood preservation, chemical pulping, and matchmaking (Streslis and Green 1962; Du Toit 1964; Murmanis 1975; Panshin and DeZeeuw 1980; Wilson and White 1986) since they cause resistance to penetration of solutions.

The experimental induction of tylosis formation under controlled culturing conditions was achieved by Leitch and colleagues (1999). Stem chips (explants) from several-year-old *Eucalyptus globulus* Labill. were incubated in a defined liquid medium (Leitch et al. 1999). In the water-only treatment (control), tylosis formation in the sapwood region adjacent to the cambium occurred in 3.7 percent of vessels. When incubated in the nutriment medium, tyloses occurred in 69.5 percent of the vessels. Chips cultured on the NAA + nutriment medium produced tyloses in 64.4 percent of the vessels. Removal of the phloem layer prior to culturing increased all treatment values. For example, tylosis formation increased to 76.1 percent on the nutriment medium. Tylosis formation occurred in a similar manner to that described in the literature (Murmanis 1975; Desch 1981) (see Figure 8.1). In earlier work, Murmanis (1975) was able to determine that under favorable conditions tylosis could form after only 2.5 hours. The synchronized induction of tylosis formation provides an opportunity for the discovery of genes and other molecular studies related to heartwood formation and cell death.

Based on these and similar developments, current and future studies on heartwood formation in vitro will strongly focus on molecular techniques to elucidate processes and isolate genes specific to both tylosis formation and programmed cell death. One particularly promising development for investigating the molecular basis of heartwood and tylosis formation was reported by Wilson (2002) based on the principles described here. This author was able to produce liquid cultures of ray cell stacks, which could be induced to synchronous tylosis formation.

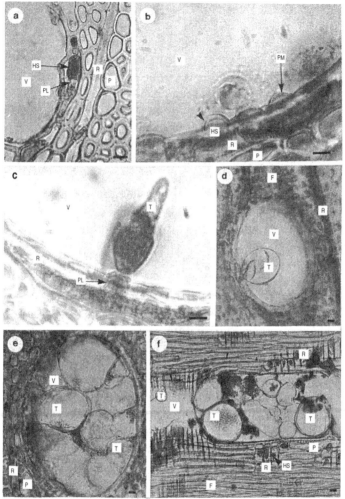

FIGURE 8.1a-f. Transverse sections of *E. globulus* MSE displaying stages of tylosis development in vitro on a defined basal medium with plant growth regulators (PGRs). (a) early development of the hydrolytic sac prior to pit membrane swelling (nutriment); (b) ray to vessel pit membrane swelling (arrowhead) as tylosis development into the vessel begins (1.0 mg·1^{-1} NAA); (c-d) increase in tylosis size as hydrolytic sac size decreases (c, nutriment; d, same as b); (e-f) vessel lumen filled with tyloses at several stages of development (e, 0.1 2,4-D + 2.0 L-methionine; f, 2.0 mg·1^{-1} GA$_3$ + 1.0 mg·1^{-1} 2,4-D + 1.0 mg·1^{-1} L-methionine, radial section). R = ray; V = vessel; F = fiber; T = tylosis; P = parenchyma; HS = hydrolytic sac; PL = protective layer; PM = pit membrane. Scale bar is 10 μm (a and e), 5 μm (b and c), and 20 μm (d).

Protoplast and Cell Suspension Cultures

With similar goals in mind (i.e., synchronized responses), protoplasts have been isolated from several plant tissues (e.g., cotyledons, roots, needles, pollen, and leaves) and woody plants (Duhoux 1980; Smith and McCown 1983; Faye and David 1983; Ahuja 1984; Hurwitz and Agrios 1984; Kirby and David 1988; Leitch and Savidge 2000). In culture, protoplasts have been used to determine, for example, the competence of certain cell types for distinct developmental pathways (Leitch and Savidge 2000).

Leinhos and Savidge (1993) reported a method to isolate protoplasts from the cambial zone and developing xylem of *Pinus* spp. and provided data on *E*-coniferin content indicating that the approach could provide insight into biochemistry and physiology of cellular differentiation. The method could also serve as a means to investigate cambial commitment and competence for totipotent expression as was suggested by David (1987). However, very few totipotency successes, in terms of regenerating whole trees from protoplasts, have yet been reported (Vardi et al. 1975; Kobayashi et al. 1983; Sticklen et al. 1985). Nevertheless, these systems provide ample opportunity for biochemical, physiological, and, potentially, molecular studies in relation to cellular differentiation and wood formation. To our knowledge, in woody tree species little if any molecular work has been done with protoplast cultures.

IN VITRO WOOD FORMATION SYSTEMS

Several wood formation systems claim to mimic *in arbor* wood formation in vitro (see Gautheret 1934, 1983; Jacquiot 1964; Zajaczkowski 1969, 1973; Savidge 1983; Zhong and Savidge 1995). However, specific differences to "normal" *in arbor* growth appear, and prolonged growth and wood formation leading to large radial files of fully differentiated tracheary elements in these systems is still wanting. More recently it has been discovered that explant competence in vitro is dependent on several factors, for example, cambial age, position in the tree, date of harvest, and the physiological status of the explant when cultured (McComb and Wroth 1986; Leitch and Savidge 1995a,b; Savidge 1993a, 1996; Leitch 1999). This thinking led to the important discovery that temporal windows during the year exist during which improved in vitro responses could be achieved (McComb and Wroth 1986). With this in mind, similar results were obtained and utilized in two particular systems for which temporal windows were charted (see Leitch 1995, 1999). Results led to much improved tracheary element formation in vitro. The two systems described here as possible wood formation model systems are a *Larix laricina* system (Savidge 1993a; Leitch 1995; Leitch and Savidge 1995a,b) and a *Eucalyptus globulus* system (Leitch 1999; Leitch et al. 1999; Leitch and Savidge 2000; Leitch 2001).

Conifers

The *Larix* system (Savidge 1993a) is an organ culture rather than a single cell culture. This system uses static basal media based on earlier work by Savidge (1983) using *Pinus contorta* explants. Here cambial cells undergo periclinal divisions (anticlinal were also noted) and produce radial files of secondary-walled, lignified, bordered-pitted, and autolyzed earlywood tracheids (Savidge 1993a; Leitch 1995; Leitch and Savidge 1995a,b) (compare Figure 8.2 and see Table 8.1). The work published by Savidge (1993a) displayed the ability of the *Larix* system to culture several-year-old cambial material (eight years old) from mature trees (30 to 40 years old) leading to organized growth very similar to that produced *in arbor*. Using this *Larix* system, Leitch and Savidge (1995a,b) studied bordered-pit formation in secondary cell walls. Features that could be manipulated through media modification included pit size, location and frequency. One major finding was that bordered pits begin cell wall modifications or readjustments shortly following cell division. This is in contrast to earlier studies, which showed bordered-pit development occurring sometime around the process of radial expansion and post cell division and prior to secondary wall formation (Murmanis and Sachs 1969; Parham and Baird 1973; Barnett and Harris 1975). Areas of modified wall structuring, as described by Wardrop (1954a,b), Frey-Wyssling and colleagues (1956), Barnett and Harris (1975), and Uehara and Hogetsu (1993) were related to early bordered-pit development (Leitch 1995; Leitch and Savidge 1995a). For this reason, some authors argued that the timing of bordered-pit formation and location is a genetically triggered process that may change developmentally if the cambium is wounded (see Savidge and Farrar 1984; Savidge 1993a,b). To date, this system has not been used for molecular studies.

Eucalypts

The experimental induction of tracheary element formation under controlled culturing conditions in young and mature aged portions of several-year-old *Eucalyptus globulus* trees was achieved by Leitch (1999). Following defined sterilization and explanting procedures these materials were grown on a defined sterile culture medium.

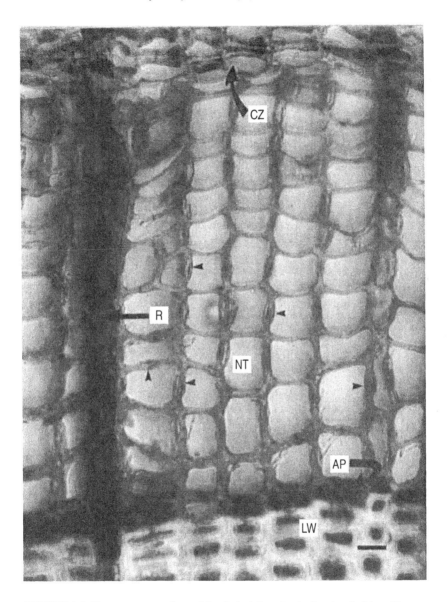

FIGURE 8.2. Transverse section of *Larix laricina* displaying tracheids with several bordered pits (arrowheads) in each radial file produced in vitro on defined basal medium with PGRs (1.4 mg·1^{-1} NAA). LW = latewood; CZ = cambial zone; NT = new tracheids; AP = axial parenchyma; R = ray. Scale bar is 10 μm.

TABLE 8.1. Staged classification of wood formation in cultured stem explants[a] (hardwood and softwood)

Stage	Description	Days[b]
0	Equilibrium: static explanted materials adjust to media conditions	3-7
1[c]	External wound response: visible callus formation on cut surfaces, mainly parenchyma cells	5-10
2[c]	Initial wound response: CZ4[d] cells de-differentiate into axial parenchyma (see Figure 8.2)	7-14
3[a]	Organized CZ activity: division of CZ cells (anti- and periclinal) leading to increases in radial file cell numbers followed by radial expansion or elongation	10-14
3[b]	*Increase in external callus leading to an equilibrium in internal pressures; brownish, random, and friable callus or green, organized, and compact callus*	10 ->
4	Continued CZ activity and beginning of secondary wall formation and lignification in radially expanded or elongated cells (see Figure 8.3)	16-20 ->
5	Continued CZ activity and differentiation of cells, and autolysis of cells completing secondary wall formation and lignification (see Figure 8.3)	20-30
6[e]	Cracking of outer surface as a result of a larger xylem response compared to phloem and bark responses; cells formed in the outer xylem may lose organization and develop microdomains of opposed cellular orientations (see Figure 8.4)	35-50 ->

[a]Based on standard culturing conditions (temperature of 25°C, 16 hours light minimum)

[b]Relate to approximate days only and are dependent on factors such as species, explant age, season, and other indogenous and exogenous factors

[c]System becomes competent for agrobacterium-mediated transformation (variable with species)

[d]CZ contains cells in the phases of cell division and radial expansion or elongation

[e]This feature not always present but can be common in certain species

Italic font: cellular response in the callus portion of the explant

Normal font: cellular response in the xylem portion of the explant as a result of CZ activity

In vitro, wood formation occurred in both apical stem segments (ASS) and in main stem explants (MSE) under several hormonal treatments. Initial responses include the formation of peripheral callus followed by cambial differentiation within the explants (see Table 8.1). Cambial divisions were followed by elongation of cambial derivatives with various amounts of secondary wall formation and lignification occurring based on material (ASS or MSE) and media (see Figure 8.3 and Table 8.1). In both systems fibers, vessels, parenchyma, and rays were produced with variations in size, presence, and frequency between treatments and material location (see Figure 8.3). Also, in both systems media were discovered which resulted in tension wood formation, kino formation (apical stem segments only), the production of microdomains (loss of axial polarity) (see also Harris 1981; Savidge and Farrar 1984), and vessel-free wood (see Figure 8.4). Significant differences in hormonal and nutritional requirements were displayed between MSE and ASS.

This research has provided the first evidence that earlywood formation in vitro using young and merchantable stem regions of *Eucalyptus globulus* is possible. This system has been successfully tried on a number of other *Eucalyptus* species *(E. delegatensis, E. nitens, E. camaldulensis)* (Leitch, unpublished).

In our hands, this system (ASS) has proven to be susceptible to *Agrobacterium*-mediated transformation as a system to test and screen genes for function and specificity (Bossinger and Leitch 2000). To the authors' knowledge, comparable in vitro systems do not exist for loblolly pine or poplar. It seems likely, however, that in time similar systems will be used for large-scale molecular investigation, complementing the ever-expanding EST databases (Allona et al. 1998; Sterky et al. 1998) and microarray technology (Hertzberg et al. 2001).

FUTURE PERSPECTIVES

The vascular cambium and the formation of woody tissue were among the first objects of in vitro research in plants (Gautheret 1934; White 1943, 1954; Ball 1950). It therefore seems surprising that the molecular revolution at this stage has not taken advantage of the benefits that in vitro xylogenesis has to offer. One reason might be that generalizations derived from molecular studies in weedy and agricultural crop species are not readily transferable to long-lived woody tree species. Molecular studies, in turn, could shed light on general morphogenetic events that are underlying gymnosperm and perennial angiosperm stem development and wood formation (Bossinger 2001). In this respect, occasionally appearing reports on wood formation in

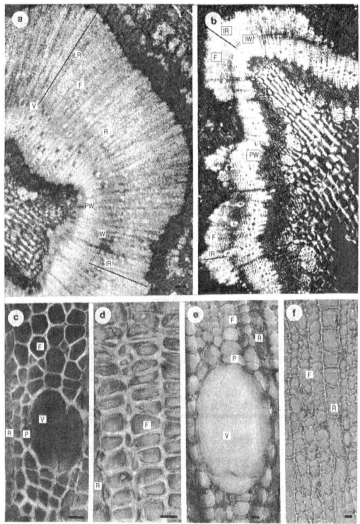

FIGURE 8.3a-f. Transverse sections of *E. globulus* ASS and MSE displaying in vitro differentiation responses on a defined basal medium with PGRs. (a-b) a large (a, 1.0 mg·1^{-1} IAA + 1.0 mg·1^{-1} GA$_4$) and smaller, variable (b, 1.0 mg·1^{-1} GA$_3$ + 0.5 mg·1^{-1} 2,4-D) response in ASS (CPO, crossed polarizing optics); (c-d) production of differentiated fibers (c, same as a) and vessel elements (d, same as b) in ASS; (e-f) production of differentiated fibers (e, 1.0 mg·1^{-1}IAA) and vessel elements (f, 2.5 mg·1^{-1} NAA + 0.5 mg·1^{-1} GA$_3$) in MSE. R = ray; V = vessel; F = fiber; P = parenchyma; IR = in vitro response; PW = wood formed prior to culture; IW = initial wound. Scale bar is 71 µm (a and b) and 10 µm (c, d, e, and f).

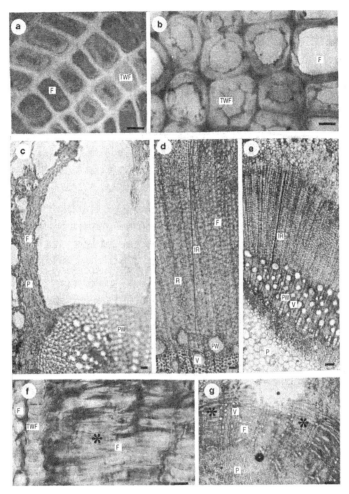

FIGURE 8.4a-g. Transverse sections of in vitro responses from ASS and MSE of *E. globulus* grown on defined basal medium displaying variability from the normal wood formation processes that occur in trees. (a-b) induction of tension wood formation in ASS (a, 1.0 mg·1^{-1} GA$_3$ + 0.5 mg·1^{-1} 2,4-D) and MSE (b, 0.5 mg·1^{-1} NAA + 2.5 GA$_3$ mg·1^{-1}); (c) induction of kino formation in ASS (1.0 mg·1^{-1} GA$_3$ + 2.0 mg·1^{-1} 2,4-D + 0.5 mg·1^{-1} L-methionine); (d-e) induction of vessel-free wood in ASS (d-e, same as a); (f-g) the production of microdomains (tangential expansion of cells) in ASS (f, 0.1 mg·1^{-1} 2,4-D; g, 2.5 mg·1^{-1} NAA + 0.5 GA$_3$ mg·1^{-1}). R = ray; V = vessel; F = fiber; P = parenchyma; IR = in vitro response; PW = wood formed prior to culture; TWF = tension wood fibers; * = microdomains. Scale bar is 5 μm (a and b), 20 μm (c and d), 71 μm (e), 10 μm (f), and 142 μm (g).

Arabidopsis (Regan et al. 1999) together with all systems described here hold much promise for shedding light on the molecular basis of wood formation in the not too distant future.

REFERENCES

Ahuja, M.R. (1984). Protoplast research in woody plants. *Silvae Genetica* 33: 32-37.

Allona, I., Quinn, M., Shoop, E., Swope, K., St. Cyr, S., Carlis, J., Riedl, J., Retzel, E., Campbell, M.M., Sederoff, R., and Whetten, R.W. (1998). Analysis of xylem formation in pine by cDNA sequencing. *Proc. Natl. Acad. Sci.* 95: 9693-9686.

Aloni, R., Feigenbaum, P., Kalev, N., and Rozovsky, S. (2000), Hormonal control of vascular differentiation in plants: The physiological basis of cambium ontogeny and xylem evolution. In R.A. Savidge, J.R. Barnett, and R. Napier (Eds.), *Cell and Molecular Biology of Wood Formation* (pp. 223-236). Oxford, UK: BIOS Scientific Publishing.

Aloni, R., Pradel, K.S., and Ullrich, C.I. (1995). The three-dimensional structure of vascular tissues in *Agrobacterium tumefaciens*-induced crown galls and in the host stems of *Ricinus communis* L. *Planta* 196: 597-605.

Ball, E. (1950). Differentiation in a callus culture of *Sequoia sempervirens*. *Growth* 14: 295-325.

Bamber, R.K. (1985). The wood anatomy of eucalypts and paper making. *Appita* 38(3): 210-216.

Barnett, J.R. and Harris, J.M. (1975). Early stages of bordered-pit formation in *Radiata* pine. *Wood Sci. Techn.* 9: 233-241.

Bolton, A.J. and Petty, J.A. (1977). Variation of susceptibility to aspiration of bordered pits in conifer wood. *J. Exp. Bot.* 28(105): 935-941.

Bossinger, G. (2001). Segments (phytomers). In *Encyclopedia of Life Sciences*, Volume 17 (pp. 61-64). London, UK: Macmillan.

Bossinger, G. and Leitch, M.A. (2000). Isolation of cambium-specific genes from *Eucalyptus globulus* Labill. In R.A. Savidge, J.R. Barnett, and R. Napier (Eds.), *Cell and Molecular Biology of Wood Formation* (pp. 203-207). Oxford, UK: BIOS Scientific Publishing.

Brett, C. and Waldron, K. (1990). *Physiology and Biochemistry of Plant Cell Walls*. London: Unwin Hyman.

Brown, C.L. (1971). Secondary growth. In M.H. Zimmermann and C.L. Brown (Eds.), *Trees: Structure and Function* (pp. 67-123). New York: Springer-Verlag.

Cui, K.M., Zhang, Z.M., Li, J.H., Li, S.W., and Li, Z.L. (1995). Changes of peroxidase, esterase isozyme activities and some cell inclusions in regenerated vascular tissues after girdling in *Broussonetia papyrifera* (L.) Vent. *Trees* 9: 165-170.

David, A. (1987). Conifer protoplasts. In J.M. Bonga and D.J. Durzan (Eds.), *Cell and Tissue Culture in Forestry*, Volume 2, *Specific Principles and Methods: Growth and Developments* (pp. 2-15). Dordrecht, The Netherlands: Martinus Nijhoff Publ.

Demura, T. and Fukuda, H. (1993). Molecular cloning and characterization of cDNAs associated with tracheary element differentiation in cultured *Zinnia* cells. *Plant Physiology* 103: 815-821.

Demura, T. and Fukuda, H. (1994). Novel vascular cell-specific genes whose expression is regulated temporally and spatially during vascular system development. *Plant Cell* 6: 967-981.

Desch, H.E. (1981). *Timber, Its Structure, Properties and Utilization.* Forest Grove, OR: Timber Press.

Domingo, C., Roberts, K., Stacey, N.J., Connerton, I., Ruíz-Teran, F., and McCann, M.C. (1998). A pectate lyase from *Zinnia elegans* is auxin inducible. *Plant J.* 13: 17-28.

Du Toit, A.J. (1964). The influence of tyloses on the manufacture of matches. *S. Afr. For. J.* 50: 27-38.

Duhoux, E. (1980). Protoplast isolation of gymnosperm pollen. *Z. Pflanzenphysiol.* 99: 207-214.

Esau, K. (1965). *Plant Anatomy,* Second Edition. New York: John Wiley and Sons.

Faye, M. and David, A. (1983). Isolation and culture of gymnosperm root protoplasts *(Pinus pinaster). Physiol. Plant.* 59: 359-362.

Freshour, G., Clay, R.P., Fuller, M.S., Albershiem, P., Darvill, A.G., and Hahn, M. (1996). Developmental and tissue-specific structural alterations of the cell-wall polysaccharides of *Arabidopsis thaliana* roots. *Plant Physiology* 110: 1413-1429.

Frey-Wyssling, A., Bosshard, H.H., and Mühlethaler, K. (1956). Die submikroskopische Entwicklung der Hoftüpfel. [The sub-microscopical development of bordered pits.] *Planta* 47: 115-126.

Fukuda, H. (1992). Tracheary element formation as a model system of cell differentiation. *Inter. Rev. Cytol.* 136: 289-332.

Fukuda, H. (1994). Redifferentiation of single mesophyll cells into tracheary elements. *Int. J. Plant Sci.* 155(3): 262-271.

Fukuda, H. (1997). Tracheary element differentiation. *Plant Cell* 9: 1147-1156.

Fukuda, H. and Komamine, A. (1980). Establishment of an experimental system for the tracheary element differentiation from single cells isolated from the mesophyll of *Zinnia elegans. Plant Physiology* 52: 57-60.

Gautheret, R.J. (1934). Culture du tissu cambial. [Culture of cambial tissue.] *C.R. Acad. Sci.* 198: 2195-2196.

Gautheret, R.J. (1983). Plant tissue culture: A history. *Bot. Mag. Tokyo* 96: 393-410.

Harris, J.M. (1981). Spiral grain formation. In R. Barnett (Ed.), *Xylem Cell Development* (pp. 256-274). Kent, England: J. Castle House Publishers Tunbridge Wells.

Harrison, M.J., Delmer, D.P., Amor, Y., Grimson, M.J., Johnson, S., and Haigler, C.H. (1997). Localization of sucrose synthase in differentiating tracheary elements of *Zinnia elegans. Plant Physiology* 114(3) (suppl): 349.

Hertzberg, M., Aspeborg, H., Schrader, J., Andersson, A., Erlandsson, R., Blomqvist, K., Bhalerao, R., Uhlén, M., Teeri, T.T., Lundeberg, J., et al. (2001). A transcriptional roadmap to wood formation. *Proc. Natl. Acad. Sci.* 98(25): 14732-14737.

Hurwitz, C.D. and Agrios, G.N. (1984). Isolation and culture of protoplasts from apple callus and cell suspension cultures. *J. Amer. Soc. Hortic. Sci.* 109: 348-350.

Iqbal, M. (1990). *The Vascular Cambium.* Somerset, UK: Research Studies Press Ltd.

Jacquiot, C. (1964). Application de la technique de culture des tissus végétaux l'étude de quelques problèmes de la physiologie du l'arbre. [Application of tissue culture techniques to trees to solve questions associated with physiology of trees.] *Ann. Sci. For.* 21: 317-473.

Kirby, E.G. and David, A. (1988). Use of protoplasts and cell cultures for physiological and genetic studies of conifers. In J.W. Hanover and D.E. Keathley (Eds.), *Genetic Manipulation of Woody Plants* (pp. 185-197). New York and London: Plenum Press.

Kobayashi, S., Uchimiya, H., and Ikeda, I. (1983). Plant regeneration from Trovita orange protoplasts. *Jpn. J. Breed.* 33: 119-122.

Kohlenbach, H.W. and Schmidt, B. (1975). Cytodifferenzierung in Form einer direktenUmwandlung isolierter Mesophyll-Zellen zu Tracheiden. [Cyto-differentiation in form of direct conversion of mesophyll cells to tracheids.] Z. *Pflanzenphysiologie* 75: 369-374.

Kozlowski, T.T. (1981). *Water Deficits and Plant Growth,* Volume VI: *Woody plant communities.* New York: Academic Press.

Larson, P.R. (1994). *The Vascular Cambium: Development and Structure.* Berlin: Springer-Verlag.

Leinhos, V. and Savidge, R.A. (1993). Isolation of protoplasts from developing xylem of *Pinus banksiana* and *Pinus strobus. Can. J. For. Res.* 23: 343-348.

Leitch, M.A. (1995). Bordered-pit development in *Larix laricina* (DuRoi) K. Koch. MScF Thesis, Univ. of New Brunswick, Faculty of Forestry, Fredericton, NB, Canada.

Leitch, M.A. (1999). The development of tissue culture techniques to study wood formation in *Eucalyptus globulus* Labill. PhD Thesis, University of Melbourne, School of Forestry, Creswick, Victoria, Australia.

Leitch, M.A. (2001). Wood formation studies utilizing tissue culture techniques. *NZ Crop and Food Research Report 16.* ISSN 1171-7564, ISBN 0-478-10822-2.

Leitch, M.A. and Savidge, R.A. (1995a). Auxin regulation of bordered-pit development in stem explants from Tamarack [*Larix laricina* (Du Roi) K. Koch]. *PGRSA Quarterly* 23(2): 97.

Leitch, M.A. and Savidge, R.A. (1995b). Evidence for auxin regulation of bordered-pit positioning during tracheid differentiation in *Larix laricina. IAWA J.* 16(3): 289-297.

Leitch, M.A. and Savidge, R.A. (2000). Tissue culture for the study of cambial activity and wood formation—A resurgence of interest in an old technique. In R.A. Savidge, J.R. Barnett, and R. Napier (Eds.), *Cell and Molecular Biology of Wood Formation* (pp. 493-512). Oxford, UK: BIOS Scientific Publishing.

Leitch, M.A., Savidge, R.A., Downes, G.M., and Hudson, I.L. (1999). Induction of tyloses in *Eucalyptus globulus* "chips." *IAWA J.* 20(2): 193-201.

Lev-Yadun, S. and Aloni, R. (1993). Effect of wounding on the relations between vascular rays and vessels in *Melia azedarach* L. *New Phytol.* 124: 339-344.

McCann, M.C. (1997). Tracheary element formation: Building up to a dead end. *Trends in Plant Sci. Rev.* 2(9): 333-338.

McCann, M.C., Domingo, C., Stacey, N.J., Milioni, D., and Roberts, K. (2000). Tracheary element formation in an *in vitro* system. In R.A. Savidge, J.R. Barnett, and R. Napier (Eds.), *Cell and Molecular Biology of Wood Formation* (pp. 457-470). Oxford, UK: BIOS Scientific Publishing.

McComb, J.A. and Wroth, M. (1986). Vegetative propagation of *Eucalyptus resinifera* and *Eucalyptus maculata* using coppice cuttings and micropropagation. *Aust. For. Res.* 16: 231-242.

Meyer, R.W. (1967). Tyloses development in white oak. *For. Prod. J.* 17(12): 50-56.

Motose, H., Fukuda, H., and Sugiyama, M. (2001). Involvement of local intercellular communication in the differentiation of *Zinnia* mesophyll cells into tracheary elements. *Planta* 213: 121-131.

Murmanis, L. (1975). Formation of tyloses in felled *Quercus rubra* L. *Wood Sci. Technol.* 9: 3-14.

Murmanis, L. and Sachs, I.B. (1969). Seasonal development of secondary xylem in *Pinus strobus* L. *Wood Sci. Techn.* 3: 177-193.

Newman, I.W. (1956). Pattern in meristems of vascular plants. I. Cell partition in living apices and in the cambial zone in relation to the concepts of initial cells and apical cells. *Phytomorphology* 6: 1-19.

Nobuchi, T., Ogata, Y., and Siripatanadilok, S. (1995). Seasonal characteristics of wood formation in *Hopea odorata* and *Shorea henryana*. *IAWA J.* 16(4): 361-369.

Ohdaira, Y., Kakegawa, K., Amino, S., Sugiyama, M., and Fukuda, H. (2002). Activity of cell-wall degradation associated with differentiation of isolated mesophyll cells of *Zinnia elegans* into tracheary elements. *Planta* 215: 177-184.

Panshin, A.J. and DeZeeuw, C. (1980). *Textbook of Wood Technology,* Fourth Edition. New York: McGraw-Hill.

Parham, R.A. and Baird, W.M. (1973). The bordered-pit membrane in differentiating Balsam fir. *Wood and Fiber* 5(1): 80-87.

Phillips, I.D.J. (1976). The cambium. In M.M. Yeoman (Ed.), *Cell Division in Higher Plants* (pp. 348-390). London: Academic Press.

Potikha, T. and Delmer, D.P. (1995). A mutant of *Arabidopsis thaliana* displaying altered patterns of cellulose deposition. *Plant J.* 7(3): 453-460.

Regan, S., Chaffey, N.J., and Sundberg, B. (1999). Exploring cambial growth with *Arabidopsis* and *Populus*. *J. Exp. Bot.* 50 (suppl.): 33.

Reiter, W.D. (1998). *Arabidopsis thaliana* as a model system to study synthesis, structure, and function of the plant cell wall. *Plant Physiol. Biochem.* 36(1-2): 167-176.

Reiter, W.D., Chapple, C.S.S., and Somerville, C.R. (1993). Altered growth and cell walls in a fucose-deficient mutant of *Arabidopsis*. *Science* 261: 1032-1035.

Roberts, A.W., Donovan, S.G., and Haigler, C.H. (1997). A secreted factor induces cell expansion and formation of metaxylem-like tracheary elements in xylogenic suspension cultures of *Zinnia*. *Plant Physiology* 115: 683-692.

Roberts, A.W. and Haigler, C.H. (1994). Cell expansion and tracheary element differentiation are regulated by extracellular pH in mesophyll cultures of *Zinnia elegans* L. *Plant Physiology* 105: 699-706.

Savidge, R.A. (1983). The role of plant hormones in higher plant cellular differentiation. II. Experiments with the vascular cambium, and sclereid and tracheid differentiation in the pine, *Pinus contorta. Histochem. J.* 15: 447-466.

Savidge, R.A. (1993a). Formation of annual rings in trees. In L. Rensing (Ed.), *Oscillations and Morphogenesis* (pp. 343-363). New York: Marcel Dekker Inc.

Savidge, R.A. (1993b). *In vitro* wood formation in chips from merchantable stem regions of *Larix laricina. IAWA J.* 14(1): 3-11.

Savidge, R.A. (1996). Xylogenesis, genetic and environmental regulation—A review. *IAWA J.* 17(3): 269-310.

Savidge, R.A. and Farrar, J.L. (1984). Cellular adjustments in the vascular cambium leading to spiral grain formation in conifers. *Can. J. Bot.* 62: 2872-2879.

Smith, M.A.L. and McCown, B.H. (1983). A comparison of source tissue for protoplast isolation from three woody plant species. *Plant Sci. Lett.* 28: 149-156.

Sterky, F., Regan, S., Karlsson, J., Hertzberg, M., Rohde, A., Holmberg, A., Amini, B., Bhalerao, R., Larsson, M., Villarroel, R., et al. (1998). Gene discovery in the wood-forming tissues of poplar: Analysis of 5,692 expressed sequence tags. *Proc. Natl. Acad. Sci. USA* 95: 13330-13335.

Sticklen, M.B., Lineberger, R.D., and Domir, S.C. (1985). Isolation and culture of *Ulmus* x Homestead protoplasts. *Hortscience* 20: 117-120.

Strelis, I. and Green, H.V. (1962). Tyloses and their detection. *Pulp and Paper Magazine of Canada* 63: T307-T310.

Uehara, K. and Hogetsu, T. (1993). Arrangement of cortical microtubules during formation of bordered-pit in the tracheids of *Taxus. Protoplasma* 172: 145-153.

Vardi, A., Speigel-Roy, P., and Galun, E. (1975). Citrus cell culture: Isolation of protoplasts, plating densities, effect of mutagens and regeneration of embryos. *Plant Cell Lett.* 4: 231-236.

Wardrop, A.B. (1954a). The mechanisms of surface growth involved in the differentiation of fibers and tracheids. *Aust. J. Bot.* 2: 165-175.

Wardrop, A.B. (1954b). Observations on crossed lamellar structures in the cell walls of higher plants. *Aust. J. Bot.* 2: 154-164.

White, P.R. (1943). *A Handbook of Plant Tissue Culture.* Lancaster, PA: J. Cattell.

White, P.R. (1954). *The Cultivation of Animal and Plant Cells.* New York: Ronald Press.

Wilson, K. and White, D.J.B. (1986). *The Anatomy of Wood: Its Diversity and Variability.* London: Stobart and Sons Ltd.

Wilson, L. (2002). Cell biology of eucalypt heartwood formation. PhD Thesis, University of Melbourne, School of Forestry, Creswick, Victoria, Australia.

Yamamoto, R., Demura, T., and Fukuda, H. (1997). Brassinosteroids induce entry into the final stage of tracheary element differentiation in cultured *Zinnia* cells. *Plant Cell Physiol.* 38(8): 980-983.

Ye, Z.H. and Droste, D.L. (1996). Isolation and characterization of cDNAs encoding xylogenesis-associated and wounding-induced ribonucleases in *Zinnia elegans. Plant Mol. Biol.* 30: 697-709.

Ye, Z.H. and Varner, J.E. (1993). Gene expression patterns associated with in vitro tracheary element formation in isolated single mesophyll cells of *Zinnia elegans. Plant Physiology* 103: 805-813.

Yoshimura, T., Demura, T., Igarashi, M., and Fukuda, H. (1996). Differential expression of three genes for different ß-tubulin isotypes during the initial culture of *Zinnia* mesophyll cells that divide and differentiate into tracheary elements. *Plant Cell Physiol.* 37: 1167-1176.

Zajaczkowski, S. (1969). Xylem formation in isolated stem segments of *Pinus silvestris* L. *Acta. Soc. Bot. Pol.* 38: 671-675.

Zajaczkowski, S. (1973). Auxin stimulation of cambial activity in *Pinus sylvestris*. I. The differential cambial response. *Physiol. Plant.* 29: 281-287.

Zhong, Y. and Savidge, R.A. (1995). Effects of IAA and GA3 on *in vitro* wood formation in merchantable stems of white ash [*Fraxinus americana* (L.)]. In *Proceedings of the Plant Growth Regulation Society of America, 22nd Annual Meeting* (pp. 231-236). C. Fritz (Ed.), P.O. Box 12014, Research Triangle Park, NC.

Zimmermann, M.H. and Brown, C.L. (1971). *Trees: Structure and Function.* New York: Springer-Verlag.

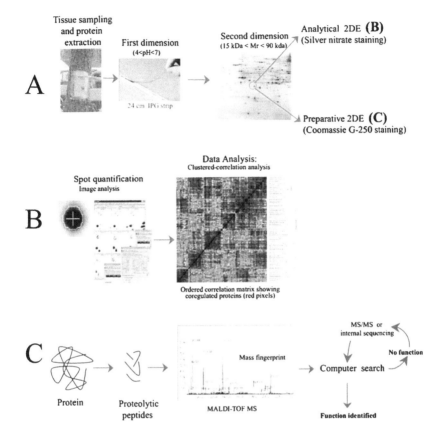

FIGURE 3.1. Schematic illustration of standard proteome analysis by two-dimensional gel electrophoresis (2DE) and mass spectrometry (MS). (A) Protein are extracted and separated by 2DE and stained. (B) Silver nitrate staining is used for analytical 2DE. Images are scanned and spot intensity analyzed using a computer-assisted system. Data are then submitted to statistical analysis (clustered correlation, analysis of variance, principal components analysis, etc.). (C) Sypro Ruby or Coomassie brilliant blue G-250 are used for preparative 2DE. Spots are excised from the gel, subjected to in-gel digestion with trypsin, and the resulting peptides are analyzed by MS.

FIGURE 4.1. The ectomycorrhizal symbiosis. (A) A basidiomycetous bolet able to form a mutualistic symbiosis with conifers and hardwood trees (photograph courtesy of F. Martin). (B) A seedling of Douglas fir *(Pseudotsuga menziesii)* colonized by the ectomycorrhizal basidiomycete *Laccaria bicolor.* The fungal mycelium has developed ectomycorrhiza on the root system and has produced a basidiocarp aboveground (photograph courtesy of P. Klett-Frey). (C) Short roots ensheathed by an ectomycorrhizal tomentelloid fungus. The orange mantle covers the root tip of *Eucalyptus globulus* (photograph courtesy of A. Jambois). (D) Transverse section of a *Eucalyptus/Pisolithus* ectomycorrhiza showing the external (EM) and internal (IM) mantles: the fungal hyphae have begun to penetrate between the epidermal cells of the root cortex (RC) to form the Hartig net (HN). Epidermal cells are radially enlarged. Extramatrical hyphae (EH) are exploring the medium (photograph courtesy of B. Dell).

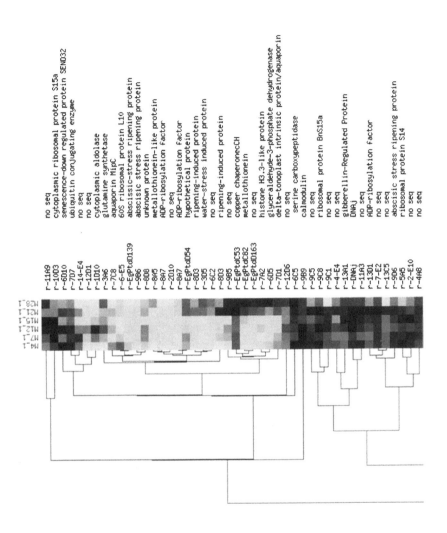

FIGURE 4.2. Cluster analysis of a set of 200 genes expressed in *Eucalyptus globulus* roots during ectomycorrhiza development. Clustering was calculated by a hierarchical method based on Euclidean distance measurements and the data represented using the method of Eisen et al. (1998). Color intensity reflects the magnitude of gene expression ratio for each gene, red indicates genes expressed at a higher level in roots colonized by *Pisolithus microcarpus* (com-

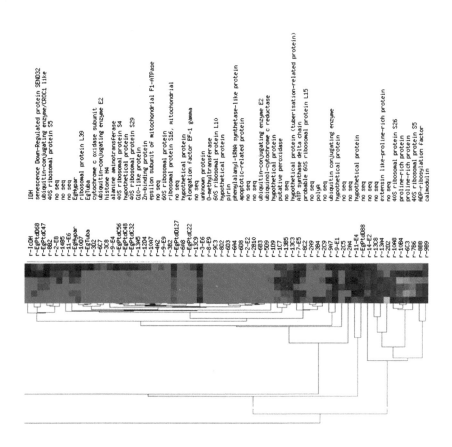

pared to nonmycorrhizal roots), green indicates genes with a down-regulated expression in symbiotic tissues, and black indicates gene expression ratio near 1. Each column represents gene expression ratios from a different developmental stage (between 0 and 28 days postcontact). No seq = sequence not available; IDH = NADP-dependent isocitrate dehydrogenase; polyA = polyA-binding protein.

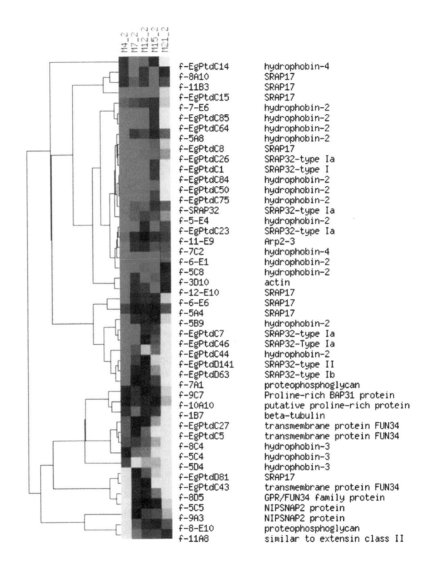

FIGURE 4.3. A group of *Pisolithus microcarpus* genes coding for structural proteins and clustering with the hydrophobines, a family of cell wall polypeptides up-regulated during *Eucalyptus globulus* ectomycorrhiza formation. For details on the hierarchical clustering, see Figure 4.2. SRAP = symbiosis-regulated acidic polypeptides; Arp = Actin-related protein.

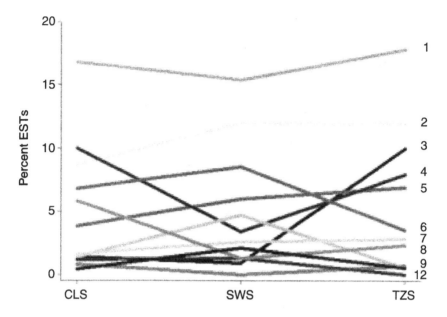

FIGURE 5.1. Percent of ESTs from 12 different functional categories represented in the three cDNA libraries: CLS (cambial region), SWS (sapwood), and TZS (transition zone). The functional categories used are (1) protein synthesis, (2) primary metabolism, (3) secondary metabolism, (4) defense, (5) signal transduction, (6) gene expression, (7) chromatin and DNA metabolism, (8) membrane transport, (9) cell division, (10) cell wall structure, (11) vesicular trafficking, and (12) cytoskeleton.

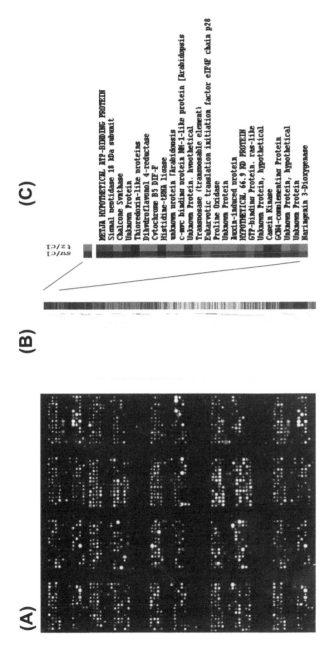

FIGURE 5.2. DNA microarray analysis of black locust. (A) Two-color overlaid scanning fluorimetric image of black locust cDNA microarray. Two hybridization data sets from the same microarray were converted into pseudo-color images and superimposed to visualize differential gene expression between the cambial region (cy3, green) and the transition zone (cy5, red). (B) Cluster analysis of microarray. (C) Exposed cluster of high-expressed clones in sapwood or transition zone compared to cambial zone. Red color indicates up-regulation in the transition zone, and green color means up-regulation in the cambial zone.

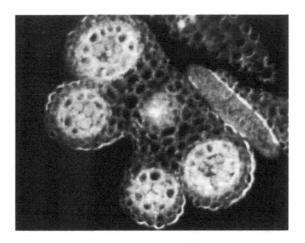

FIGURE 9.1. Expression of the *Pinus radiata* chalcone synthase (CHS) promoter fused to the *uid*A reporter gene, in developing microspores of transgenic *Arabidopsis*. In this dark-field microscopical image of transsection through an *Arabidopsis* anther, the β-glucuronidase (GUS) stain appears pink rather than blue.

PART III:
FOREST TREE TRANSGENESIS

Chapter 9

Genetic Modification in Conifer Forestry: State of the Art and Future Potential— A Case Study with *Pinus radiata*

Christian Walter
Julia Charity
Lloyd Donaldson
Lynette Grace
Armando McDonald
Ralf Möller
Armin Wagner

INTRODUCTION

Conifers have been very successful and competitive in occupying various ecological niches around the world. They appear highly flexible with regard to their ability to grow in various and often harsh environmental conditions, and among other characteristics, this makes them suitable to play a major role in the commercial exploitation of forests. Conifer forestry contributes to the economic and social well-being of many countries, and the use of plantations or sustainable managed forests has a significant positive impact on natural forests, in particular on those that make major contributions to the world climate health. The world demand for forest products, such as pulp, paper, and timber, is ever increasing, and the wood for these products will in the future have to come from highly managed plantations if the world's population chooses to leave natural forests alone.

The agricultural industry is in a strong position to evaluate and capture the benefits from conventional breeding and the application of modern biotechnology. Crops of higher value and of higher manageability are used commercially in many countries. The world's total area covered with transgenic crops rose to more than 100 million acres in the year 2000 (James, 2000). However, even with current production rates an inadequate volume of agricultural products are produced to adequately feed the world population, and our ability to feed a growing population would be significantly reduced if modern breeding and biotechnology were not avilable in modern agriculture (Krattiger, 1998).

With these facts in mind, today's commercial forestry is in a unique position. Conventional breeding has already demonstrated enormous potential, and genetically improved trees are now grown in plantation forestry in many countries (Pandy, 1995). Considering that this represents value gained from only a few breeding and genetic improvement cycles, one can assume with reasonable confidence that more potential to enhance gain exists (Burdon et al., 1997). This may very well be achieved by further conventional breeding efforts supported by molecular techniques. Recently developed molecular protocols allow the testing of genotypes and provide economical and reliable options for quality assurance, thereby significantly improving breeding efforts (Carson et al., 1996; Kumar et al., 2000).

However, not every desirable trait is readily available in the breeding populations. Many traits may be available in other organisms, and genetic engineering (GE) now offers the potential to transfer and integrate these traits into breeding populations. GE technologies have become available to the main forest tree species including conifers, eucalypts, and poplars (McCown et al., 1991; Jouanin et al., 1993; Walter, Carson, et al., 1998; MacRae and Van Staden, 1999).

This chapter concentrates on the application of biotechnology, and in particular GE in conifer trees. We will discuss tissue culture and transformation technologies, show evidence for the expression of foreign genes in regenerated transgenic conifers, and provide data on conifer promoter function. In addition, we address the assessment of novel traits using microanalytical techniques.

TISSUE CULTURE AS A TOOL
FOR MOLECULAR RESEARCH

Tissue culture plays a crucial role in many applications related to modern plantation forestry. It provides for the clonal multiplication of superior genotypes, may enable the production of artificial seed, and serves as a basis for genetic modification. In this section, embryogenesis in conifers, with specific examples drawn from research with *Pinus radiata,* will be discussed.

Somatic embryogenesis is defined as a tissue culture method in which proliferative embryonal suspensor masses (ESM) are established from somatic cells. By manipulating components of the medium and culture conditions, the ESM are induced to form somatic embryos and subsequently plants. In the case of conifers, ESM have usually been initiated from juvenile explant tissues, especially from immature zygotic embryos either within the intact megagametophyte or excised from the seed (reviewed in Becwar, 1993). As the embryos within the seed become more developed, competency for somatic embryogenesis appears to decline. This has especially been observed with *Pinus* species in which precotyledonary zygotic embryos are most responsive (Becwar et al., 1990; Garin et al., 1998; Lelu et al., 1999; Jain et al., 1989; Li et al., 1998). In *Picea glauca* (Moench) Voss × *P. engelmannii* Parry ex Englem. (interior spruce), competence to form embryogenic tissue is limited to the cotyledonary stage of development prior to accumulation of storage proteins (Roberts et al., 1989). Recently it has been reported that embryogenic cultures of both *P. radiata* (Smith, personal communication) and *Picea abies* (Paques, personal communication) have been initiated from trees up to 25 years old, and plants with juvenile characteristics were regenerated from these cultures.

Somatic embryogenesis has been demonstrated for a wide range of conifers including *Picea* sp. (Attree et al., 1990; Sutton et al., 1993), *Pinus* spp. (Gupta and Durzan, 1986; Klimaszewska and Smith, 1997; Li et al., 1998; Garin et al., 1998; Lelu et al., 1999; Jain et al., 1989; Becwar et al., 1990; Haggman et al., 1991; Jones and van Staden, 1995), *Larix* spp. (Nagmani and Bonga, 1985; von Aderkas et al., 1990; Kim et al., 1999), *Abies* sp. (Schuller et al., 1989; Salajova et al., 1996), and *Pseudotsuga menziesii* (Durzan and Gupta, 1987).

A detailed description of somatic embryogenesis (SE) from *P. radiata* seed has already been published (Smith, 1996; Walter and Grace, 2000). The protocols have also been successfully used to produce plants of *P. taeda, P. elliottii, P. strobus,* and *Picea abies* (Walter and Smith, 1999).

Advantages of Somatic Embryogenesis for Plantation Forestry

Genetic improvement of forest trees can be accelerated using SE technology integrated with both conventional breeding and molecular biology technologies, such as genetic engineering.

A major advantage of SE is that a sample of each clone can be cryostored in a juvenile state indefinitely while field trials are being carried out. The chosen superior clones can then be retrieved from storage and propagated. Depending on the cost-effectiveness, millions of plants could be produced using SE technology, or smaller numbers of somatic plantlets could be established as hedges from which large numbers of cuttings could be produced for field planting. SE techniques have been sufficiently refined to be applied in commercial tree improvement programs for *Pseudotsuga menziesii* (Gupta, personal communication), *Pinus taeda* (Handley, personal communication), *P. radiata* (Aitken-Christie, personal communication), and *Picea glauca* × *P. engelmannii* (Grossnickle et al., 1997).

In a recent review of SE technology of conifer tree species, Timmis (1998) stated that it might be possible to establish six ha of forest from a one-liter flask of embryogenic tissue. Although automation has been studied quite extensively, it has usually been successfully applied only at the maintenance and proliferation steps, and to date no published SE system is fully automated. Despite recent advances in artificial seed production, this technology is not yet available for operational application.

Application of Somatic Embryogenesis to Genetic Modification

Genetic engineering is dependent upon the availability of both tissue culture techniques allowing regeneration of a plant from a few cells and on DNA transfer methods enabling the introduction of foreign genes into the genome of such cells. Embryogenic tissue is a suitable target tissue for gene transfer, and SE may be the most effective protocol for genetic engineering in conifers. The advantages of SE for transformation include the following: (1) tissues can be proliferated very quickly, either in liquid or on a solid medium; (2) tissues can be temporarily suspended in liquid and plated on a medium in a thin layer which allows for easier selection of resistant cells after the tissue has been transformed; (3) plants can be regenerated from single cells, potentially resulting in lower frequencies of chimera formation; (4) individual transformed lines (transclones) can be cryopreserved while plants are regenerated and tested in containment; and (5) embryogenic suspensions provide a source of rapidly dividing cells which is a prerequisite for transformation (Sangwan et al., 1992; Iida et al., 1991).

GENETIC ENGINEERING IN CONIFERS

Genetic engineering enables the generation of genotypes with additional traits, or the modification of existing ones. New traits are usually introduced through overexpression of heterologous genes, whereas existing traits can be modified by suppression of endogenous genes. Genetic engineering is a relatively new technology for forest trees, and due to long growth cycles, only limited data are currently available characterizing the expression of novel genes over the life span of mature trees.

Genetically engineered conifers became a reality in the early 1990s (Huang et al., 1991), and since then many protocols have been published for a range of conifer species (Table 9.1). In this section Biolistic transformation technologies are briefly reviewed, and more recent developments with *Agrobacterium*-mediated transformation are included.

Biolistic Transformation

Artificial gene transfer technologies for plants involve various approaches to directly introduce genes in the form of pure DNA into plant cells. For example, electroporation of cells has been reported for a range of species (for example, maize: D'Halluin et al., 1992; rice: Battraw and Hall, 1992). Another direct DNA transfer method is Biolistics, whereby small (1 μm diameter) spherical bullets are coated with DNA and shot into target cells. These bullets are usually made of gold or tungsten, and DNA may integrate into the genome of the cells if they are in a competent physiological state and the physical conditions for delivery are appropriate for the species under investigation (Klein et al., 1987). Biolistic techniques have successfully been applied to conifer transformation followed by the regeneration of transgenic plants (Table 9.1).

Selection of transgenic lines (transclones) usually involves an antibiotic selection gene (for example *nptII* [neomycin phototransferase II] or *aphIV* [aminocyclitol phosphotransferase]) to differentiate transformed cells from nontransformed cells. In addition to antibiotic resistance, other genes of different function have also been transferred to conifers and expressed in transgenic tissue and plants. This includes the *uidA* reporter gene (for example, Walter et al., 1994; Walter, Grace, et al., 1998), genes involved in reproductive development (Mellerowicz, personal communication) and lignin formation (Wagner, unpublished), and genes for insect resistance (Grace, unpublished). The introduction and expression of a herbicide resistance gene into *P. radiata* and *Picea abies* (Bishop-Hurley et al., 2001) has successfully demonstrated the correct function of this gene in transgenic regenerated plants. Transgenic plants with the *bar* gene (De Block et al., 1987) integrated in their genome were resistant against operational doses of the herbicide Buster (the active ingredient is phosphinothricin).

Field trials of genetically modified conifers are in progress to investigate long-term gene expression, performance of engineered plants, and aspects of environmental risk and effects on nontarget organisms. Initial results from transgenic conifer trials confirm the continued expression of transgenes in the field over a longer period of time (Ellis, personal communication; Walter, unpublished). However, data from field trials with other transgenic

TABLE 9.1. Genetic Transformation of Conifers

Species	Explant	Method	Stage	Reference
Pinus radiata	Et	Biolistic	Plants	Walter et al., 1994; Walter, Carson, et al., 1998; Walter, Grace, et al., 1998
	Et	Biolistic	?	Wagner et al., 1997
	Et	At	Plants	Charity, unpublished
	Cotyledons	At	Plants	Charity et al., 2001
	Axillary shoots	At	Plants	Grant, personal communication
P. taeda	Et	At	Stable tissue	Wenck et al.,1999
	Et	Biolistic	Stable tissue	Grace, unpublished Tang et al., 2001
P. pinaster	Et	At	Plants	Trontin, personal communication
	Et	Biolistic	Plants	Trontin, personal communication
P. strobus	Et	At	Plants	Levee et al., 1999
Larix spp.	?	At	Plants	Levee et al., 1997
	Sh	Ar	Plants	Huang et al., 1991
	Sh	Ar	Plants	Shin et al., 1994
Picea sitchensis	Immature and mature embryos	At	?	Drake et al., 1997
P. glauca	Mature embryos	Biolistic	Plants	Ellis et al., 1993
	Et	At	Plants	Klimaszewska et al., 2001
P. mariana	Et	Biolistic	Plants	Charest et al., 1993
	Et	At	Plants	Klimaszewska et al., 2001
P. abies	Et	Biolistic	Plants	Walter et al., 1999
	Et	Biolistic	Plants	Clapham et al., 2000
	Et	At	Plants	Klimaszewska et al., 2001
	Et	At	Plants	Wenck et al., 1999

Abies nordmanniana	Et	Biolistic	Plants	Find, personal communication
	Et	At	Plants	Rahmat, personal communication

Ar = *Agrobacterium rhizogenes;* At = *Agrobacterium tumefaciens;* Et = embryogenic tissue; Sh = shoots; ? = not known.

tree species also suggest that complex integrations are present and that gene expression can be highly variable (see Chapter 12; Kumar and Fladung, 2000; Wagner, unpublished). Reversions to wild type, based on loss of the transgene in some transgenic plants, have also been reported (Fladung, 1999). The data from further trials are expected to contribute to a better understanding of gene expression in conifers and to provide the facts required for an informed debate on potential or perceived risks associated with genetically modified trees.

Agrobacterium-*Mediated Transformation of* Pinus Radiata

Recently there has been an increasing trend to develop transformation protocols for conifers using the soil bacterium *Agrobacterium tumefaciens* or *A. rhizogenes* as an alternative to Biolistic techniques (see also Table 9.1). The reasons underlying this trend are most likely related to the perceived or observed advantages, such as lower copy number, less fragmentation of the transgenes, and the precision of gene integration (Hadi et al., 1996; Pawlowski and Somers, 1996; Kumar and Fladung, 2000).

Huang and colleagues (1991) developed a transformation system for seedling hypocotyls of *Larix* spp. using *A. rhizogenes,* and transgenic plants were regenerated. Since then, other protocols for organogenic tissue of various conifer species have also been investigated. For example, a transformation protocol was developed for detached cotyledons of zygotic embryos of *P. radiata* using the reporter gene *uidA* and the selectable marker *nptII* (Holland et al., 1997). The frequency of gene transfer in detached cotyledons, as evidenced by histochemical staining of *uidA,* varied depending upon treatments such as wounding (Holland et al., 1997) and vacuum infiltration (Charity et al., 2001). Although success in developing an *Agrobacterium-*mediated transformation protocol for organogenic tissue has been difficult for most conifer species, Tang and colleagues (2001) recently demonstrated stable transformation of *P. taeda* using tissue derived from mature zygotic embryos.

The transformation of embryogenic tissue is an alternative to *Agrobacterium*-mediated transformation of organogenic tissue. Regeneration of transgenic plants via somatic embryogenesis after *Agrobacterium* transformation has been established in a number of conifers including *Larix kaempferi* × *L. decidua* (Levée et al., 1997), *P. strobus* (Levée et al., 1999), *Picea glauca, P. mariana,* and *P. abies* (Klimaszewska et al., 2001).

An *Agrobacterium*-mediated transformation protocol for radiata pine, using embryogenic cell suspensions, was developed, and 29 putative transgenic lines were produced. Molecular analysis including *nptII* ELISA (enzyme-linked immunoabsorbent assay) and Southern hybridization confirmed the transgenic nature of the lines (Charity, unpublished). Embryos and plantlets were regenerated from ten transclones, which were phenotypically similar to nontransformed plants. Functional analysis of the cotransformed *bar* gene was tested by painting in vitro plantlets with high concentrations of glufosinate—the active ingredient in the herbicide Basta. Only transgenic plants survived while all control plants either died or showed signs of browning.

MODIFICATION OF TRAITS BY SUPPRESSION OF GENE EXPRESSION

Investigation of gene function and modification of existing traits *in planta* are often accomplished by suppressing the transcription or translation process of a targeted endogenous gene. Recombinant vectors, containing sense, antisense or, more recently, RNAi constructs are generally used for this purpose, and a substantial amount of scientific data is available on the effectiveness of different constructs in suppressing endogenous genes for angiosperm species. However, to our best knowledge no scientific data are available on this subject for gymnosperm species.

As an example of genes involved in a major biochemical pathway, a series of those associated with the phenylpropanoid metabolism in *P. radiata* were isolated and used in stable transformation experiments to assess the effect of different constructs in suppressing the corresponding endogenous genes in this conifer.

Here we summarize data on parameters affecting the suppression level of *CAD,* a gene encoding for cinnamyl alcohol dehydrogenase, which is involved in the biosynthesis of monolignols in pine (MacKay et al., 1997). Approximately 80 *P. radiata* transclones containing either *CAD* sense or antisense constructs were generated using a Biolistic transformation process. In both constructs, *CAD* was controlled by the maize polyubiquitin promoter (Christensen et al., 1992). This promoter has previously been analyzed in transient Biolistic transformation experiments, in which it confirmed strong expression of the attached *uidA* reporter gene in both embryogenic and xylogenic *P. radiata* tissue (Wagner, unpublished). The effects of the different constructs on the endogenous CAD activity, as well as the development of *CAD* suppression levels over time and in different tissue types were monitored.

CAD measurements in all the transclones revealed that neither construct was able to substantially suppress the endogenous *CAD* gene in embryogenic tissue (Table 9.2). Conversely, 86 percent of all transclones containing a *CAD* sense construct revealed an overexpression phenotype at this developmental stage, indicating that they contained a functional copy of the transgene. Interestingly, the distribution of the CAD activity found in embryogenic tissue was considerably different in six-month-old regenerated transgenic seedlings, compared to embryogenic tissue. In seedlings, none of the transclones containing a *CAD* sense construct showed a significant overexpression phenotype (Table 9.2). Instead, transclones exhibiting substantial cosuppression of the endogenous *CAD* were identified. Interestingly, only transclones with a substantial overexpression phenotype at the embryogenic state also showed suppression at the seedling level, indicating that the expression level of the transgene is important in triggering suppression in *P. radiata*. Additional experiments revealed that the suppression level in transclones was strongest in tissue types with a high endogenous CAD activity such as developing xylem, which might indicate that the level of endogenous gene expression is also important for the establishment of

TABLE 9.2. Development of CAD activity in transgenic *P. radiata* lines containing CAD antisense or sense constructs, over plant development (embryo to seedling), with numbers of plants showing specific CAD activity indicated

Percent of wt CAD activity	Embryogenic tissue		Seedlings	
	Antisense	Sense	Antisense	Sense
> 200	–	21	–	–
150-200	–	6	–	–
100-150	6	9	1	5
50-100	30	6	7	2
10-50	1	–	1	2
0-10	–	–	–	1
Total	37	42	9	10

suppression mechanisms based on sense constructs in *P. radiata*. This could also partially explain why no substantial suppression of *CAD* in transclones containing a *CAD* sense construct was identified at the embryogenic state, in which the endogenous CAD activity is low.

The *CAD* antisense construct was less effective in suppressing the endogenous CAD activity at all developmental stages and in all tissue types tested, when compared to the sense construct. More experiments with different constructs are needed to investigate whether this is a general phenomenon or just a trend based on the constructs used in this study.

The dramatic differences in the *CAD* suppression levels in different tissue types, the changes of its suppression level over time, as well as the differences in effectiveness of the constructs used in this study clearly indicate more research is necessary to better understand factors affecting suppression mechanisms in gymnosperms.

ANALYSIS OF PROMOTERS CONTROLLING GENE EXPRESSION IN CONIFERS

Promoters are very attractive objects to study for several reasons: (1) linked to reporter genes, their expression pattern can be assessed; (2) they reveal information on potential *cis* active elements and therefore help to understand signal cascades involved in the regulation of the gene; (3) they are essential tools for the isolation of transcription factors involved in gene regulation; and (4) they play an important role in practical applications, e.g., by controlling the temporal and spatial expression patterns of transgenes.

Specific promoters of genes involved in different developmental processes, including the formation of male cones and xylogenesis, were isolated from genomic *P. radiata* DNA, based on cDNA sequence information. Further characterization was achieved by fusions to the *uidA* reporter gene and histochemical analysis of expression in transgenic model plants such as *Arabidopsis thaliana* (Höfig, personal communication) or *Nicotiana tabacum* (Moyle et al., 2002).

Promoters Associated with Male Cone Formation

Reproductive development in gymnosperms is of great scientific and commercial interest. However, only limited information is available on the genes and promoters involved and on homologies with sequences and reproductive pathways in angiosperms (Tandre et al., 1995; Mouradov et al., 1997; Mellerowicz et al., 1998). Special attention is focused on genes and promoters involved in male cone formation (Walden et al., 1999). Using differential screening, a series of different cDNAs specifically expressed in male cone tissues were isolated from radiata pine. Northern hybridization confirmed the expression of the respective genes in male cone tissue and the absence of expression in other tissue types. Analysis of DNA sequences and homologies between the sequences grouped one of the cDNAs isolated (PrCHS1) with plant chalcone synthetase genes and another sequence (PrLTP2) with a clade of lipid transfer protein genes. In situ expression analysis confirmed the expression of PrCHS1 mainly in the tapetum of developing male reproductive structures, whereas PrLTP2 was confirmed to express mainly in developing microspores (Walden et al., 1999).

This sequence information formed the basis for the isolation of the genomic upstream sequences of PrCHS1 and PrLTP2 which were subsequently fused to a *uidA* reporter gene. *Arabidopsis thaliana* plants were transformed with these constructs and heterozygous transgenic populations produced. Plants from the T3 generation were grown to maturity, and their *uidA* expression patterns were assessed. Microscopic studies applied a very sensitive ultra-darkfield technology that shows *uidA* expression in pink rather than blue. The results indicate that the PrCHS1 promoter directs expression toward the *Arabidopsis* tapetum and to microspores at an early stage in development (pollen mother cells) (Figure 9.1). As expected, the activity of the PrLTP2 promoter was also observed in these tissues; however, the onset

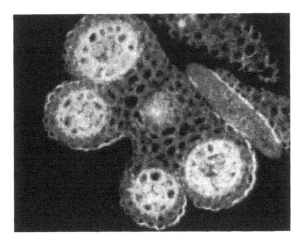

FIGURE 9.1. Expression of the *Pinus radiata* chalcone synthase (CHS) promoter fused to the *uidA* reporter gene, in developing microspores of transgenic *Arabidopsis*. In this dark-field microscopical image of transsection through an *Arabidopsis* anther, the β-glucuronidase (GUS) stain appears pink rather than blue. (See also color photo section.)

of expression was observed later, at the tetrad stage of microspore development.

The results of this study illustrate that sequences upstream of differentially expressed conifer genes are highly conserved with respect to their function in two such distant species as *A. thaliana* and *P. radiata*. These species are considered to be approximately 300 million years apart from each other in evolutionary terms.

TISSUE CULTURE PROTOCOLS
FOR THE EARLY ASSESSMENT OF GENE FUNCTION

The chemical composition and structure of the primary and/or secondary cell walls of the tracheids influence the wood properties of gymnosperms. It should be possible to modify these properties by genetic modification (Chapple and Carpita, 1998); however, the identification and assessment of promising candidate genes for desired modifications is potentially a difficult and time-consuming endeavor. This is particularly true for conifer species, in which the production of transgenic trees and their analysis can take years to complete.

Although this functional analysis of genetic information remains a cumbersome task, an ever-increasing amount of DNA sequence information is becoming available in electronic databases (Sommerville and Sommerville, 1999). For example, the first genome of a higher plant, *Arabidopsis thaliana,* has recently been completely sequenced (Arabidopsis Genome Initiative, 2000). More than 15 percent of the 25,498 *Arabidopsis* genes are potentially involved in the synthesis or modification of cell walls (Carpita et al., 2001). In the area of tree genomics, projects aimed at the production of expressed sequence tags (ESTs) from wood-forming tissues of economically important tree species, such as loblolly pine (*P. taeda* L.) or hybrid aspen (*Populus tremula* L. × *tremuloides* Michx.) have been initiated (Allona et al., 1998; Sterky et al., 1998; <http://web.ahc.umn.edu/biodata/doepine/>; <http://web.ahc.umn.edu/biodata/nsfpine/>).

The potential function of a DNA sequence can often be deduced from existing databases; however, experimental confirmation of gene function is required in most cases (Bouchez and Höfte, 1998; Sommerville and Sommerville, 1999). Overexpression or suppression of a candidate gene in transgenic plants and the analysis of the resulting phenotype may help to reveal its exact function. Furthermore, spatial and temporal expression patterns assessed by in situ hybridization can provide additional insights.

Some information about the function of genes of gymnosperms can be gained from differential screening of cDNA databases created from the zone of differentiating wood and from a dormant cambial zone. These sequences could be analyzed using computer programs, and subsequently, the candidate genes would have to be cloned and introduced into plants to experimentally confirm their function. However, the creation of a transgenic plant is a time consuming task, and to assign functions to a large number of genes would be uneconomical.

In an attempt to address functional gene analysis at high throughput and in economically viable time frames, an in vitro system for early assay of gene function using parenchymatous callus cultures of *P. radiata* has been developed. Callus cultures were induced from xylem strips of one- to four-year-old radiata pine and from hypocotyls of eight-day-old radiata pine seedlings. The callus cells formed primary cell walls, and differentiation was induced by the transfer of cells to induction media, supplemented with various concentrations of auxins and cytokinins. After ten and twenty days the callus cultures were assessed for the presence of tracheary elements and for the presence of cells developing a lignified secondary cell wall. A significant induction of cell differentiation was noted on induction medium. Within twenty days after cell transfer, an average of 10 to 20 percent of the xylem callus cells formed a reticulately thickened and lignified secondary cell wall (Figure 9.2). These cells differentiated as single cells or in nodules of up to hundreds of cells. They resembled parenchyma cells that differentiated in response to wounding (Kuroda and Shimaji, 1984). Differential interference contrast microscopy, confocal microscopy, and electron microscopy did not detect any bordered pits. Using microanalysis techniques, cell wall properties were evaluated.

After transfer of cells from the hypocotyl-derived calli to an induction medium, tracheary elements differentiated. They showed helical, scalariform, or reticulated secondary cell wall thickenings with integrated bordered pits (Figure 9.3A). Some tracheary elements displayed a completely closed secondary wall with integrated bordered pits (Figure 9.3B).

In an additional experiment, prior to the induction of secondary cell wall formation, cells of the xylem calli and the hypocotyl calli were genetically transformed using Biolistics. Transient expression of the *uidA* gene has been observed in cell lines three days after bombardment, and preliminary results indicate stable transformation. The investigation of chemical composition and structure of the cell walls using microanalysis techniques may enable the assessment of gene function at high throughput and in a comparatively short time period. In addition to studying aspects of primary and sec-

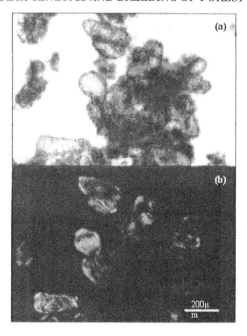

FIGURE 9.2. (a) Light micrograph showing cells cultured on induction media viewed in bright field; (b) light micrograph showing the same cells, but in polarized light. Cell walls of some cells are birefringent, indicating deposition of cell wall material in a crystalline order.

ondary cell wall formation, the system might also prove useful to study the function of genes involved in defense mechanisms, cell division, cell growth, or starch biosynthesis.

However, the value of this tissue culture system as a research tool for developmental biology of wood formation in a mature tree still needs further evaluation. Research needs to correlate characteristics of cell walls in culture with traits displayed in trees at a later age. The system might prove useful to test larger numbers of candidate genes in tissue culture and subsequently continue with the analysis of a much smaller number of promising genes in transgenic plants.

To fully capture the advantages of such a secondary cell wall producing tissue culture technology, specific microanalysis tools were developed to assess the chemical, biochemical, anatomical, and physical properties of cell walls.

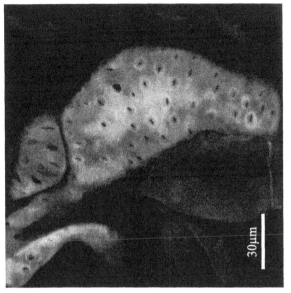

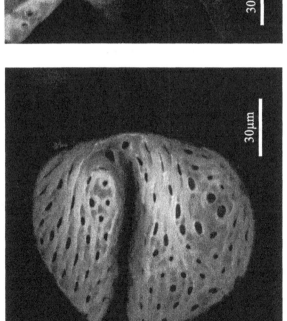

FIGURE 9.3. Confocal light scanning micrograph of tracheary elements: (A) secondary cell wall with scalariform-reticulate pattern and bordered pits; (B) tracheary element with secondary cell wall and bordered pits

229

TOOLS TO ASSESS GENE FUNCTION
AT THE MICRO LEVEL

The analysis of gene function by evaluation of phenotype in transformed xylogenic cell lines or in young seedlings presents unique problems in comparison to the evaluation of large mature trees—the most obvious of which is the small amount of material available for analysis. The juvenile features of young trees also require knowledge of how features in juvenile wood are related (or not related) to the properties of mature wood. Special techniques have been and are continuing to be developed to deal with small sample quantity and to relate the features of juvenile wood to the expectation in a mature crop.

Chemical Analysis

Wood from conifers is a cellular lignocellulosic material consisting of cellulose microfibrils embedded in an acetylglucomannan, 4-*O*-methyl-glucuronoarabinoxylan and lignin matrix (McDonald and Donaldson, 2001). The composition of wood varies at all levels, from tree to tree, from species to species, among cell types, and within the cell wall itself, and is governed by the expression of specific genes that control macromolecular synthesis and are triggered by internal and external factors. The chemical composition of wood is highly variable, and this variability has a significant effect on wood properties (Fengal and Wegner, 1989). Targeted breeding or genetic engineering offers the possibility to alter or control wood's chemical composition and quality by introducing desired genes into the genome. In vitro methodology for conifer plant regeneration from breeding programs or genetically transformed cells offers a rapid method for the generation of tissue for wood-quality screening purposes, which can be analyzed chemically for specific traits.

Because only small amounts of tissue are generated from culture systems, microanalytical techniques are required to determine the chemical composition of the in vitro developed cell walls and to quantify any changes. In addition, rapid and economical methods are required for analyzing large sample sets in reasonable time frames required as a breeding-screening tool. The analytical techniques are targeted at the main polymeric constituents such as lignin and polysaccharides.

Generally, lignin is made up of monolignol precursors, which are *trans* p-coumaryl, coniferyl, and sinapyl alcohols, while in conifers lignin is made up primarily of coniferyl alcohol that undergoes dehydrogenative polymerization to form macromolecular lignin (Fengal and Wegner, 1989; see Chapter 7). To establish the composition of lignin and to estimate content, pyroly-sis gas chromatography-mass spectrometry (GC-MS) offers a simple and reliable technique at the 1-100 µg level in wood samples, and with minimal sample preparation (Fullerton and Franich, 1983; McDonald et al., 2000; Meier and Faix, 1992; Ralph and Hatfield, 1991). Pyrolysis is a high-tem-perature treatment (500 to 600°C) which thermally cracks nonvolatile wood (polymeric components) into a complex mixture of low molecular weight volatile degradation products that can be analyzed by GC-MS. Conifer lignin degradation results in characteristic products, mainly guaiacol and phenol derivatives. This technique produces degradation products, and only limited information can be gained on the actual structure of the lignin pres-ent in the cell wall. However, information about the type of units that make up the lignin can provide evidence of changes resulting from the expression of transgenes or natural variation.

Pyrolysis GC-MS was employed to identify the presence of lignin in induced *P. radiata* callus tissue and softwood lignin pyrolysis products (guaiacol, methyl guaiacol, vinyl guaiacol, eugenol, *cis*-isoeugenol, and *trans*-isoeugenol) together with phenol and methyl phenol were found (Figure 9.4). This suggests that the induction of cell differentiation had occurred, giving rise to lignified secondary cell walls. In contrast, unin-duced callus tissue showed only phenol and methyl phenol in the pyrolysis chromatogram, which probably arose from cell protein. As an indication of softwood lignin content, the phenol to guaiacol ratio can be used qualita-tively. The guaiacol/phenol ratios for callus tissue, differentiated tissue, and isolated lignin (Figure 9.4c) were 0.05, 1.1, and 3.3, respectively.

Solid state ^{13}C cross-polarization magic angle spinning (CP-MAS) NMR (nuclear magnetic resonance) spectroscopy is a nondestructive technique for the chemical analysis of wood with a required sample size of 100 mg. For wood (Figure 9.5c) the ^{13}C NMR signals at δ-104 and 107 (C-1), 74 and 77 (C-2,-3,-5), 86 and 91 (C-4), and 64 (C-6) are readily assigned to that of cellulose (crystalline and amorphous) and hemicellulose (Gil and Pascoal Neto, 1999; Jarvis et al., 1996). In addition, the signal at δ-66 is attributable to C-5 of xylan, and the signals at δ-174 and 23 are associated with an *O*-acetyl group. Softwood lignin can be easily identified by ^{13}C NMR which gives characteristic signals for guaiacyl (G) units (Figure 9.5c) at about δ-150 (C-3, -5), δ-145 (C-4), 134 (C-1), 110-118 (C-2,-6), and 56 (OCH$_3$) (Sosanwo et al., 1995; Leary and Newman, 1992; Gil and Pascoal Neto, 1999). In ad-dition to the identity of the major cell wall components in the sample, an es-timate of lignin content can be obtained by integrating the guaiacyl C-3 and C-4 signals between δ-140 and 160 relative to the whole spectrum (Leary and Newman, 1992).

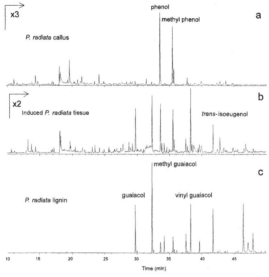

FIGURE 9.4. Pyrolysis GC-MS chromatograms of *Pinus radiata* (a) callus tissue, (b) induced callus tissue, and (c) milled wood lignin

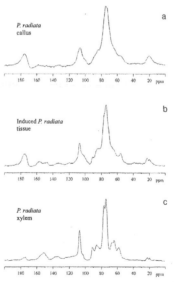

FIGURE 9.5. ^{13}C CP-MAS NMR spectra of *Pinus radiata* (a) callus tissue, (b) induced callus tissue, and (c) xylem

The ^{13}C CP-MAS NMR spectrum of *P. radiata* callus tissue (Figure 9.5a) shows signals attributable to galacturonan pectic polysaccharide at δ-175 (C-6), 101 (C-1), 82 (C-4), and 54 (methoxyl) and those of cellulose (crystalline and amorphous). Furthermore, the signal at δ-111 was assignable to C-1 of an α-linked arabinan (Carnachan, 2001). Additional signals to those of primary wall polysaccharides at δ-150 (C-3, -5), 145 (C-4), 134 (C-1), 110-118 (C-2, -6), and 56 (OCH$_3$) are assignable to lignin that were observed in the induced callus tissue sample (Figure 9.5b), which suggests that the tissue was differentiated and having a lignified secondary wall. These results support those obtained from pyrolysis GC-MS that the induced callus tissue was lignified.

The pyrolysis GC-MS and ^{13}C CP-MAS NMR techniques outlined in this chapter were selected based on ease of sample preparation with minimal wet chemistry to give a rapid assessment of chemical content/composition on woody tissue as a first screen. In our laboratory, these techniques have been successfully employed to screen the extent of lignification in differentiated callus tissue. However, there are some drawbacks to being semiquantitative. Other techniques such as carbohydrate analysis, Klason lignin content, and thioacidolysis are more selective in obtaining detailed chemical composition and can be used at the final screening stage.

Microscopic

Although chemical analysis can describe in great detail many aspects of an altered cell wall "phenotype," it is often useful and sometimes essential to characterize the same material microscopically. Transformational changes that lead to altered chemistry can also lead to, or be the result of, altered anatomy and/or ultrastructure. In the case of lignin content, for example, important anatomical considerations in understanding the exact nature of the altered phenotype in a transclone or transgenic plant exist. Consider the case of a reduction in lignin content in transgenic trees with secondary xylem. Reduced lignin content can potentially be achieved in three ways: (1) reduced lignification of the cell wall; (2) an increase in the thickness of the secondary cell wall, leading to dilution of the high lignin concentration in the middle lamella region (this leads to a moderate correlation between lignin content and basic density: correlation coefficient of 0.75); (3) an increase in nonlignified tissues within the xylem such as rays or resin canals. Microscopic analysis can thus lead to an understanding of the exact nature of the phenotype when considering altered composition in quantitative terms, while chemistry is well suited to defining the phenotype in qualitative terms (altered chemical structure).

Confocal Microscopy

Confocal laser scanning microscopy (CLSM) allows the evaluation of transgenic cell cultures or secondary xylem in terms of altered morphology and topochemistry with relatively high resolution and often very simple sample preparation allowing the evaluation of large numbers of samples (Knebel and Schnepf, 1991).

Confocal microscopy has also been used to provide images for the automated analysis of cell dimensions (Figure 9.6) (Donaldson and Lausberg, 1998; Moëll and Donaldson, 2001). The advantages are simple preparation (no embedding) and high quality, high contrast optical sections from thick microtome sections that can be processed automatically without operator intervention, thus eliminating bias. Other technologies, such as SilviScan, are more suited to rapid analysis of wood properties from large trees in that they are designed to work with increment cores (nondestructive) or strips cut from disks (destructive) (Evans, 1994).

Transmission Electron Microscopy (TEM)

Electron microscopy allows the imaging of ultrastructural features in cell walls from either cell cultures or from secondary xylem. The advantage of this technique is its very high resolution and dynamic range from single cells down to the macromolecule level. Although specimen preparation is

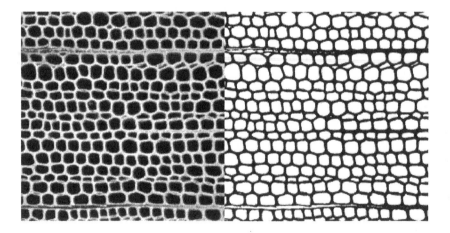

FIGURE 9.6. A confocal image of radiata pine secondary xylem used for cell dimension assessment and its binary segmentation

much more involved and time consuming than other methods, the information provided can be very valuable when applied to a small set of interesting samples identified by other techniques. In conifers, TEM offers high-resolution information on lignin distribution that can confirm observations by, for example, confocal microscopy while providing more detailed information (Donaldson, 2001). The use of $KMnO_4$ as a lignin stain for guaiacyl lignin (Bland et al., 1971) has provided a wealth of information on the structural organization of tracheid cell walls and on the lignification process (Donaldson, personal communication). Unfortunately, this technique does not work well with angiosperms because of the predominance of syringyl lignins in fiber secondary walls. Work on hardwood lignin distribution by TEM has involved the use of specific antibody probes to a variety of chemical features of lignin to evaluate transgenic phenotypes (Ruel et al., 1999).

CONCLUSIONS

Genetic modification of crops including forest trees is a new technology that will most probably have a significant influence on silvicultural practice in the future and that will change specific quality traits in plantation forests. Food crops with genetic modifications are already covering agricultural lands of substantial size, and forest trees will most likely experience a similar destiny. The most obvious advantage of this new technology will be the availability of choice for our society: the choice to leave natural forests alone while satisfying the ever-increasing requirement for fiber and solid wood products.

In this chapter, we have summarized current technologies used to genetically transform conifers and have given examples of successful expression of new genes in forest trees. Microanalysis tools including chemical, biochemical, and ultrastructural methods will help to characterize modified varieties and to evaluate the effect of specific genes in the future. The challenge for our society will be in finding the right balance between risk management and the economical use of genetically modified material. If the right decisions are made, based on solid scientific results rather than emotion, genetic engineering will deliver on the positive side and will provide us with fiber and solid wood products of improved quality, produced in an environmentally sustainable manner.

REFERENCES

Allona, I., Quinn, M., Shoop, E., Swope, K., St. Cyr, S., Carlis, J., Riedl, J., Retzel, E., Campbell, M.M., Sederoff, R.R., and Whetten, R.W. (1998). Analysis of xylem formation in pine by cDNA sequencing. *Proc. Natl. Acad. Sci. USA* 95: 9693-9698.

Arabidopsis Genome Initiative (2000). Analysis of the genome sequence of the flowering plant *Arabidopsis thaliana. Nature* 408: 796-815.

Attree, S.M., Tautorus, T.E., Dunstan, D.I., and Fowke, L.C. (1990). Somatic embryo maturation, germination and soil establishment of plants of black and white spruce (*Picea mariana* and *Picea glauca*). *Can. J. Botany* 68: 2583-2589.

Battraw, M. and Hall, T.C. (1992). Expression of a chimeric neomycin-phosphotransferase II gene in first and second generation transgenic rice plants. *Plant Science* 86: 191-202.

Becwar, M.R. (1993). Conifer somatic embryogenesis and clonal forestry. In M.R. Ahuja and W.J. Libby (Eds.), *Clonal Forestry I Genetics and Biotechnology I* (pp. 200-203). Berlin, Heidelberg: Springer-Verlag.

Becwar, M.R., Nagmani, R., and Wann, S.R. (1990). Initiation of embryogenic cultures and somatic embryo development in loblolly pine *(Pinus taeda). Can. J. Forest Res.* 20: 810.

Bishop-Hurley, S.L.R., Zabkievicz, J., Grace, L.J., Gardner, R.C., and Walter, C. (2001). Conifer genetic engineering: Transgenic *Pinus radiata* (D Don) and *Picea abies* (Karst) plants are resistant to the herbicide Buster. *Plant Cell Rep.* 20: 235-243.

Bland, D.E., Foster, R.C., and Logan, A.F. (1971). The mechanism of permanganate and osmium tetroxide fixation and distribution of lignin in the cell wall of *Pinus radiata. Holzforschung* 25: 137-143.

Bouchez, D. and Höfte, H. (1998). Functional genomics in plants. *Plant Physiology* 118: 725-732.

Burdon, R.D., Firth, A., Low, C.B., and Miller, M.A. (1997). Native provenances of *Pinus radiata* in New Zealand: Performance and potential. *New Zealand Forestry* 41: 32-36.

Carnachan, S.M. (2001). Structural organization and compositions of the cell walls of selected gymnosperms. PhD Thesis, University of Auckland, Auckland, New Zealand.

Carpita, N.C., Tierney, M., and Campbell, M.M. (2001). Molecular biology of the plant cell wall: Searching for the genes that define structure, architecture and dynamics. *Plant Mol. Biol.* 47: 1-5.

Carson, M.J., Walter, C., Grace, L.J., Menzies, M.I., Richardson, T., Carson, S.D., Burdon, R.D., and Gardner, R.C. (1996). Genetic modification of forest trees in New Zealand—Science and public perception. *Proceedings of the Royal Society of New Zealand Conference "Gene technology: Benefits and risks"* (pp. 29-36). Wellington: Royal Society of New Zealand.

Chapple, C. and Carpita, N. (1998). Plant cell walls as targets for biotechnology. *Curr. Opin. Plant. Biol.* 1: 179-185.

Charest, P.J., Calero, N., Lachance, D., Datla, R.S.S., Duchesne, L.C., and Tsang, E.W.T. (1993). Microprojectile-DNA delivery in conifer species: Factors affecting assessment of transient gene expression using the ß-glucuronidase reporter gene. *Plant Cell Rep.* 12: 189-193.

Charity, J.A., Holland, L., Donaldson, S.S., Grace, L.J., and Walter, C. (2001). *Agrobacterium*-mediated transformation of *Pinus radiata* organogenic tissue using vacuum infiltration. *Plant Cell Tiss. Org. Cult.* 70: 51-60.

Christensen, A.H., Sharrock, R.A., and Quail, P.H. (1992). Maize polyubiquitin genes: Structure, thermal perturbation of expression and transcript splicing, and promoter activity following transfer to protoplast by electroporation. *Plant Mol. Biol.* 18: 675-689.

Clapham, D., Demel, P., Elfstrand, M., Koop, H-U., Sabala, I., and van Arnold, S. (2000). Gene transfer by particle bombardment of embryogenic cultures of *Picea abies* and the production of transgenic plantlets. *Scand. J. Forest. Res.* 15: 151-160.

De Block, M., Botterman, J., Vandewiele, M., Dockx, J., Thoen, C., Gosselé, V., Rao Movva, N., Thompson, C., van Montagu, M., and Leemans, J. (1987). Engineering herbicide resistance in plants by expression of a detoxifying enzyme. *EMBO J.* 6: 2513-2518.

D'Halluin, K., Bonne, E., Bossut, M., De Beuckeleer, M., and Leemans, J. (1992). Transgenic maize plants by tissue electroporation. *Plant Cell* 4: 1495-1505.

Donaldson, L.A. (2001). Lignification and lignin topochemistry—An ultrastructural view. *Phytochemistry* 57: 859-873.

Donaldson, L.A. and Lausberg, M.J.F. (1998). Comparison of conventional transmitted light and confocal microscopy for measuring wood cell dimensions by image analysis. *IAWA J.* 19: 321-336.

Drake, M.W., John, A., Power, J.B., and Davey, M.R. (1997). Expression of the gus A gene in embryogenic cell lines of Sitka spruce following *Agrobacterium*-mediated transformation. *J. Exp. Bot.* 48: 151-155.

Durzan, D.J. and Gupta, P.K. (1987). Somatic embryogenesis and polyembryogenesis in Douglas-fir cell suspension cultures. *Plant Science* 52: 229-235.

Ellis, D.D., McCabe, D.E., McInnis, S., Ramachandran, R., Russell, D.R., Wallace, K.M., Martinell, B.J., Roberts, D.R., Raffa, K.F., and McCown, B.H. (1993). Stable transformation of *Picea glauca* by particle acceleration. *Nat. Biotechnol.* 11: 84-89.

Evans, R. (1994). Rapid measurement of the transverse dimensions of tracheids in radial wood sections from *Pinus radiata*. *Holzforschung* 48: 168-172.

Fengal, D. and Wegner, G. (1989). *Wood: Chemistry Ultrastructure, Reactions.* Berlin: De Gruyter.

Fladung, M. (1999). Gene stability in transgenic aspen *(Populus)*. I. Flanking DNA sequences and T-DNA structure. *Mol. Gen. Genet.* 260: 574-581.

Fullerton, T.J. and Franich, R.A. (1983). Lignin analysis by pyrolysis-GC-MS. *Holzforschung* 37: 267-269.

Garin, E., Isabel, N., and Plourde, A. (1998). Screening of large numbers of seed families of *Pinus strobus* L. for somatic embryogenesis from immature and mature zygotic embryos. *Plant Cell. Rep.* 18: 37-43.

Gil, A.M. and Pascoal Neto, C. (1999). Solid state NMR studies of wood and other lignocellulosic materials. *Ann. R. NMR S.* 37: 75-117.

Grossnickle, S.C., Cry, D.R., and Polonenko, D.R. (1997). Somatic embryogenesis tissue culture for the propagation of conifer seedlings: A technology comes of age. *Tree Planters Notes* 47: 48-57.

Gupta, P.K. and Durzan, D.J. (1986). Somatic polyembryogenesis from callus of mature sugar pine embryos. *BioTechnology* 4: 643-645.

Hadi, M.Z., McMullen, M.D., and Finer, J.J. (1996). Transformation of 12 different plasmids into soybean via particle bombardment. *Plant Cell. Rep.* 15: 500-505.

Haggman, H., Jokela, A., Krajnakova, J., Kauppi, A., Niemi, K., and Aronen, T. (1991). Somatic embryogenesis of Scots pine: Cold treatment and characteristics of explants affecting induction. *J. Exp. Bot.* 50 (341) 1769-1778.

Holland, L., Gemmell, J.E., Charity, J.A., and Walter, C. (1997). Foreign gene transfer into *Pinus radiata* cotyledons by *Agrobacterium tumefaciens*. *New Zeal. J. For. Sci.* 27: 289-304.

Huang, Y., Diner, A.M., and Karnosky, D.F. (1991). *Agrobacterium rhizogenes* mediated genetic transformation and regeneration of a conifer: *Larix decidua. In Vitro Cell Dev. Biol.* 27: 201-207.

Iida, A., Yamashita, T., Yamada, Y., and Morikawa, H. (1991). Efficiency of particle-bombardment-mediated transformation is influenced by cell stage in synchronised cultured cells of tobacco. *Plant Physiology* 97: 1585-1587.

Jain, S.M., Dong, N., and Newton, R.J. (1989). Somatic embryogenesis in slash pine *(Pinus elliottii)* from immature embryos cultured in vitro. *Plant Science* 65: 233-241.

James, C. (2000). Global status of commercialized transgenic crops: 2000. *ISAAA Briefs No. 21.* Ithaca, NY: ISAAA.

Jarvis, M.C., Fenwick, K.M., and Apperley, D.C. (1996). Cross-polarization kinetics and proton NMR relaxation in polymers of *Citrus* cell walls. *Carbohyd. Res.* 288: 1-14.

Jones, N.B. and van Staden, J. (1995). Plantlet production from somatic embryos of *Pinus patula. J. Plant Phys.* 145: 519-525.

Jouanin, L., Brasileiro, A.C.M., Leple, J.C., Pilate, G., and Cornu, D. (1993). Genetic transformation: A short review of methods and their applications, results and perspectives for forest trees. *Ann. Sci. For.* 50: 325-336.

Kim, Y.W., Youn, Y., Noh, E.R., and Kim, J.C. (1999). Somatic embryogenesis and plant regeneration from immature zygotic embryos of Japanese larch *(Larix leptolepis). Plant Cell Tiss. Org. Cult.* 55: 95-101.

Klein, T.M., Wolf, E.D., Wu, R., and Sanford, J.C. (1987). High-velocity microprojectiles for delivering nucleic acids into living cells. *Nature* 327: 70-73.

Klimaszewska, K., Lachance, D., Pelletier, G., Lelu, A.M., and Seguin, A. (2001). Regeneration of transgenic *Picea glauca, P. mariana,* and *P. abies* after cocultivation of embryogenic tissue with *Agrobacterium tumefaciens. In Vitro Cell Dev. Biol.* 37: 748-755.

Klimaszewska, K. and Smith, D.R. (1997). Maturation of somatic embryos of *Pinus strobus* is promoted by a high concentration of gellan gum. *Physiol. Plant.* 100: 949-957.

Knebel, W. and Schnepf, E. (1991). Confocal laser scanning microscopy of fluor-escently stained wood cells: A new method for three-dimensional imaging of xylem elements. *Trees* 5: 1-4.

Krattiger, A.F. (1998). The importance of ag-biotechnology for global prosperity. *ISAAA Briefs No. 6.* Ithaca, NY: ISAAA.

Kumar, S., Carson, S.D., and Garrick, D.J. (2000). Detecting linkage between a fully informative marker locus and a trait locus in outbred populations using analysis of variance. *Forest Genetics* 7: 47-56.

Kumar, S. and Fladung, M. (2000). Determination of transgene repeat formation and promoter methylation in transgenic plants. *Biotechniques* 28: 1128-1134.

Kuroda, K. and Shimaji, K. (1984). Wound effects on xylem cell differentiation in a conifer. *IAWA Bulletin* 5: 295-305.

Leary, G.J. and Newman, R.H. (1992). Cross polarization/magic angle spinning nuclear magnetic resonance (CP/MAS NMR) spectroscopy. In S.Y. Lin and C.W. Dence (Eds.), *Methods in Lignin Chemistry* (pp. 147-161). *Springer Series in Wood Science*. Berlin: Springer-Verlag.

Lelu, M.A., Bastien, C., Drugeault, A., Gouez, M.L., and Klimaszewska, K. (1999). Somatic embryogenesis and plantlet development in *Pinus sylvestris* and *Pinus pinaster* on medium with and without growth regulators. *Physiol. Plant.* 105: 719-728.

Levée, V., Garin, E., Klimaszewska, K., and Seguin, A. (1999). Stable genetic transformation of white pine (*Pinus strobus* L.) after cocultivation of embryogenic tissues with *Agrobacterium tumefaciens. Mol. Breeding* 5: 429-440.

Levée, V., Jouanin, L., Dornu, D., and Pilate, G. (1997). *Agrobacterium tumefaciens* mediated transformation of hybrid larch (*Larix kaempferi* x *L. decidua*) and transgenic plant regeneration. *Plant Cell Rep.* 16: 680-685.

Li, X.Y., Huang, F.H., and Gbur, E.E. Jr. (1998). Effect of basal medium, growth regulators and phytagel concentration on initiation of embryogenic cultures from immature zygotic embryos of loblolly pine (*Pinus taeda* L.). *Plant Cell Rep.* 17: 298-301.

MacKay, J.J., O'Malley, D.M., Presnell, T., Booker, F.L., Campbell, M.M., Whetten, R.W., and Sederoff, R.R. (1997). Inheritance, gene expression, and lignin characterisation in a mutant pine deficient in cinnamyl alcohol dehydrogenase. *Proc. Natl. Acad. Sci. USA* 94: 8255-8260.

MacRae, S. and van Staden, J. (1999). Transgenic eucalyptus. In Y.P.S. Bajaj (Ed.), *Biotechnology in Agriculture and Forestry* 44 (pp. 88-112). Berlin, Heidelberg: Springer-Verlag.

McCown, B.H., McCabe, D.E., Russell, D.R., Robinson, D.J., Barton, K.A., and Raffa, K.F. (1991). Stable transformation of *Populus* and incorporation of pest resistance by electric discharge particle acceleration. *Plant Cell Rep.* 9: 590-594.

McDonald, A.G. and Donaldson, L.A. (2001). Constituents of wood. *The Encyclopedia of Materials: Science and Technology* (pp. 9612-9616). Dordrecht, The Netherlands: Elsevier Science.

McDonald, A.G., Fernandez, M., Kreber, B., and Laytner, F. (2000). The chemical nature of kiln brown stain in radiata pine. *Holzforschung* 54: 12-22.

Meier, D. and Faix, O. (1992). Pyrolysis-gas chromatography-mass spectrometry. In S.Y. Lin and C.W. Dence (Eds.), *Methods in Lignin Chemistry* (pp. 177-199). *Springer Series in Wood Science*. Berlin: Springer-Verlag.

Mellerowicz, E., Horgan, K., Walden, A.R., Coker, A., and Walter, C. (1998). PRFLL-a *Pinus radiata* homologue of *FLORICAULA* and *LEAFY* is expressed in buds containing vegetative shoot and undifferentiated male cone primordia. *Planta* 206: 619-629.

Moëll, M. and Donaldson, L.A. (2001). Comparison of segmentation methods for digital image analysis of confocal microscope images to measure tracheid cell dimensions. *IAWA J.* 22: 267-288.

Mouradov, A., Glassik, T., and Teasdale, R. (1997). Isolation and characterisation of a FLORICAULA/LEAFY-like cDNA from *Pinus radiata*. *Plant Physiology* 113: 664-665.

Moyle, R., Moody, J., Phillips, L., Walter, C., and Wagner, A. (2002). Isolation and characterisation of a *Pinus radiata* lignin biosynthesis related *O*-methyltransferase promoter. *Plant Cell Rep.* 20: 1052-1060.

Nagmani, R. and Bonga, J.M. (1985). Embryogenesis in subcultured callus of *Larix decidua*. *Can. J. Forest Res.* 15: 1088-1091.

Pandy, D. (1995). Forest Resources Assessment 1990. Tropical Forest Plantation Resources. *FAO Forestry Paper* 128. ISBN 9251037302.

Pawlowski, W.P. and Somers, D.A. (1996). Transgene inheritance in plants genetically engineered by microprojectile bombardment. *Mol. Biotechnol.* 6: 17-30.

Ralph, J. and Hatfield, R.D. (1991). Pyrolysis-GC-MS characterisation of forage materials. *J. Agr. Food. Chem.* 39: 1426-1437.

Roberts, D.R., Flinn, B.S., Webb, D.T., Webster, F.B., and Sutton, B.C.S. (1989). Characterisation of immature embryos of interior spruce by SDS-PAGE and microscopy in relation to their competence for somatic embryogenesis. *Plant Cell Rep.* 8: 285-288.

Ruel, K., Burlat, V., and Joseleau, J.P. (1999). Relationship between ultrastructural topochemistry of lignin and wood properties. *IAWA J.* 20: 203-211.

Salajova, T., Jasik, J., Kormutak, A., Salaj, J., and Hakman, I. (1996). Embryogenic culture initiation and somatic embryo development in hybrid firs (*Abies alba* x *Abies cephalonica*, and *Abies alba* x *Abies numidica*). *Plant Cell Rep.* 15: 527-530.

Sangwan, R.S., Bourgeois, Y., Brown, S., Vasseur, G., and Sangwan-Norrell, B. (1992). Characterisation of competent cells and early events of *Agrobacterium*-mediated genetic transformation in *Arabidopsis thaliana*. *Planta* 188: 439-456.

Schuller, A., Reuther, G., and Geier, T. (1989). Somatic embryogenesis from seed explants of *Abies alba*. *Plant Cell Tiss Org. Cult.* 17: 53-58.

Shin, D-I., Podila, G.K., Huang, Y., and Karnosky, D.F. (1994). Transgenic larch expressing genes for herbicide and insect resistance. *Can. J. For. Resistance.* 24: 2059-2067.

Smith, D.R. (1996). Growth Medium. U.S. Patent 08-219879.

Somerville, C. and Somerville, S. (1999). Plant functional genomics. *Science* 285: 380-383.

Sosanwo, O.A., Fawcett, A.H., and Apperley, D. (1995). ^{13}C CP-MAS NMR spectra of tropical hardwoods. *Polym. Int.* 36: 247-259.

Sterky, F., Reagan, S., Karlsson, J., Hertzberg, M., Rohde, A., Holmberg, A., Armini, B., Bhalerao, R., Larsson, M., Villarroel, R., et al. (1998). Gene discovery in the wood-forming tissues of poplar: Analysis of 5,692 expressed sequence tags. *Proc. Natl. Acad. Sci. USA* 95: 13330-13335.

Sutton, B.C.S., Grossnickle, S.C., Roberts, D.R., Russell, J.H., and Kiss, G.K. (1993). Somatic embryogenesis and tree improvement in interior spruce. *Forestry* 91: 34-38.

Tandre, K., Albert, V.A., Sundas, A., and Engstrom, P. (1995). Conifer homologues to genes that control floral development in angiosperms. *Plant Mol. Biol.* 27: 69-78.

Tang, W., Tian, Y., Ouyang, F., and Guo, Z. (2001). *Agrobacterium tumefaciens* mediated transformation of loblolly pine and transgenic plant regeneration. *Planta* 213: 981-989.

Timmis, R. (1998). Bioprocessing for tree production in the forest industry: Conifer somatic embryogenesis. *Biotechnol. Prog.* 14: 156-166.

von Aderkas, P., Klimaszewska, K., and Bonga, J.M. (1990). Diploid and haploid embryogenesis in *Larix leptolepsis, L. decidua,* and their reciprocal hybrids. *Can. J. Forest. Res.* 20: 9-14.

Wagner, A., Moody, J., Grace, L.J., and Walter, C. (1997). Stable transformation of *Pinus radiata* based on selection with Hygromycin B. *New Zeal. J. Forest. Sci.* 27: 280-288.

Walden, A.R., Walter, C., and Gardner, R.C. (1999). Genes expressed in *Pinus radiata* male cones include homologues to anther-specific and pathogenesis response genes. *Plant Physiology* 121: 1103-1108.

Walter, C., Carson, S.D., Menzies, M.I., Richardson, T., and Carson, M. (1998). Review: Application of biotechnology to forestry—Molecular biology of conifers. *World J. Microb. Biot.* 14: 321-330.

Walter, C. and Grace, L.J. (2000). Genetic engineering of conifers for plantation forestry: *Pinus radiata* transformation. In S.M. Jain and S.C. Minocha (Eds.), *Molecular Biology of Woody Plants,* Volume 2 (pp. 79-104). Dordrecht, The Netherlands: Kluwer.

Walter, C., Grace, L.J., Donaldson, S.S., Moody, J., Gemmell, J.E., van der Maas, S., Kwaalen, H., and Loenneborg, A. (1999). An efficient biolistic transformation protocol for *Picea abies* (L.) Karst embryogenic tissue and regeneration of transgenic plants. *Can. J. Forest. Res.* 29: 1539-1546.

Walter, C., Grace, L.J., Wagner, A., Walden, A.R., White, D.W.R., Donaldson, S.S., Hinton, H.H., Gardner, R.C., and Smith, D.R. (1998). Stable transformation and regeneration of transgenic plants of *Pinus radiata* D. Don. *Plant Cell Rep.* 17: 460-468.

Walter, C. and Smith, D.R. (1999). Transformation of *Pinus radiata.* In Y.P.S. Bajaj (Ed.), *Biotechnology in Agriculture and Forestry* 44 (pp. 193-211). Berlin, Heidelberg: Springer-Verlag.

Walter, C., Smith, D.R., Connett, M.B., Grace, L.J., and White, D.W.R. (1994). A biolistic approach for the transfer and expression of a *gus* reporter gene in embryogenic cultures of *Pinus radiata*. *Plant Cell Rep.* 14: 69-74.

Wenck, A.R., Quinn, M., Whetten, R.W., Pullman, G., and Sederoff, R. (1999). High efficiency *Agrobacterium*-mediated transformation of Norway spruce *(Picea abies)* and loblolly pine *(Pinus taeda)*. *Plant Mol. Biol.* 39: 407-416.

Chapter 10

Transgenic Forest Trees
for Insect Resistance

Wang Lida
Han Yifan
Hu Jianjun

INTRODUCTION

Insects are responsible for substantial losses in forest tree species, and their damage can sometimes be a limiting factor for tree growth and survival. In practice, the use of insecticides is rather limited in forestry, owing in part to the extensive forest areas, large size of trees, the development of resistance by insects, and environmental impacts of the insecticides. Thus, insecticide application is usually restricted to nurseries and small plantations with young trees (Tzvi et al., 1998).

Modern molecular biology, especially molecular breeding by transferring an insect-resistant gene into the plants (termed as gene technology), can help to avoid or minimize the damage by providing novel methods for insect control. Compared to the traditional methods of insect control, methods based on gene technology have many merits. For example, transgenic plants harboring resistance genes eliminate the need for repeated insecticide application and thus avoid associated hazards. Also, it is possible to induce insecticidal activity when insect attack occurs and/or in the necessary plant organs. Furthermore, the use of transgenics can have some environmental advantages and requires relatively small investments.

Historically, insect pests have caused very serious problems in China because of the continental climate. However, recent research on insect-resistant forest tree breeding shows considerable promise for alleviating these problems. In the following section we review tree species that have been made transgenic using different insect-resistant genes, field performance of the transgenic trees, and their biosafety.

WOODY PLANT SPECIES TRANSGENIC
TO INSECT-RESISTANT GENES

Poplars

Due to its small genome and ease of vegetative propagation, poplar (*Populus* species and hybrids) represents a useful model system for investigation of the genetics and molecular biology of woody species. Moreover, because of their desirable characteristics for biomass production, *Populus* species and hybrids are among the leading candidates for short-rotation energy and fiber forest plantations. In many countries within temperate regions, poplar trees represent the primary source of wood. However, increased planting of poplar could contribute to the development of poplar disease and insect problems, which in turn may threaten the development of afforestation. An effective approach for forest pest control would be the implementation of insect-resistant transgenic trees. The first insect-resistant, transgenic poplar plant (*Populus alba* × *P. grandidentata*) was produced by McCown and colleagues (1991) using electric discharge particle acceleration with a plasmid containing a *Cry1Aa* gene from *Bacillus thuringiensis (Bt)*. Subsequently, many insect-resistant transgenic *Populus* species and hybrids have been produced (Table 10.1).

Conifers

Insect feeding experiments with *Bt*-transgenic European larch trees (Shin et al., 1994) showed a decrease in the average weight of larvae fed on transgenic needles compared to those fed on untransformed plants. Furthermore, needle consumption was considerably lower in transgenic trees, although no increase in the larval mortality rate was observed. In another attempt to use *Bt* in conifers, Ellis and colleagues (1993) introduced the *Cry1A Bt* toxin gene into white spruce plants. Although spruce budworm larvae fed on transformed callus exhibited reduced body weight, no pronounced increase in mortality of the larvae was observed. Another research group also used a gene from *Bacillus thuringiensis (pBinBt)* to transform larch by a particle gun (Ewald et al., 1999). Approximately 250 somatic embryos of *Larix decidua* appeared after about 42 days. The somatic embryos were transferred to solid medium for germination. One hundred plants were obtained in red light conditions, and 94 plants were planted into vermiculite and subsequently planted into the nursery. Approximately 30 plants were screened by polymerase chain reaction (PCR) to verify presence of the gene, and 15 plants showed a positive PCR reaction (Ewald et al., unpublished).

TABLE 10.1. Research on transgenic insect-resistant forest trees

Species	Gene	Method	Insect bioassay	Confirmation of transformation	Reference
Populus alba × P. grandidentata; P. nigra × P. trichocarpa	Bt	Electric discharge particle acceleration	Malacosoma disstria Lymantria dispar		McCown et al. (1991)
Populus nigra	Bt	Agrobacterium tumefaciens-mediated	—	Dot blot, Southern blot	Wu and Fan (1991)
Populus nigra	Bt	Agrobacterium tumefaciens-mediated	Apocheima cinerarius Erschoff	PCR, Southern blot, Western blot	Tian et al. (1993)
			Lymantria dispar Linn		Chen et al. (1996)
Populus deltoides	Bt	Agrobacterium tumefaciens-mediated	—	PCR	Chen et al. (1995)
Populus deltoides; P. nigra; P. ×euramericana	Lcl	Agrobacterium tumefaciens-mediated	—	PCR	Li et al. (1996)
Populus euramericana	Bt	Agrobacterium tumefaciens-mediated	—	PCR	Wang, Bian, et al. (1997)

TABLE 10.1 (continued)

Species	Gene	Method	Insect bioassay	Confirmation of transformation	Reference
Populus ×euramericana 'Branbatic'; Populus ×euramericana 'Robusta'; 82182-9 (P.×euramericana ×yunnanensis)	Bt	Agrobacterium tumefaciens-mediated	Lymantria dispar Linn	PCR, Southern blot	Wang, Han, et al. (1997)
NL-80106 (Populus deltoides × P. simonii)	Bt	Agrobacterium tumefaciens-mediated	Lymantria dispar Linn	PCR, PCR-Southern blot	Rao et al. (2000)
Populus tremula × P. tremuloides	OCI	Agrobacterium tumefaciens-mediated	Chrysomela tremulae	–	Leple et al. (1995)
Populus alba L. × P. grandidentata; P. ×euramericana	PIN2	Agrobacterium tumefaciens-mediated	–	Northern blot	Klopfenstein et al. (1993)
Populus deltoides × P. nigra	PIN2	–	Chrysomela scripta	–	Kang (1997)
Populus euramericana	PIN2	–	Chrysomela scripta	–	Kang et al. (1997)
Populus alba × P. grandidentata	PIN2	Agrobacterium tumefaciens-mediated	Plagiodera versicolora	PCR	Klopfenstein et al. (1997)

Plant species	Gene	Transformation method	Target insect	Analysis	Reference
Populus tomentosa; (Populus tomentosa × P. bolleana) × P. tomentosa	CpTI	Agrobacterium tumefaciens-mediated	–	PCR, PCR-Southern blot	Hao et al. (1999, 2000)
Populus nigra	KTi$_3$	Agrobacterium tumefaciens-mediated	Lymantria dispar Linn	Southern blot	Massimo et al. (1998)
			Clostera anastomosis	Northern blot	
Populus alba L. × (P. davidiana Dode + P. simonii Carr.) × P. tomentosa Carr.	Bt	Agrobacterium tumefaciens-mediated	Clostera anachoreta	PCR, Southern blot	Tian et al. (2000)
	API		Lymantria dispar Linn		
Populus tremula × P. tremuloides	Bt OCI	–	Chrysomela tremulae	–	Cornu et al. (1996)
Populus deltoides × P. simonii	AaIT	Agrobacterium tumefaciens-mediated	Lymantria dispar	PCR, PCR-Southern blot	Wu et al. (2000)
Larix decidua	Bt	Agrobacterium rhizogenes-mediated	–	Southern blot, Western blot, Northern blot	Shin et al. (1994)
Picea glauca	Bt	Particle acceleration	Choristoneura fumiferana	PCR, Southern blot	Ellis et al. (1993)
Diospyros kaki	Bt	Agrobacterium tumefaciens-mediated	Plodia interpunctella; Monema flavescens	PCR, Southern blot, Western blot	Tao et al. (1997)
Eucalyptus camaldulensis	Bt	Agrobacterium tumefaciens-mediated	Chrysophtharta bimaculata; C. agricola; C. variicolis	Southern blot, Western blot	Harcourt et al. (2000)

Eucalypts

The genus *Eucalyptus* contains over 500 different species and is the most widely grown tree genus in the world. Members of this genus are grown primarily for their hardwood timber and pulp for paper. However, repeated defoliation of plantation trees by insect pests, such as the Tasmanian eucalypt leaf beetle *(Chrysophtharta bimaculata)* significantly reduces wood production and increases plantation rotation times. Harcourt and colleagues (2000) have introduced the *Cry3A* gene into *E. camaldulensis* (river red gum). Transgenic plants from two lines tested were resistant to first instars of chrysomelid beetles.

INSECT-RESISTANT GENES

Bacillus thuringiensis (Bt) *Genes*

The first report on inducing insect resistance describes the transfer of the *Bt Cry1Ab* gene into tobacco plants. The transgenic tobacco plants obtained showed resistance to *Manduca sexta* (Barton et al., 1987). Since 1989, transgenic conifers and poplars containing *Bt* transgene have also been successfully produced (Table 10.1). In China, genetic transformation experiments on forest trees with a *Bt* gene were initiated in 1990. Subsequently, insect-resistant transgenic poplar (*P. nigra* L.) plants expressing *Bt Cry1Ac* were obtained in 1993 by *Agrobacterium*-mediated transformation (Tian et al., 1993).

Bacterial Gene LcI

Li and colleagues (1996) used leaves of in vitro-cultured *Populus deltoides, P. nigra,* and *P. ×euramericana* as explants for transformation with a bacterial gene, *LcI*. The gene was synthesized using the sequence of a bacterial protein from *Bacillus subtilis*. The PCR analysis indicated that the bacterial gene was present in the genome of poplar. It was further demonstrated that the bacterial protein was responsible for more than 75 percent mortality to *Anoplophora glabripennis* (Coleoptera, Cerambycidae). Seven transgenic poplar plants with resistance to *A. glabripennis* have been preliminarily selected through an artificial inoculation method (Hu et al., unpublished).

Insect-Specific Scorpion Neurotoxin Gene

Insect-specific scorpion neurotoxin AaIT is the bioactive substance that is isolated from *Androctonus australis*. AaIT is mainly poisonous to Lepidopteran insects, and it appears to be safe for other living organisms. Wu and colleagues (2000) inserted an insect-specific scorpion neurotoxin *AaIT* gene into a binary vector and transformed a hybrid poplar clone N-106 *(P. deltoides × P. simonii)* using *Agrobacterium tumefaciens*. Sixty-two regenerated plants were obtained. PCR and PCR-Southern analyses showed that the *AaIT* gene was present in the genome of some of the regenerated poplar plants. One of the transformed plants, named "A5," was significantly resistant to feeding by first instar larvae of *Lymantria dispar,* compared to the untransformed control plant. The "A5" line showed a decrease in leaf consumption by larvae, a lower larval weight gain, and a higher larval mortality rate of *L. dispar.* Enzyme-linked immunoabsorbent assay (ELISA) analysis demonstrated that the *AaIT* gene was expressed in this transformed poplar plant.

Proteinase Inhibitor (PI) Gene

Successful protection against insect pests has been observed in transgenic plants that contain genes coding for proteinase inhibitors (Duan et al., 1996; Hilder et al., 1987; Johnson et al., 1990; Leple et al., 1995). Proteinase-inhibitor proteins are among the most prevalent proteins in nature. They occur in all living organisms and are a natural insect-resistant compound in plants. Cowpea trypsin inhibitor (CpTI) is a serine proteinase inhibitor. Hidler and colleagues (1987) first transferred a *CpTI* gene into tobacco *(Nicotiana tabacum)*. The *CpTI* gene was expressed in tobacco, with the CpTI content approaching 0.9 percent of the total soluble protein in the transgenic plant. It inhibited the growth of *Heliothis virescens* and caused their mortality after feeding with transgenic tissue.

In forest trees, Hao and colleagues (1999, 2000) successfully introduced the *CpTI* gene into two *Populus tomentosa* clones by *Agrobacterium*-mediated gene transformation. The transformed shoots were rooted in a selection medium containing 50 mg/L kanamycin and transferred to the greenhouse. The presence of *CpTI* genes in transgenic plants was confirmed by PCR and PCR-Southern blot.

Massimo and colleagues (1998) have reported the production of transgenic *P. nigra* containing the soybean Kunitz proteinase inhibitor gene. Southern blot analysis demonstrated the presence of the *KTi$_3$* gene in the poplar genome. Northern blot analysis of different kanamycin-resistant plantlets confirmed the accumulation of KTi_3 mRNA but revealed different levels of its expression. However, the expression of the KTi_3 gene in transgenic poplars did not show any inhibitory effect on the growth and development of the tested insects.

Leple and colleagues (1995) analyzed the toxicity to *Chrysomela tremulae* of transgenic poplars expressing a cysteine proteinase inhibitor. As a first step, cysteine proteinases were determined to be the major digestive proteinases of *C. tremulae,* and oryzacystatin (OCI), a cysteine proteinase inhibitor, was shown to inhibit this activity in vitro. The gene encoding OCI was introduced into *Populus tremula* × *P. tremuloides* via *Agrobacterium*-mediated transformation, and transgenic plants expressing OCI at a high level were selected. Feeding tests on these transgenic plants demonstrated the toxicity of OCI-producing poplar leaves against *C. tremulae* larvae. The highest mortality rates were noted among third instar larvae and the pupae, ranging from 29.8 to 43.5 percent among the three transgenic lines tested (control mortality rate was 4.5 percent) at day 25.

Combinations of Different Genes

The expression of two or more insect-resistant genes having different insecticidal mechanisms may be helpful to increase the insecticidal activity of transgenic plants and also reduce the evolutionary probability of insects becoming more tolerant to insect-resistant transgenic plants (Zhu and Zhu, 1997). We have studied insect-resistant transgenic poplar plants containing both *Bt* and *PI* genes. The explants of transgenic *Populus nigra* containing a *Bt* gene were transformed with the *PI* gene using *Agrobacterium tumefaciens.* The presence of the *PI* gene in the transgenic poplar was confirmed by PCR and Southern blot analyses. Larvae of *Lymantria dispar* were fed with leaves of the transgenic plants containing the *Bt* and *PI* transgenes, transgenic plants containing only the *Bt* transgene, and nontransgenic plants. The bioassay showed that the transgenic plants containing both *Bt* and *PI* transgenes had enhanced toxicity (insect mortality was 100 percent after ten days) to larvae compared to the plants containing only *Bt* gene (insect mortality was 10 percent after ten days) (Han et al., unpublished). Similarly, a partially modified *Bt CryIAc* gene and the arrowhead inhibitor *(API)* gene have been used to construct a plant transformation vector pBtiA. This vector was used (Tian et al., 2000) to transform the hybrid poplar "741" (*Populus* x*aldatomentosa* cl. "741") by *Agrobacterium*-mediated transformation, and ten kanamycin-resistant plants were regenerated. A subsequent bioassay demonstrated that three of these transgenic plant lines were highly resistant to *Clostera anachoreta* (Fabricius). The mortality of insect larvae on one plant was higher than 90 percent at six days after infestation, and the growth of the surviving larvae was seriously inhibited. Results of PCR and Southern blot analyses using the *Cry1Ac* gene fragment as the probe indicated that both the *Bt Cry1Ac* and *API* transgenes were integrated as a single copy into the genomes of each of these three plants. Protein dot-blot immunoassay and ELISA analyses revealed that at least the Cry1Ac protein was produced in these three transgenic plants, and the expression levels were estimated to be approximately 0.015 percent of the leaf total soluble protein.

FIELD PERFORMANCE OF INSECT-RESISTANT TRANSGENIC TREES

Growth and Morphology

Field observations on the growth and morphology of clonal cuttings of *Populus tomentosa* transgenic to the *Bt* transgene were performed by Zheng and colleagues (1995). Only 44.4 percent of the clones planted in the field displayed no reduction in productivity and no change in morphological characteristics. For the remaining number of clones, the height growth of one-year-old cutting plants was decreased about 30 percent. In addition, total dry weight of stems, branches, and leaves was 50 percent lower than control plants. A large change in leaf morphology was also apparent. The branch angles were narrower and the growth increment of branches was higher compared to the control plants. Therefore, their economic value for timber production was reduced. The plants of these clones also appeared to have more buds and relatively higher net leaf weight. However, somaclonal or position effect variegation occurring during in vitro stages could be a possible reason for the morphological abnormalities observed later in the field. Our results on the growth and morphology of *Bt*-transgenic *P. nigra* (Wang et al., 1996) are substantially in agreement with previous reports on transgenic hybrid *P. alba × P. grandidentata* (McCown et al., 1991; Robinson et al., 1994). Similar to these reports, we also observed changes in leaf morphology in many transgenic plants. Leaf morphology of transgenic plants was grouped into three discrete classes (see Table 10.2): similar to those of *P. nigra* (type a); similar to those of the section *P. ×euramericana* (type b); and similar to those of the section Tacamahaca (type c). Changes in leaf morphology apparently followed discrete patterns suggested to be controlled by transcription factors. It is possible that these controlling elements have been affected by somaclonal or position effect variation (Wang et al., 1996). The vegetative period of transgenic plants with type b leaf shape is

TABLE 10.2. Growth and phenotype of six-year-old poplar clones *(P. nigra)* transgenic to *Bt* gene

Clones (No.)	Phenotype	Average diameter (cm)	Average height (m)
153 (12)	b	20.1	16.7
172 (13)	b	20.7	16.8
12 (12)	b	22.0	17.3
192 (3)	c	14.4	12.3
P. nigra (control) Ck2 (17)	a	16.5	16.3
P. nigra (control) Ck3 (11)	a	14.8	13.3

Note: Phenotypic leaf morphology of the plants: type a = similar to those of *P. nigra;* type b = similar to those of the section *P.* ×*euramericana;* type c = similar to those of the section Tacamahaca.

longer than that of nontransgenic plants, and the growth of these transgenic plants is different from the control (Table 10.2).

Field Tests on Insect Resistance

Kleiner and colleagues (1995) performed evaluations of field-grown transgenic poplar against forest tent caterpillar *(Malacosoma disstria)* and gypsy moth *(Lymantria dispar)* following winter dormancy. Hybrid *Populus* plants genetically engineered with a *Bacillus thuringiensis CryIA(a)* δ-endotoxin gene were field planted in summer 1993 and evaluated for insect resistance three times during the 1994 growing season. Foliage of plants containing the transgene encoding the δ-endotoxin displayed reduced feeding in bioassays conducted during the months of June, July, and August of each year. Larvae of the forest tent caterpillar and gypsy moth consumed significantly less foliage and showed reduced wet-weight gains when fed transgenic leaves than larvae fed with control foliage. In addition, mortality of early third-instar forest tent caterpillar and gypsy moth was greater when transgenic foliage was fed than the control foliage. Mortality of late third-instar gypsy moth feeding on transgenic foliage did not differ from the control larvae (Kleiner et al., 1995).

In China, the first field test on insect resistance of *Populus nigra* transgenic to the *Bt* gene was conducted by Hu and colleagues (1999) at Manas Forest Station, Xinjiang Uygur Autonomous Region, China. They evaluated the resistance of transgenic *Populus nigra* to *Apocheima cinerarius* and *Orthosia incerta* based on the rate of damaged leaves within the transgenic poplar plantation and larval and pupal density in the soil. The rate of damaged leaves within the transgenic stand was 10 percent, considerably lower than the 40 percent displayed by the control plants. In addition, the rates of damaged leaves of both *P. nigra* controls and nontransgenic plants *P. ×euramericana* cv. Robusta in the transgenic plantation were lower than the plants from surrounding stands. Thus, nontransgenic trees within the transgenic plantation appear to benefit from the decreased growth and survival of the larvae within the transgenic plantation. As a result, the pupae per square meter in the transgenic plantation soil was 18, much lower than 88 pupae in *P. ×euramericana* cv. Robusta stands and 73 pupae in *P. nigra* stands outside of the transgenic plantation. In 1997 and 1999, some clones of genetically modified (GM) *P. nigra* were approved for release to the environment by the Chinese Agricultural Genetic Engineering Safety Community of the Chinese Agriculture Ministry.

Presence of Bt *Gene in the Primary Transgenic* P. nigra *Transgenic Lines*

In 1997 and 2001, samples were collected from four-year-old (data from Prof. Sala lab, University of Milano, Italy) and eight-year-old transgenic *P. nigra,* which were planted in Manas Forest Station, Xinjiang Uygur Autonomous Region, China. The presence of *Bt* gene was tested by PCR analysis. Our results showed that both *npt II* and *Bt* gene could be PCR amplified from DNA isolated from transgenic *P. nigra* plants.

Inheritance of the Bt *Gene in Progenies*

Cross-pollinated seeds were collected from transgenic *P. nigra* clone 12 (GM-12) and *P. nigra* clone 153 (GM-153), and progenies were tested for the inheritance of the *Bt* transgene by PCR analysis. Out of 24 progenies of *P. nigra* clone 12, seventeen tested positive for the *Bt* gene. This result indicates that the *Bt* transgene segregates with 3:1 ($\alpha = 0.05$) ratio in the progenies. Southern blot analysis revealed that two copies of the *Bt* transgene were present in *P. nigra* GM-12. Similarly, out of 29 seeds of *P. nigra* clone GM-153 tested, 14 were found positive, whereas 15 tested negative for the *Bt* transgene. The ratio in the progenies indicates a 1:1 segregation, which was also consistent with the result of Southern blot analysis that showed one copy of the *Bt* transgene in the *P. nigra* GM-153 genome. Thus, the *Bt* transgene appears to follow a Mendelian distribution in our experiment.

BIOSAFETY OF INSECT-RESISTANT TRANSGENIC PLANTS

Worldwide, the biosafety of the transgenic plants has aroused a general public concern. Before transgenic plants are commercially field planted, questions need to be addressed that are related to public health, environmental pollution, and ecological balance of natural forest ecosystems. Based on these considerations, Li and colleagues (2000) conducted a study on the introduction of male sterility genes into insect-resistant transgenic *Populus nigra* by using the *TA29-Barnase* gene. A chimaeric *TA29-Barnase* gene (see Chapter 11) was introduced into insect-resistant transgenic poplar, and insect-resistant poplar containing the *Barnase* transgene were obtained. If male sterility is achieved, this approach may provide a means to prevent the pollen pollution from transgenic plants to nearby nontransgenic plantations. However, a system leading to female sterility is also needed to prevent vertical gene transfer via seed.

The work has also been initiated in our lab to address the effects of insect-resistant transgenic plants on natural enemies of insects and soil microorganisms (Han et al., unpublished). For this study, insect-resistant transgenic poplar *(Populus nigra)* were used which contained a *Bt* transgene and showed a high level of insecticidal activity in the field.

Effect of Insect-Resistant Transgenic Poplars on Untargeted Insects

The results on the effect of insect-resistant transgenic poplars on nontarget insects are shown in Tables 10.3 and 10.4. The data suggest that transgenic insect-resistant poplars have not negatively affected the natural enemies of targeted insects. Furthermore, the species, number, and parasitic rate of the natural enemies in the transgenic plantation were higher than in the control plantation. This result indicates that insect-resistant transgenic *Populus nigra* can limit the targeted insects in the transgenic plantation while protecting the biodiversity of the surrounding environment.

TABLE 10.3. Survey results of the nontargeted insects in the transgenic poplar plantation

Investigation items	Trees or area investigated	Control plantation	Neighboring plantation of TP	Transgenic plantation (TP)
Ending date of *Orthosia incerta* Hufnagel pupation	1 ha	May 10	May 13	May 14
Pupation rate on May 16, 2001	1 ha	95%	30%	30%
Rate of leaves damaged	10 trees	90%	10%	10%
Pupal density of *Orthosia incerta* Hufnagel in the soil	10 m^2	150 heads/m^2	10 heads/m^2	8 heads/m^2
Nontargeted insects	1 ha	One kind of Carabidae larvae 20 heads/m^2	Two kinds of parasitic flies	43% parasitic rate
			Two kinds of parasitic wasps	10% parasitic rate
			Two kinds of Carabidae adults	0.5 heads/m^2

Effect of Insect-Resistant Transgenic Poplars on Soil Microorganisms

The comparison of soil microorganisms of transgenic, nontransgenic, and neighboring plantations are shown in Tables 10.5 and 10.6. The results indicate that soil microorganisms are not significantly different between transgenic and nontransgenic plantations. Thus, these transgenic poplars have apparently not influenced the microorganisms in the soil.

CONCLUSION AND OUTLOOK

Our results demonstrate that transgenic poplars have exhibited substantial insect resistance during two years of growth in the field. In contrast, the poplar leaves were heavily consumed in the nontransgenic plantation. However, no significant difference in leaf loss was observed between the transgenic and nontransgenic poplars grown together for two years within a mixed plantation. This response is likely attributable to the extensive reduc-

TABLE 10.4. Results of *Orthosia incerta* Hufnagel pupae fed on transgenic poplar inoculated with *Chouioia cunea* Yang

Items	Control pupae	Pupae from neighboring plantation	Pupae from transgenic plantation
Inoculation date—eclosion	May 25-June 11-12	May 25-June 11-12	May 25-June 11-12
Size of pupae inoculated (mm)	6.04×18.29	6.68×19.22	6.63×19.39
Number of wasp (heads)	100	36	80
Eclosion rate (%)	100	100	100
Eclosion number (heads/pupa)	492.9	539	550.7
Male wasp proportion	6.1	6.4	8.5

tion in pupal density by the transgenic poplars. In addition, tree growth in the nursery was remarkably improved, which may be related to the protection against insect attacks. Thus, the transgenic poplar can be planted together with other nontransgenic poplar species to protect them from damage by insects.

Through small-scale field tests, transgenic poplar have demonstrated an obvious improvement in insect resistance and other growth parameters. Such rapid improvements cannot be achieved by traditional breeding methods to enhance insect resistance. However, the application of genetic engineering for tree production has raised various concerns. These concerns are being addressed at our lab in which research on environmental release of transgenic trees is ongoing and scientific research is reinforced to prevent genetic pollution.

In China, the prospects of tree genetic engineering for commercial purposes has imminent applications. Since China began protecting natural forests, the lack of wood products has become more serious. To overcome this shortage, the first step is to raise plantations using fast-growing species, such as poplar and larch. However, this approach is threatened by serious insect damage of poplar and larch as the annual damage rate by leaf insects and borers in poplar plantations approaches 40 to 80 percent. Recently, the Chinese government emphasized the commercialization of products from genetic engineering, while being cautious to minimize risks to the environment. In 1993, China enacted genetic engineering regulations. The State

TABLE 10.5. Comparison of soil microorganisms of transgenic and nontransgenic plants

Clone	Bacteria (total sum)		Actinomycetes		Mold	
	mean	SD	mean	SD	mean	SD
12	1.06E+08	1.01E+08	1.24E+07	2.02E+06	2.23E+05	1.86E+05
153	3.80E+07	1.74E+07	3.62E+07	4.78E+07	4.19E+05	4.57E+05
172	3.30E+07	1.33E+07	1.34E+07	3.31E+06	1.04E+05	6.92E+04
P. ×eur.	2.81E+07	9.78E+06	1.00E+07	4.70E+06	2.21E+05	2.72E+05
Ck2	3.83E+07	2.98E+06	1.16E+07	1.20E+06	6.00E+04	1.50E+04
Ck3	3.98E+07	2.50E+07	1.27E+07	3.50E+06	6.01E+04	2.34E+04

12, 153, 172 = *Populus nigra* transgenic plants; CK2, CK3 = *Populus nigra* nontransgenic plants; SD = standard deviation

TABLE 10.6. Comparison of soil microorganisms of transgenic, nontransgenic, and neighboring plantations

Plantation	Bacteria (total sum)		Actinomycetes		Mold	
	mean	SD	mean	SD	mean	SD
TP	1.99E+07	4.78E+06	5.23E+06	1.70E+06	3.39E+04	2.41E+04
NP	2.65E+07	4.61E+06	9.55E+06	6.16E+06	4.24E+04	8.53E+03
EP	2.18E+07	4.58E+06	5.88E+06	1.75E+06	5.20E+04	4.04E+04

TP = transgenic plantation; NP = neighboring plantation; EP = *Populus ×euramericana* cv. Robusta nontransgenic plantation; SD = standard deviation

Science and Technology Commission of the People's Republic of China published the "Safety Administration Regulation on Genetic Engineering," and these regulations were enacted on December 24, 1993. Subsequently, the Ministry of Agriculture published the "Safety Administration Implementation Regulation on Agricultural Biological Genetic Engineering," and these regulations have been enforced since July 10, 1996. Based on these regulations, the Chinese Agriculture Ministry approved the release of transgenic plants to the field at six sites (ca. 300 ha) in 1997 and 1999. The commercial release of transgenic poplars was approved in the year 2002. The released plants will be female, and seeds are almost impossible to germinate in the arid soil in which most of the poplars are planted. Furthermore, we are continuing the work in various other research areas, such as transferring male sterility genes, transformation of poplar species with double transgenes (e.g., *PI* and *Bt*), and safety assessments of transgenic poplars in the field.

REFERENCES

Barton, K.A., Whiteley, H.R., and Ying, N.S. (1987). *Bacillus thuringiensis* δ-endotoxin expressed in transgenic *Nicotiana tabacum* provides resistance to lepidopteran insects. *Plant Physiology* 85: 1103-1109.

Chen, Y., Han, Y.F., Tian, Y.C., Li, L., and Nie, S.J. (1995). Study on the plant regeneration from *Populus deltoides* explant transformed with Bt toxin gene. *Sci. Silvae Sin.* 31: 77-82.

Chen, Y., Li, Q., Li, L., and Han, Y.F. (1996). Western blot analysis of transgenic *Populus nigra* plants transformed with *Bacillus thuringiensis* toxin gene. *Sci. Silvae Sin.* 32: 274-276.

Cornu, D., Leple, J.C., Bonade, B.M., Ross, A., Augustin, S., Delplanque, A., Jouanin, L., and Pilate, G. (1996). Expression of a proteinase inhibitor and a *Bacillus thuringiensis* delta-endotoxin in transgenic poplars. In M.R. Ahuja, W. Boerjan, and D.B. Neale (Eds.), *Somatic Cell Genetics and Molecular Genetics of Trees* (pp. 131-136). Dordrecht, Netherlands: Kluwer Academic.

Duan, X., Li, X., Xue, Q., Abo-El-Saad, M., Xu, D., and Wu, R. (1996). Transgenic rice plants harboring an introduced potato proteinase inhibitor II gene are insect resistant. *Nat. Biotechnol.* 14: 494-498.

Ellis, D.D., McCabe, D.E., McInnis, S., Ramachandran, R., Russell, D.R., Wallace, K.M., Martinell, B.J., Roberts, D.R., Raffa, K.F., and McCown, B.H. (1993). Stable transformation of *Picea glauca* by particle acceleration. *Bio-Technology* 11: 84-89.

Ewald, D., Li, L., Li, M.L., and Han, Y.F. (1999). Attempts to transform hybrid larch with a *Bacillus thuringensis* gene. *Forest Biotechnology* 99: Poster 45.

Hao, G.X., Zhu, Z., and Zhu, Z.T. (1999). Transformation of *Populus tomentosa* with insecticidal cowpea proteinase inhibitor gene. *Acta Bot Sin.* 41: 1276-1282.

Hao, G.X., Zhu, Z., and Zhu, Z.T. (2000). Obtaining of cowpea proteinase inhibitor transgenic *Populus tomentosa. Sci. Silvae Sin.* 36: 116-119.

Harcourt, R.L., Kyozuka, J., Floyd, R.B., Bateman, K.S., Tanaka, H., Decroocq, V., Llewellyn, D.J., Zhu, X., Peacock, W.J., and Dennis, E.S. (2000). Insect- and herbicide-resistant transgenic eucalypts. *Mol Breeding* 6: 307-315.

Hilder, V.A., Gatehouse, A.M.R., Sherman, S.E., Barker, R.F., and Boulter, D. (1987). A novel mechanism of insect resistance engineered in tobacco. *Nature* 330: 160-163.

Hu, J.J., Liu, Q.Y., Wang, K.S., Zhang, B.E., Tian, Y.C., and Han, Y.F. (1999). Field test on insect-resistance of transgenic plants *(Populus nigra)* transformed with Bt toxin gene. *Forest Res.* 12: 202-205.

Johnson, R., Narvaez, J., An, G., and Ryan, C. (1990). Expression of proteinase inhibitor I and II in transgenic tobacco plants: Effects on natural defense against *Manduca sexta* larvae. *Proc. Natl. Acad. Sci. USA* 86: 9871-9875.

Kang, H.D. (1997). Gene manipulation of PIN2 (proteinase inhibitor II) to the cottonwood leaf beetle (Coleoptera: Chrysomelidae) in transgenic poplar (*Populus deltoides* x *P. nigra*). *J. Korean Forest. Soc.* 86: 407-414.

Kang, H.D., Hall, R.B., Heuchelin, S.A., McNabb, H.S. Jr, Mize, C.W., and Hart, E.R. (1997). Transgenic *Populus:* In vitro screening for resistance to cottonwood leaf beetle (Coleoptera: Chrysomelidae). *Can. J. Forest. Res.* 27: 943-944.

Kleiner, K.W., Ellis, D.D., McCown, B.H., and Raffa, K.F. (1995). Field evaluation of transgenic poplar expressing a *Bacillus thuringiensis* cry1A(a) δ-endotoxin gene against forest tent caterpillar (Lepidoptera: Lasiocampidae) and gypsy moth (Lepidoptera: Lymantriidae) following winter dormancy. *Environ. Entomol.* 24: 1358-1364.

Klopfenstein, N.B., Allen, K.K., Avila, F.J., Heuchelin, S.A., Martinez, J., Carman, R.C., Hall, R.B., Hart, E.R., and McNabb, H.S. Jr. (1997). Proteinase inhibitor II gene in transgenic poplar: Chemical and biological assays. *Biomass Bioenerg.* 12: 299-311.

Klopfenstein, N.B., McNabb, H.S. Jr., Hart, E.R., Hall, R.B., Hanna, R.D., Heuchelin, S.A., Allen, K.K., Shi, N.Q., and Thornburg, R.W. (1993). Transformation of *Populus* hybrids to study and improve pest resistance. *Silvae Genetica* 42: 86-90.

Leple, J.C., Bonade, B.M., Augustin, S., Pilate, G., Dumanois, L.T.V., Delplanque, A., Cornu, D.M., and Jouanin, L. (1995). Toxicity to *Chrysomela tremulae* (Coleoptera: Chrysomelidae) of transgenic poplars expressing a cysteine proteinase inhibitor. *Mol. Breeding* 1: 319-328.

Li, L., Qi, L.W., Han, Y.F., Wang, Y.C., and Li, W.B. (2000). A study on the introduction of male sterility of anti-insect transgenic *Populus nigra* by TA29-Barnase gene. *Sci. Silvae Sin.* 36: 28-32.

Li, Y., Chen, Y., Li, L., and Han, Y.F. (1996). Transformation of antibacterial gene LcI into poplar species. *Forest Res.* 9: 561-567.

Massimo, C., Gianni, A., Alma, B., Corrado, F., and Massimo, D. (1998). Regeneration of *Populus nigra* transgenic plants expressing a Kunitz proteinase inhibitor (KTi_3) gene. *Mol. Breeding* 4: 137-145.

McCown, B.H., McCabe, D.E., Russell, D.R., Robinson, D.J., Barton, K.A., and Raffa, K.F. (1991). Stable transformation of *Populus* and incorporation of pest resistance by electric discharge particle acceleration. *Plant Cell Rep.* 9: 590-594.

Rao, H.M., Chen, Y., Huang, M.R., Wang, M.X., Wu, N.F., and Fan, Y.L. (2000). Genetic transformation of poplar NL-80106 transferred by Bt gene and its insect-resistance. *J. Plant Resources and Environ.* 9: 1-5.

Robinson, D.J., McCown, B.H., and Raffa, K.F. (1994). Responses to gypsy moth (Lepidoptera: Lymantriidae) and forest tent caterpillar (Lepidoptera: Lasiocampidae) to transgenic poplar, *Populus* spp., containing a *Bacillus thuringiensis* δ-endotoxin gene. *Environ. Entomol.* 23: 1030-1041.

Shin, D.I., Podila, G.K., Huang, Y., and Karnosky, D.F. (1994). Transgenic larch expressing genes for herbicide and insect resistance. *Can. J. Forest Res.* 24: 2059-2067.

Tao, R., Dandekar, A.M., Uratsu, S.L., Vail, P.V., and Tebbets, J.S. (1997). Engineering genetic resistance against insects in Japanese persimmon using the cryIA(c) gene of *Bacillus thuringiensis. J. Am. Soc. Hortic. Sci.* 122: 764-771.

Tian, Y.C., Li, T.Y., Mang, K.Q., Han, Y.F., Li, L., Wang, X.P., Lu, M.Z., Dai, L.Y., Han, Y.N., and Yan, J.J. (1993). Insect tolerance of transgenic *Populus nigra* plant transformed with *Bacillus thuringiensis* toxin gene. *Chin. J. Biotechnol.* 9: 291-298.

Tian, Y.C., Zheng, J.B., Yu, H.M., Liang, H.Y., Li, C.Q., and Wang, J.M. (2000). Studies of transgenic hybrid poplar 741 carrying two insect-resistant genes. *Acta Bot. Sin.* 42: 263-268.

Tzvi, T., Amir, Z., and Arie, A. (1998). Forest-tree biotechnology: Genetic transformation and its application to future forests. *Trends Biotechol.* 16: 439-446.

Wang, G.J., Castiglione, S., Chen, Y., Li, L., Han, Y.F., Tian, Y.C., Gabriel, D.W., Han, Y.N., Mang, K.Q., and Sala, F. (1996). Poplar (*Populus nigra* L.) plants transformed with a *Bacillus thuringiensis* toxin gene: Insecticidal activity and genomic analysis. *Transgenic Res.* 5: 289-301.

Wang, X.P., Bian, Z.X., Zhang, X.H., and Lu, M.Z. (1997). PCR analysis of transgenic (*P. x euramericana*) plants. *Sci. Silvae Sin.* 33: 380-383.

Wang, X.P., Han, Y.F., Dai, L.Y., Li, L., and Tian, Y.C. (1997). Studies on insect-resistant transgenic (*P. x euramericana*) plants. *Sci. Silvae Sin.* 33: 69-74.

Wu, N.F. and Fan, Y.L. (1991). Transgenic poplar plants transformed with *Bacillus thuringiensis* toxin gene. *Chinese Science Bulletin* 9: 705-708.

Wu, N.F., Sun, Q., Yao, B., Fan, Y.L., Rao, H.Y., Huang, M.R., and Wang, M.X. (2000). Insect-resistant transgenic poplar expressing AaIT gene. *Chinese J. Biotechnol.* 16: 129-133.

Zheng, J.B., Wang, F., and Liu, Q.L. (1995). Studies on the clones of gene-transformed *Populus tomentosa. Sci. Silvae Sin.* 31:181-184.

Zhu, X.S. and Zhu, Y.X. (1997). Proceeding in plant anti-insect genetic engineering. *Acta Bot. Sin.* 39: 282-288.

Chapter 11

Modification of Flowering in Forest Trees

Juha Lemmetyinen
Tuomas Sopanen

INTRODUCTION

Flowering and seed production represent the main mode of reproduction and dispersal in most plants, including forest trees. Trees, together with perennial herbaceous plants, differ from annual and biennial plants in their ability to flower over a long period of years. The flowering of trees starts only after a juvenile phase, which may last from a few years to as long as ten to twenty years (Clark, 1983; Poethig, 1990). During this phase, the trees do not respond to the environmental or internal factors which, in the mature phase, induce flowering (Longman, 1976). Sometimes certain structural features, for example the form or phyllotaxy of the leaves, are also different in the two phases. The genetic basis of the change from juvenility to maturity is poorly understood and may vary in different species.

Recent advances in understanding the genetic basis of flowering induction and flower development provide us with potential tools to regulate different aspects of flowering, the two main aims being either the prevention or the acceleration of flowering. Prevention of flowering would be especially important for improving the environmental safety of testing and cultivating transgenic trees, by preventing the escape of pollen or seed-borne transgenes from them into native species. The acceleration of flowering would be extremely useful for speeding up breeding programs.

This chapter provides a brief overview of the genetic regulation of flowering time and flower development in the model plant *Arabidopsis* and the corresponding genes in forest trees. We then describe attempts that have been made to prevent or accelerate flowering using genetic modification. The modification of flowering in trees has earlier been reviewed in Brunner and colleagues (1998), Meilan and Strauss (1997), Nilsson and Weigel (1997), Rishi and colleagues (2001), Skinner and colleagues (2000), Strauss and colleagues (1995, 2001), and Tzfira and colleagues (1998).

REGULATION OF FLOWERING TIME AND FLOWER DEVELOPMENT IN ARABIDOPSIS

Most of what is known about the genes regulating flowering time or flower development has been gained from studies in *Arabidopsis* (for reviews, see, e.g., Araki, 2001; Jack, 2001a,b; Koornneef et al., 1998; Levy and Dean, 1998; Lin, 2000; Simpson et al., 1999; Yanofsky, 1995). The first phase in flowering is floral induction, after which the apical meristem of the shoot turns into an inflorescence meristem and starts to form initials of flowers on its flanks. *Arabidopsis* is a facultative long-day plant, but vernalization (cold treatment) also accelerates flowering in the winter annual ecotypes. However, the plants can also flower without long days or vernalization, but the flowering is then delayed. Nutrients also have an effect on the time of flowering (Meilan and Strauss, 1997; Ohto et al., 2001). One internal factor controlling flowering is the age of the plant.

Flowering time is regulated through several distinct pathways, which interact in various ways (Figure 11.1). The operation of these pathways is triggered by the detection of external or internal signals, which then leads to a chain of events that ultimately result in activation of the floral meristem genes and the formation of floral meristems. Of these genes, *LEAFY (LFY)* and *APETALA1 (AP1)* are the most important (Yanofsky, 1995). The exact details of how the various genes and factors interact in this network are not yet clear.

One of the main principles established in *Arabidopsis* is that during vegetative growth flowering is prevented by the products of floral repressor genes, and the initiation of flowering is due, at least partly, to reversal of this inhibition (Koornneef et al., 1998). Important floral repressor genes are *EMBRYONIC FLOWER (emf)* 1 and 2 (Aubert et al., 2001; Yoshida et al., 2001). In *emf1* and *emf2* mutants, flowering starts immediately after germination. So far, the exact positions of these genes in the regulatory pathways are not known, but somehow they repress the expression of genes such as *LFY* and *AP1*. Other genes that delay flowering are *TERMINAL FLOWER (TFL)* 1 and 2 (Ratcliffe et al., 1998; Sundås-Larsson et al., 1998). The *tfl* mutants flower early but not as early as the *emf* mutants, and the inflorescence meristem is consumed in the formation of a terminal flower. On the other hand, plants overexpressing *TFL1* flower late (Ratcliffe et al., 1998). *TFL1* may exert its effect by repressing the expression of genes that accelerate flowering, such as *FCA*, but it is also important in preventing the formation of a flower at the apex of the inflorescence meristem by delaying the expression of *LFY* and *AP1* (Ratcliffe et al., 1998; Page et al., 1999). Another gene that delays flowering is *HASTY*, which controls the change from the juvenile to the adult phase (Telfer and Poethig, 1998).

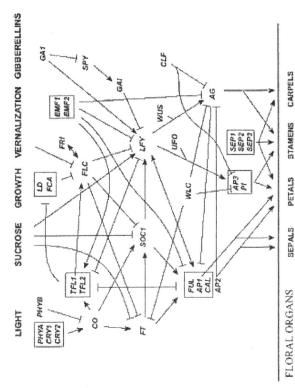

FIGURE 11.1. *Arabidopsis* genes regulating flowering time and flower development. The figure is modified from Blázquez (2000) and the recent results have also been incorporated, based on the references cited. The arrowheads indicate activation, whereas the bars indicate repression. Genes involved, at least partly, in the same functions are boxed. It should be noted that in many cases the effects may be indirect and that a line connecting two genes may represent a chain of events. It should also be noted that for clarity many genes not mentioned in the text have been omitted. The figure shows how several external and internal factors lead to activation of the floral meristem genes and how these then participate in the activation of genes determining the identity of the floral organs. The ABC model is depicted at the bottom of the figure. It shows how the A, B, and C functions, representing various combinations of organ identity genes, determine the identities of the floral organs.

Genes that repress flowering also include *FLOWERING LOCUS C (FLC), FRIGIDA (FRI), CURLY LEAF (CLF), WAVY LEAVES* and *COTYLEDONS (WLC), PHYTOCHROME B (PHYB), SPINDLY (SPY)* (Levy and Dean, 1998), *SHORT VEGETATIVE PHASE (SVP),* and *MADS Affecting Flowering 1 (MAF1)* (Hartmann et al., 2000; Ratcliffe et al., 2001). It would be extremely interesting to know whether tree genes, related to those mentioned, play a role in regulating juvenility and in determining the sites of flower or inflorescence formation. The latter phenomenon is important in determining the form of the tree, because it affects the mode of branching.

One of the pathways that accelerates flowering in *Arabidopsis* is the day-length pathway (Simpson et al., 1999; Lin, 2000). This pathway starts from the photoreceptors PHYTOCHROME (PHY) A, CRYPTOCHROME (CRY) 1, and CRY 2, which interact with the circadian clock. In long days, a signal transduction pathway leads to the up-regulation of the *CONSTANS (CO)* gene. In turn, *CO* activates *SUPPRESSOR OF OVEREXPRESSION OF CO1 (SOC1),* a MCM1-AGAMOUS-DEFICIENS-SRF (MADS)-box gene also called *AGL20,* which is up-regulated in the apical meristem by flowering induction (Samach et al., 2000). *SOC1* expression is one of the several factors that lead to up-regulation of the floral meristem gene *LFY.* Another gene activated very early is *FRUITFULL (FUL),* which is already active in the inflorescence meristem before formation of the floral meristems (Mandel and Yanofsky, 1995a; Hempel et al., 1997). The role of *FUL* in the inflorescence meristem is not known in detail, but it has been suggested that it maintains the identity of the inflorescence meristem and promotes floral meristem specification (Ferrándiz et al., 2000). In addition to *SOC1, CO* also activates the flowering inducer gene *FLOWERING LOCUS T (FT)* that, in turn, activates *APETALA1 (AP1)* expression (Kardailsky et al., 1999).

The gibberellin (GA) pathway also accelerates flowering (Richards et al., 2001; Simpson et al., 1999). It is especially important in short days, as shown by the fact that mutants defective in GA synthesis *(ga1)* do not flower at all or flower very late in short days. This pathway also leads to the up-regulation of *LFY* (Blázquez and Weigel, 2000).

The autonomous promotion pathway stimulates flowering even in the absence of exogenous inducers (Simpson et al., 1999). The autonomous pathway includes genes such as *LUMINIDEPENDENS (LD)* and *FCA,* which accelerate flowering. They function, at least partly, by down-regulating the floral repressor gene *FLC,* which is active in noninduced conditions and prevents the up-regulation of *SOC1* (Michaels and Amasino, 2001). Vernalization accelerates flowering by suppressing *FLC* expression (Sheldon et al., 2000).

According to Jack (2001a), the long-day promotive pathway, the gibberellin pathway, and the autonomous promotion pathway all converge at *SOC1.*

The up-regulation of *SOC1* leads to the activation of the floral meristem genes *LFY, AP1, CAULIFLOWER (CAL),* and *FUL.* As a result of the activation of the floral meristem identity genes, floral meristems start to form.

Floral organs start to form on the flanks of the floral meristem. The formation starts from the outer whorl and from the sepals, and then the initials of petals, stamens, and finally carpels appear. The identity of the floral organs is partly determined by the so-called ABC genes acting in various combinations (Weigel and Meyerowitz, 1994; Yanofsky, 1995). The identity of sepals is determined when the A class genes *AP1* and *AP2* function alone in certain parts of the floral meristem. Petals are formed when the A class genes are expressed together with the B class genes *AP3* and *PISTILLATA (PI),* the products of which function as a heterodimer. Stamens are formed when the B function genes are expressed together with the C function gene *AGAMOUS (AG). AG* alone determines the identity of the carpels and ensures determinate development of the flower.

In addition to the classical ABC genes, other genes are also needed for the identity of the floral organs (Jack, 2001b). For example, at least one of the three *SEPALLATA (SEP)* genes is needed for the formation of petals, stamens, and carpels (Pelaz et al., 2001). The SEP3 polypeptide interacts with AG, AP3/PI, or AP1 polypeptides, and only these larger complexes are capable of specifying the organ identity (see also Theissen, 2001). For example, *LFY* is needed for the activation of the homeotic genes, but in the case of the B and C function genes it seems to function together with *UNUSUAL FLORAL ORGANS (UFO)* and *WUSCHEL (WUS),* respectively (Lee et al., 1997; Lohmann et al., 2001).

Many of the genes mentioned here encode transcription factors, and most of these genes, including *FLC, SOC1, FUL, AP1, AP3, PI, AG, SEP 1, 2,* and *3,* belong to the so-called MADS-box genes (Shore and Sharrocks, 1995; Alvarez-Buylla et al., 2000). The products of the MADS-box genes function as homo- or heterodimers or larger complexes (Theissen et al., 2000; Jack, 2001b). They recognize conserved *cis*-elements, termed CArG boxes [CC(A/T)$_6$GG] in the promoters of certain genes (Tilly et al., 1998). Although the MADS-box genes seem to be the most important gene family in flower development, members of other, often small, gene families are also involved (e.g., *LFY, TFL1, AP2, EMF1, EMF2*).

GENES REGULATING FLOWER DEVELOPMENT IN TREES

The main patterns of flower development seem to be rather similar among angiosperms, and MADS-box genes are also important in gymnosperms such as conifers. However, there are differences even within dicots. For example, in some cases gene duplication has led to two genes that either continue to perform their earlier functions or acquire partially or completely new functions.

Trees are a heterogenous group of plants, representing gymnosperms and dicot and monocot angiosperms. Therefore, it is obvious that different trees will show some differences in their regulation of flower development. Table 11.1 provides an overview of isolated genes apparently regulating flower development in trees. The table also shows the most similar genes or group of genes in *Arabidopsis* (except *TM3*, which is in tomato). However, in most cases a functional similarity has not been confirmed. At present, little is known about the genetic regulation of flower development in trees, but the current state of knowledge is described in the sections that follow.

Poplar

Poplars belong to the Rosids (order Malpighiales), one of the two main branches of dicots (Sheppard et al., 2000). Large areas of poplars are cultivated because of their rapid growth and use in fiber and energy production. Genetic transformation of poplars is relatively easy, and they have become very important in biotechnological and genetic research (Klopfenstein et al., 1997). In poplars, male and female inflorescences are on separate trees (Sheppard et al., 2000). The inflorescence is a catkin, bearing structurally simple flowers. The female flowers consist of a perianth cup surrounding the pistil, whereas the male flowers have 40 to 60 stamens inside a perianth cup.

Several genes participating in flower formation have been isolated from *Populus trichocarpa*, black cottonwood. One of these is the *LFY* homologue *Populus trichocarpa LEAFY (PTLF)* (Rottmann et al., 2000). *PTLF* is strongly expressed in developing inflorescences. The expression is highest in the floral meristems and developing flowers, especially in immature anthers and carpels, with some expression in the perianth cup. *PTLF* is also expressed in young bracts and weakly in leaf primordia and even in seedlings. Overexpression of *PTLF* in *Arabidopsis* results in early flowering, although not as early as that caused by overexpression of *LFY*. Two homologues of the floral meristem identity gene *AP1* have also been isolated (Skinner et al., 2000).

Another gene isolated was *Populus trichocarpa DEF (PTD)* (Sheppard et al., 2000), a homologue of the B function gene *AP3*. *PTD* expression was first detected in male flowers after differentiation of the perianth cup, the expression covering the entire floral meristem with the exception of the

TABLE 11.1. List of isolated genes apparently regulating flower development in trees

Tree genes	Plant species	Similar gene(s) in *Arabidopsis*	References
Poplar:			
PTAG1 and 2	Populus trichocarpa	AG (AGAMOUS)	Brunner et al., 2000
PTAP1-1 and 2		AP1 (APETALA1)/ FUL (FRUITFULL)	Skinner et al., 2000
PTD	Populus trichocarpa	AP3 (APETALA3)	Sheppard et al., 2000
PTLF	Populus trichocarpa	LFY (LEAFY)	Rottmann et al., 2000
two TFL1 homologues		TFL1 (TERMINAL FLOWER 1)	Dye et al., 2001
Eucalyptus:			
EAP1 and 2	Eucalyptus globulus ssp. bicostata	AP1/FUL	Kyozuka et al., 1997
ETL	Eucalyptus globulus ssp. bicostata	AGL14/ TM3	Decroocq et al., 1999
ELF1 and 2	Eucalyptus globulus ssp. globulus	LFY	Southerton, Strauss, et al., 1998
EGM1 and 3	Eucalyptus grandis W. Hill ex Maiden	SEP1, 2, and 3 (SEPALLATA1, 2, and 3)	Southerton, Marshall, et al., 1998
EGM2	Eucalyptus grandis W. Hill ex Maiden	PI (PISTILLATA)	Southerton, Marshall, et al., 1998
AG homologue		AG	Southerton et al., 2001
TFL1 homologue		TFL1	Collins and Campbell, 2001
Birch:			
BpMADS1	Betula pendula	SEP1, 2, and 3	Lemmetyinen, Pennanen, et al., 2001
BpMADS2	Betula pendula	PI	unpublished
BpMADS3-5	Betula pendula	AP1/ FUL	Elo, Lemmetyinen, Turunen, et al., 2001
BpMADS6	Betula pendula	AG	unpublished

TABLE 11.1 *(continued)*

Tree genes	Plant species	Similar gene(s) in *Arabidopsis*	References
BpMADS7	Betula pendula	AGL11	unpublished
BpMADS8	Betula pendula	AP3	unpublished
BpFLO	Betula pendula	LFY	unpublished
Apple:			
MdMADS1	Malus × domestica, Fuji	SEP1, 2, and 3	Sung and An, 1997
MdMADS2	Malus × domestica, Fuji	AP1/ FUL	Sung et al., 1999
MdMADS3 and 4	Malus × domestica, Fuji	SEP1, 2, and 3	Sung et al., 2000
MdMADS5	Malus × domestica, Granny Smith	AP1/ FUL	Yao et al., 1999
MdMADS6-9	Malus × domestica, Granny Smith	SEP1, 2, and 3	Yao et al., 1999
MdMADS10	Malus × domestica, Granny Smith	AGL11	Yao et al., 1999
MdMADS11	Malus × domestica, Granny Smith	AGL6	Yao et al., 1999
MdPI	Malus x domestica, Granny Smith	PI	Yao et al., 2001
MdH1	Malus × domestica, Granny Smith	BEL1	Dong et al., 2000
Other dicot tree species:			
CaMADS1	Corylus avellana	AG	Rigola et al., 1998
LAG	Liquidambar styraciflua	AG	Liu et al., 1999
PlaraLFY	Platanus racemosa	LFY	Frohlich and Parker, 2000
Conifers:			
CjMADS1 and 2	Cryptomeria japonica	PI/AP3	Fukui et al., 2001
DAL1	Picea abies	AGL6	Tandre et al., 1995
DAL2	Picea abies	AG	Tandre et al., 1995, 1998

DAL3	Picea abies	AGL14/ TM3	Tandre et al., 1995
DAL10	Picea abies	?	Sundström, 2001
DAL11-13	Picea abies	PI/AP3	Sundström et al., 1999
PaAP2L1 and 2	Picea abies	AP2 (APETALA2)	Vahala et al., 2001
SAG1	Picea mariana	AG	Rutledge et al., 1998
NLY (NEEDLY)	Pinus radiata	LFY	Mouradov, Glassick, Hamdorf, Murphy, Fowler, et al., 1998
PrDGL	Pinus radiata	PI/AP3	Mouradov et al., 1999
PrMADS2 and 3	Pinus radiata	AGL6	Mouradov, Glassick, Hamdorf, Murphy, Marla, et al., 1998
PrMADS4-9	Pinus radiata	AGL14/TM3	Walden et al., 1998
PRFLL	Pinus radiata	LFY	Mellerowicz et al., 1998
PMADS1	Pinus resinosa	AGL6	Liu and Podila, 1997
DFL1 and 2	Pseudotsuga menziesii	LFY	Strauss et al., 1995

Note: The list also shows the most similar genes or group of genes in *Arabidopsis* (except *TM3*, which is in tomato), but it should be noted that in most cases a functional similarity has not been confirmed. Most, but not all, genes in the list have been mentioned in the text. The presence of some genes is based on unpublished results or on conference abstracts, and therefore the sequences are not yet available in the sequence databases.

perianth cup. *PTD* was also expressed in the middle of young female flowers, resembling the expression of *PI* and *DEFICIENS (DEF)*. *Populus trichocarpa* has two very similar *AG* homologues, *PTAG1* and *PTAG2 (Populus trichocarpa AGAMOUS)* (Brunner et al., 2000). They are both expressed in the same way, in the middle of the flower and in the developing stamens and pistils, but not in the perianth cup. These C-function genes were also, rather surprisingly, expressed in all vegetative tissues tested. In addition to the homeotic and floral meristem genes, some earlier genes, such as *TFL1* and *CO,* have also been studied in poplar (Brunner et al., 2001).

Eucalyptus

Eucalyptus species are especially important in Australia, but, because of their economic importance, they are also extensively cultivated elsewhere. Similar to poplars, *Eucalyptus* also belongs to the Rosids (order Myrtales). In *Eucalyptus,* the inflorescences form in leaf axils and are determined dichasia (Southerton, Marshall, et al., 1998). The flowers are unusual: sepal primordia fuse together, as do the petal primordia, forming a covering structure called the operculum, which protects the developing inner parts of the flower but is shed when the flowers open.

Several genes regulating flower development have been isolated from *Eucalyptus globulus.* Two of these, *EAP1* and *EAP2,* are similar (Kyozuka et al., 1997) and belong to the *AP1/FUL* group of MADS-box genes (Theissen et al., 2000). *EAP1* and *EAP2* are both expressed predominantly in flowers, being active at all stages of floral development (Kyozuka et al., 1997). In mature flower buds, both genes are expressed in the operculum; *EAP1* is also expressed at low level in the anthers, but *EAP2* expression is not detectable in these organs. When *EAP1* and *EAP2* were overexpressed in *Arabidopsis,* they caused a phenotype similar to that resulting from the overexpression of *AP1,* including early flowering and the formation of apical flowers.

Eucalyptus globulus has also two genes resembling *LFY,* namely *EARLY FLOWERING 1 (ELF1)* and *ELF2,* but only *ELF1* appears to be functional (Southerton, Strauss, et al., 1998). *ELF1* is expressed in each of the floral organs, at least at some stages of their early development. *ELF1* is also expressed in vegetative tissues; weak expression is observed in leaf primordia and young leaves of juvenile trees, and stronger expression occurs in corresponding tissues from mature trees. In *Arabidopsis,* the promoter of *ELF1* functions like the promoter of *LFY,* suggesting similar regulation of activity, and ectopic expression of *ELF1* causes a similar phenotype in *Arabidopsis* as that resulting from the overexpression of *LFY,* i.e., early flowering and terminal flower (Southerton, Strauss, et al., 1998).

Eucalyptus has two relatively different genes, *EGM1* and *EGM3* (Southerton, Marshall, et al., 1998), belonging to the *AGL2/4* group of MADS-box genes, which contains the three *SEP* genes of *Arabidopsis.* Both *EGM1* and *EGM3* are expressed preferentially in flowers. In the young flower buds, EGM1 is expressed in the floral meristem and at the base of sepals. The *Eucalyptus PI* homologue *(EGM2)* is expressed, like other B function genes, in stamens and petals (Southerton, Marshall, et al., 1998). Homologues of *AG* (Southerton et al., 2001) and *TFL1* (Collins and Campbell, 2001) have also been isolated.

Birch

Birches belong to the Rosids (order Fagales) and are important broad-leaved forest tree species in temperate regions. In birch, the male and female flowers are on the same tree but on separate inflorescences, termed catkins (Atkinson, 1992). The flowers are in groups of three in the axil of three fused scales, the male flowers consisting of two to three reduced tepals and two stamens, and the female flowers consisting of a single pistil.

Three genes *(BpMADS3-5)* belonging to the *AP1/FUL* group have been isolated from silver birch *(Betula pendula)* (Elo, Lemmetyinen, Turunen, et al., 2001). *BpMADS3* seems to be the homologue of *AP1*, whereas *BpMADS4* and *BpMADS5* are more closely related to *FUL*. Our preliminary unpublished results with in situ hybridization suggest that *BpMADS4* and *BpMADS5* are both expressed in the inflorescence meristem starting at a very early stage of its development. This supports their classification as *FUL*-like genes (Hempel et al., 1997). In addition to inflorescence, *BpMADS3* also appears to be expressed weakly in vegetative tissues, and *BpMADS4* is clearly expressed in the vegetative parts including roots, whereas *BpMADS5* expression seems to be inflorescence specific (Elo, Lemmetyinen, Turunen, et al., 2001). These three genes cause very early flowering when overexpressed in tobacco.

Birch also has a homologue of *SEP3*, designated *BpMADS1*, which is expressed in both the male and female inflorescences (Lemmetyinen, Pennanen, et al., 2001). In situ hybridization analysis suggests that *BpMADS1* expression starts from the very early stages of flower development and continues even in the developing seed coat (Hassinen et al., unpublished results). Other genes isolated from birch by our group include homologues of *AG, LFY, AP3,* and *PI,* as well as the ovule-specific *AGL11* (unpublished results). According to the limited results available, these birch genes seem to play roles similar to those of their *Arabidopsis* counterparts.

Other Dicot Tree Species

Genes involved in the regulation of flowering or fruit development have also been isolated from apple (see references in Yao et al., 2001; Sung et al., 2000), from *Corylus avellana* (an *AG* homologue) (Rigola et al., 1998), and from *Platanus racemosa* (a *LFY* homologue) (Frohlich and Parker, 2000).

Conifers

The reproductive structures of conifers differ from those of angiosperms (Rutledge et al., 1998). All conifers have separate male and female cones (strobili). The pollen cones are made up of many microsporophylls having pollen-producing microsporangia on their abaxial surfaces. The female cones consist of many ovuliferous scales bearing ovules on their adaxial surface. The most extensively studied species with respect to regulation of reproductive development are Norway spruce *(Picea abies),* black spruce *(Picea mariana),* and Monterey pine *(Pinus radiata).*

In *Pinus radiata,* there are two different *LEAFY* homologues, *NEEDLY (NLY)* and *Pinus radiata FLO/LFY-like (PRFLL)* (Mellerowicz et al., 1998; Mouradov, Glassick, Hamdorf, Murphy, Fowler, et al., 1998). They differ from each other almost as much as they differ from *LFY* and do not contain a proline-rich domain at the amino terminus or an acidic region in the middle part of the proteins, which are characteristic of the angiosperm genes in the LFY family (Mellerowicz et al., 1998; Mouradov, Glassick, Hamdorf, Murphy, Fowler, et al., 1998). *NLY* is expressed in both types of cones at very early stages of development, as well as in vegetative shoot buds and needles, but not in roots (Mouradov, Glassick, Hamdorf, Murphy, Fowler, et al., 1998). Constitutive expression of *NLY* in *Arabidopsis* results in early flowering and terminal flowers, suggesting that, in spite of considerable structural differences, this conifer *LFY* homologue functions like *LFY* itself. This notion is further supported by the fact that *LFY::NLY* (an *NLY* expression construct driven by an *LFY* promoter) can complement the phenotypic defects in a severe *lfy* mutant (Mouradov, Glassick, Hamdorf, Murphy, Fowler, et al., 1998). *PRFLL* is expressed in vegetative buds but not in other vegetative tissues (Mellerowicz et al., 1998). During development, PRFLL is strongly expressed in male cones but not in the female ones. Douglas fir *(Pseudotsuga menziesii)* also has two paralogous genes resembling *LFY* (Strauss et al., 1995).

Conifers seem to have their B- and C- function gene counterparts, but the B-function genes are not divided into two distinct families like those in angiosperms (Sundström et al., 1999; Theissen et al., 2000). Three putative B-function genes, *DAL11, DAL12,* and *DAL13 (DEFICIENS AGAMOUS LIKE),* have been isolated from *Picea abies* (Sundström et al., 1999) and two genes, *CjMADS1* and *CjMADS2,* from sugi *(Cryptomeria japonica)* (Fukui et al., 2001). They are all specifically expressed in the developing male cones, but there are clear differences in their tissue specificity. It has been postulated that even at the very early stages in the evolution of vascular plants, the B-function genes were involved in the specification of the male floral organs (Theissen et al., 2000). Two *APETALA2*-like genes have been isolated from *Picea abies* (Vahala et al., 2001).

The *Pinus radiata* gene *PrDGL (Pinus radiata DEF/GLO-like)* most closely resembles *DAL13* and is also specifically expressed in the male cones (Mouradov et al., 1999). C-function genes in conifers include *DAL2* from *Picea abies* (Tandre et al., 1998) and *SAG1* from *Picea mariana* (Rutledge et al., 1998). *DAL2* and *SAG1* are both expressed in developing male and female cones but not in any vegetative tissues, except for very low expression of *SAG1* in needles and vegetative buds. When ectopically expressed in *Arabidopsis,* these genes result in conversion of petals to stamens and sepals to carpels, as well as loss of indeterminacy of the inflorescence meristem. This suggests that the basic functions of the C-class genes in the determination of the male and female reproductive organs are present already in conifers.

One conifer gene having no close relatives in angiosperms is *DAL10,* which is expressed in seed and pollen cones throughout development (Sundström, 2001). Interestingly, it is expressed in the bracts and the central part of the cone. This gene might be a candidate for determining the identity of reproductive shoots. This is supported by the observation that overexpression of *DAL10* in *Arabidopsis* activates *AG.*

Six cDNA clones belonging to the *TM3* group of MADS-box genes (Pnueli et al., 1991) have been isolated from *Pinus radiata (PrMADS4-9)* (Walden et al., 1998) and one from *Picea abies (DAL3)* (Tandre et al., 1995). It is possible that there are at least twelve different genes in the *TM3* family in *Pinus radiata* (Walden et al., 1998). However, the function of these genes is still unclear. Several genes related to the *AGL6* group have also been isolated from gymnosperm species (Tandre et al., 1995; Mouradov, Glassick, Hamdorf, Murphy, and Marla, 1998).

PREVENTION OF FLOWERING

One of the aims of genetic manipulation of flowering in trees is to produce sterile trees by preventing the formation of the inflorescences, flowers, or the sexual organs of the flowers. The advantages sought often vary (for more detail, see Strauss et al., 1995). Sterility would greatly increase the ecological safety and acceptability of transgenic trees by preventing the escape of the transgenes into native populations, which would often exist near proposed transgenic plantations. This is important, because long-term effects of the transgenes, especially various ecological effects, are difficult to predict precisely. Sterility could also be used to prevent the spread of exotic plant species. Preventing the formation of fertile flowers and therefore seeds, as well, may also allow the trees to devote more resources to vegetative growth. Therefore, the introduction of sterility might increase tree growth, performance, and yield (Brunner et al., 1998). The prevention of flowering would also help to reduce the amount of allergenic pollen, particularly if the sterile trees replace fertile varieties near centers of habitation. The pollen of birch and sugi, for example, are highly allergenic and thereby contribute to public health-care costs.

Tissue-Specific Ablation

To date, two main approaches have been employed in the prevention of flowering: tissue-specific ablation and suppression of essential gene(s). In tissue-specific ablation, genes are used which encode proteins that kill the cells in which they are synthesized. The basic idea is to ligate a cytotoxic gene to a gene promoter that only functions in the floral tissues. When the gene is turned on at a certain stage of inflorescence or flower development, the gene product kills the target cells, and this prevents further inflorescence or flower development.

The two cytotoxic genes used are the *Diphtheria toxin A (DTA)* gene and the *BARNASE* gene, from *Corynebacterium diphtheriae* and *Bacillus amyloliquefaciens,* respectively (Day et al., 1995; Hartley, 1989; Strauss et al., 1995). *DTA* encodes an enzyme that inactivates translation elongation factor 2 and therefore prevents protein synthesis, leading to cell death. The DTA protein is harmless to animals eating the plants. *BARNASE* encodes a ribonuclease that, after removal of the signal sequence, remains in the cytoplasm, where it efficiently degrades the RNAs and thereby kills the cells. *Bacillus* also synthesizes a small protein, BARSTAR, which is a very efficient inhibitor of *BARNASE* (Hartley, 1989). *BARSTAR* could possibly be used to prevent unwanted expression of *BARNASE* in selected tissues (Beals and Goldberg, 1997).

BARNASE was first used to generate sterile male tobacco and oilseed rape, using the tapetum-specific promoter *TA29* (Mariani et al., 1990). In this case, the flowers and even the anthers developed normally, but the tapetum (the innermost cell layer of the pollen sac) did not form, and therefore the pollen grains did not mature. *TA29::DTA* also resulted in male sterility in tobacco (Koltunow et al., 1990). Female sterility with normal male fertility was obtained in tobacco using a stigma-specific promoter *STIG1 (stigma-specific gene 1)* ligated to *BARNASE* (Goldman et al., 1994). Complete ablation of petals and stamens was achieved with *AP3::DTA* in *Arabidopsis* and tobacco, while sepals and pistils developed normally (Day et al., 1995). This study clearly demonstrated that the promoter of a homeotic gene of one species may function in the desired tissue-specific way in a distantly related species.

The first demonstration of complete prevention of flower formation using tissue-specific ablation was by Nilsson and colleagues (1998). They transferred an *LFY::DTA* construct into *Arabidopsis* and showed that the construct prevented the formation of inflorescences, or at least that of flowers. In most transgenic plants, however, the vegetative development was disturbed, apparently due to weak activation of the *LFY* promoter already in the vegetative tissues (Blázquez et al., 1997), but in some cases the rosette seemed to be quite normal, although no stems formed. This suggests that it might be possible to prevent flowering using an *LFY::DTA* construct by carefully selecting lines that do not express the transgene at high levels.

Our group has been testing the use of *BARNASE* in the prevention of flowering, using gene promoters isolated from silver birch. So far, we have tested three promoters that were selected because of their inflorescence-specific expression. The promoter of *BpMADS1* was first tested using promoter::GUS (β-glucuronidase) fusion in tobacco and *Arabidopsis* (Lemmetyinen, Pennanen, et al., 2001), and a strong flower-specific expression was observed. Interestingly, there was clear expression in the apex of the shoot at about the same time as the initiation of the floral inflorescence. When the *BpMADS1::BARNASE* construct was transferred into tobacco, some lines formed sterile flowers without stamens and pistils, whereas in other transgenic lines only a few sepal-like leaves formed in the place of the inflorescence. The vegetative growth before flowering time was normal in most cases, and, because of the formation of branches, the dry weights of some sterile lines were clearly higher than those of the controls.

In *Arabidopsis,* some transgenic *BpMADS1::BARNASE* transformants produced inflorescences with flowers that lacked stamens and pistils, while some transformants did not produce any inflorescences (Lemmetyinen, Pennanen, et al., 2001). Shootless transformants formed a large mass of leaves even in long days and therefore looked completely different from those containing a *LFY::DTA* construct (Nilsson et al., 1998). Thus, the formation of the *Arabidopsis* inflorescence could be completely prevented without harmful effects on vegetative growth.

To test the effects of various gene constructs on flowering in birch, we are using very early flowering birch clones BPM2 and BPM5 (Lemmetyinen et al., 1998), which start inflorescence formation in the greenhouse about four months after rooting, being about 50 cm high. Because the transformation of birch takes about six months, with these lines it is possible to detect the effects of a transgene on inflorescences within one year from the transformation, which is unique among trees.

BpMADS1::BARNASE is now being tested in the early flowering birch (Lemmetyinen, Keinonen, et al., 2001; Lemmetyinen, Pennanen, et al., 2001). Most of the lines containing the construct look normal and have normal male inflorescences (the female inflorescences stay in the buds and have not yet been studied). In one line, however, the male inflorescences do not contain stamens, although they have normal scales. In this line, all inflorescences later die. In several other lines, no inflorescences form. Some of these lines look normal, at least during their initial period of growth, but in some lines there are clear disturbances (slow, bushy growth, partly necrotic leaves, dichotomic branching). These preliminary results suggest that it may be possible to use *BpMADS1::BARNASE* to prevent the formation of inflorescences in birch without harmful effects on growth. Because *BpMADS1::BARNASE* seems to function well in three widely different dicot species, it is likely to function in other dicots as well.

Another birch promoter preliminarily tested in model plants is *BpMADS5* (Elo, Lemmetyinen, Turunen, et al., 2001; Lemmetyinen, Keinonen, et al., 2001). The *BpMADS5:: uidA*-containing tobacco or *Arabidopsis* lines do not seem to have any *uidA* expression in the vegetative tissues, but this construct is expressed early in the inflorescence meristems of *Arabidopsis* and in the flowers of both plants. *BpMADS5::BARNASE* could completely prevent the formation of inflorescences in tobacco and *Arabidopsis*. In most lines, the vegetative growth was severely disturbed, but in some lines the initial growth was quite normal. *BpMADS8::BARNASE* has so far only been tested in *Arabidopsis,* in which it results in male-sterile, female-fertile lines (Lännenpää and Sopanen, unpublished results). This was expected, because *BpMADS8* is the birch homologue of the B-function gene *AP3*.

Transgenic strategies for generating sterility have also been tested in poplar, which has the advantage of being dioecious. Although *TA29::DTA* and *TA29::BARNASE* specifically prevented the formation of the tapetum in tobacco anthers without causing any vegetative effects, these constructs resulted in decreased growth in both male and female clones of hybrid aspen (Skinner et al., 2000). Similarly, *SLG::DTA, TTS::DTA,* and *TTS::BARNASE* decreased vegetative growth. This shows that even promoters active late in flower development can result in vegetative expression in other species. One possible explanation might be that not all the necessary *cis*-elements are present in the isolated 5'-sequences but are located further upstream or in the introns, as is the case with *AG* (Sieburth and Meyerowitz, 1997). Because *PTD,* the poplar homologue of *AP3,* is not expressed in vegetative tissues but is expressed in both the male and female flowers, the use of its promoter for controlling the expression of sterility transgenes is being studied (Sheppard et al., 2000). The use of the poplar homologues of *LFY, AP1, AG,* and *AP3* to engineer sterility in poplar is also in progress (Skinner et al., 2000). In addition to tissue-specific ablation, other gene suppression approaches for generating sterile poplars are also being used.

Suppression of Expression

Another possible strategy to induce sterility would be to prevent the expression of a gene (or several genes) that is absolutely necessary for inflorescence or flower formation, or for the formation of stamens and pistils. In the antisense RNA technique, the gene to be suppressed is introduced as a cDNA in reverse (antisense) orientation under the control of a constitutive promoter (generally CaMV35S). This leads to the specific prevention of translation of the normal mRNA (Meyer and Saedler, 1996). More efficient suppression of gene expression could possibly be achieved using RNAi technique (Wesley et al., 2001; Waterhouse et al., 2001). In this technique, a construct consisting of an intron flanked by inversely oriented, target gene sequences is transformed into a plant. Expression of this construct results in the formation of at least 21 base pairs (bp) long, double-stranded RNA molecules, which direct specific degradation of the target gene transcripts. The deletion of a gene, or parts of it, using homologous recombination would be the ideal method of gene suppression. To date, this approach has been practically impossible in plants; however, there are suggestions about how this might be achieved (Kumar and Fladung, 2001a).

In several cases, the introduction of antisense (as) RNA constructs has resulted in complete or partial sterility. Antisense suppression of *TAG1* (the tomato homologue of *AG*) results in homeotic conversion of stamens into petaloid organs and prevention of ovule formation (Pnueli, Hareven, Rounsley, et al., 1994). In *Arabidopsis,* the introduction of *AG*-as also results in sterile flowers (Mizukami and Ma, 1995). The introduction of *TM5*-as (the tomato homologue of *SEP3*) into tomato also results in sterile lines (Pnueli, Hareven, Broday, et al., 1994). In potato, surprisingly, antisense suppression of citrate synthase leads to early abortion of flower buds (Landschutze et al., 1995). Transformation of *Arabidopsis* with an *AG (RNAi)* construct results in a strong sterile *ag* mutant phenotype (Chuang and Meyerowitz, 2000).

Due to the long generation times of trees, only limited data are currently available on the use of gene suppression techniques for the modification of flowering in these species. They are being tested in poplar (Brunner et al., 1998), *Eucalyptus* (Southerton et al., 2001), and birch (Elo, Lemmekyinen, Keinonen, et al., 2001; Lemmetyinen, Keinonen, et al., 2001). We have some preliminary results on the effects of antisense constructs in the early-flowering birch clones. In most cases, the effects of various antisense constructs made using MADS-box genes do not give very clear phenotypes, but

BpMADS4-as delayed flowering and/or decreased the number of inflorescences in most of the transgenic lines tested. Even more important, two lines have not produced any inflorescences after more than 600 days, although in all control plants inflorescences emerged after about 100 days of cultivation. So far, only one close relative of *BpMADS4* has been isolated from other plants, *DEFH28* from *Antirrhinum* (Müller et al., 2001). In poplar, the effect of antisense *PTLF* is being studied (Rottmann et al., 2000). Putative candidates to be tested for suppression are the floral meristem identity genes (e.g., homologues of *LFY, AP1*), the C-function genes (homologues of *AG*) and the B-function genes in male clones of dioecious trees. As mentioned previously, several floral genes have been isolated from *Eucalyptus*. The effects of down-regulating their expression using RNAi are currently being studied, but due to the long juvenile period, results are not yet available (Southerton et al., 2001).

Other Techniques

Other techniques for inducing sterility primarily include transfer of a construct conferring a dominant negative mutation, as shown with a truncated version of *AG* in *Arabidopsis* (Mizukami et al., 1996), but so far there are no reports of the use of this approach in trees. In *Arabidopsis,* many genes when overexpressed prevent or delay flowering. Overexpression of these types of genes, alone or in combination, might be one approach to the prevention of flowering in trees. The use of one such gene, namely *FLC,* is currently being examined in *Eucalyptus* (Southerton et al., 2001).

Possible Problems

All of the approaches described, except those based on homologous recombination, have the disadvantage that mutation of the transgene may lead to the restoration of fertility. With the cytotoxin approach, the cytotoxin gene may mutate so that the cytotoxin is no longer active, or a change in the promoter (methylation or mutation) may lead either to decreased expression or to expression in vegetative tissues, which might have a negative impact on plant growth or vigor. Although the possibilities of such mutations are low, they cannot be completely eliminated. Changes in environmental conditions may trigger the activation of the gene in vegetative tissues. Another reason for loss of gene function could be silencing. Generally, gene silencing takes place when the levels of expression are high (Waterhouse et al., 2001). Because the cells expressing cytotoxin genes die, high levels are never achieved, and therefore silencing is not likely. For antisense constructs, however, the probability of silencing is much greater, especially if multiple copies of the transgene are present. Gene silencing has been studied in aspen, and it has been suggested that transgene inactivation generally depends on the presence of transgene repeats but probably on the type of flanking regions as well (Kumar and Fladung, 2001b; see Chapter 12).

ACCELERATION OF FLOWERING

Forest tree breeding is extremely slow and tedious, largely because of their long generation times. Breeding programs could be shortened by accelerating the flowering transiently during the period of the breeding. This would be especially useful if the traits to be combined could be tested in young plants. Such properties might be disease or pest resistance, or modification of cellular metabolism or cell-wall properties. If the interesting traits are linked to molecular markers, then the presence of genes for such traits, which may only be evident in older plants, could be detected and selected in young plants that could then be used in accelerated breeding programs.

In some cases, the flowering of trees can be accelerated using specific treatments (e.g., GAs in conifers or high CO_2 in birch; Meilan and Strauss, 1997; Lemmetyinen et al., 1998), but these methods are not applicable to all species. Therefore, there is a great interest in the use of transgenic techniques to accelerate the flowering of trees. In this approach, the unlinked fast-flowering transgene would be eliminated by screening for transgene segregation during the last selection step. Alternatively, after completion of the program, the transgene could be eliminated by other means, for example, using an inducible gene excision system such as Cre/loxP (Zuo et al., 2001). Therefore, the resulting trees would not be transgenic.

With transgenic techniques, acceleration of flowering has been achieved in *Arabidopsis* using floral meristem genes. Overexpression of *LFY* or *AP1* with the *CaMV 35S* promoter resulted in very early flowering and the formation of terminal flowers (Weigel and Nilsson, 1995; Mandel and Yanofsky, 1995b). Also, overexpression of several flowering time genes such as *CO* or *FT* (Simon et al., 1996; Kardailsky et al., 1999), as well as MADS-box genes such as *SOC1, AG,* and *FUL* have resulted in early flowering (Samach et al., 2000; Mizukami and Ma, 1992; Ferrándiz et al., 2000). On the other hand, repression of genes inhibiting flowering (e.g., *FLC* or *EMF*) using the antisense RNA technique has led to the acceleration of flowering (Sheldon et al., 2000; Aubert et al., 2001). These results indicate that there should be many possibilities for transgenically modifying flowering time.

The first demonstration of the transgenic acceleration of flowering in trees was achieved by the overexpression of *LFY* which caused flower formation of hybrid aspen after only a few months in tissue culture and after six to seven months of growth in soil (Weigel and Nilsson, 1995). This was remarkable, because aspen generally starts flowering when 8 to 20 years old. In this case, however, the plants did not form normal catkins but formed single male flowers at the ends of the shoots as well as in the axils of the leaves. These flowers were sterile, because the anthers did not open. In addition, the rooting of these plants was poor.

When *PTLF,* the poplar homologue of *LFY,* was overexpressed in poplar, only one of the many tested lines flowered precociously (Rottmann et al., 2000). Competence to react to overexpression of *LFY* also varied widely among the different genotypes tested. However, consistent early flowering was observed in one male genotype. Interestingly, in another genotype the gender of the flowers changed. Again, the anthers did not open. Therefore, it seems that in poplar, *LFY* may not be the best gene to target for accelerated flowering.

Another example of the use of floral meristem genes in trees is in citrus (Peña et al., 2001), which normally flowers at the age of six to ten years. In citrus, overexpression of *AP1* or *LFY* resulted in the production of normal, fertile flowers within one year, and the flowers also produced normal fruits. The seeds of these fruits gave rise to early flowering seedlings, clearly demonstrating the usefulness of this approach in breeding. Interestingly, the annual cycle of flowering did not change: the plants did not flower continuously but started to flower a second time in the next spring.

Three of the birch MADS box genes, *BpMADS3* (a homologue of *AP1*), *BpMADS4,* and *BpMADS5* (homologues of *FUL*), greatly accelerate flowering when overexpressed in tobacco (Elo, Lemmetyinen, Turunen, et al., 2001). In transgenic lines carrying these constructs, the flowers are quite normal and they produce fertile seeds. The progeny are also very early-flowering. When overexpressed in the early-flowering birch line BPM2, *BpMADS4* also caused very early flowering (Elo, Lemmetyinen, Keinonen, et al., 2001). In the earliest flowering line, the plantlets formed inflorescences after about two weeks in soil, at a height of about 2 cm, whereas the controls start inflorescence formation when four months old, at a height of about 50 cm. In this case, all of the meristems turned into inflorescences, and these were much smaller than the normal ones. In a somewhat weaker phenotype, flower formation started about two months after rooting, the plants being about 5 cm high and the inflorescences having normal appearance. The promising feature of the *BpMADS4* is that the overexpression results in normal inflorescences.

CONCLUSIONS

Current strategies for the modification of flowering in trees are primarily based on existing knowledge of the genes that affect flowering time and flower development in model plants, especially *Arabidopsis*. Using this knowledge, many genes involved in flowering have already been isolated and characterized from a number of tree species. Recent programs aiming at characterization of expressed sequence tags (ESTs) will certainly greatly increase the number of genes known in forest trees.

The transformation of trees and phenotypic analysis of the resultant transgenic lines are relatively slow processes. Nevertheless, the results achieved to date with respect to the prevention and the acceleration of tree flowering are encouraging. Consequently, there is good reason to believe that transgenic fast-flowering and sterile lines of economically important tree species will be available in the near future.

The fast-flowering varieties will expedite conventional breeding programs to develop superior lines for plantation forestry. Such superior lines could be precisely analyzed to ensure that the fast-flowering transgene, used to expedite their development, had been completely removed during the process of natural genetic segregation. The variety and precision of current molecular biological techniques allow us to determine whether the smallest unit (i.e., a single nucleotide) of extraneous (e.g., transgenic) DNA is present.

Transgenic sterile tree varieties will also be of considerable value to the future sustainability of the plantation forestry industries. These transgenic sterile lines could be further transformed with other gene constructs, conferring desirable traits, and the resultant progeny could be properly assessed in the natural environment. Such transgenic trees will probably be designed to produce either no flowers, or even no inflorescences, and hence the likelihood of transgene outcrossing to related native trees would be eliminated.

REFERENCES

Alvarez-Buylla, E.R., Liljegren, S.J., Pelaz, S., Gold, S.E., Burgeff, C., Ditta, G.S., Vergara-Silva, F., and Yanofsky, M.F. (2000). MADS-box gene evolution beyond flowers: Expression in pollen, endosperm, guard cells, roots and trichomes. *Plant J.* 24: 457-466.

Araki, T. (2001). Transition from vegetative to reproductive phase. *Curr. Opin. Plant Biol.* 4: 63-68.

Atkinson, M.D. (1992). *Betula pendula* Roth (*B. verrucosa* Ehrh.) and *B. pubescens* Ehrh. *J. Ecol.* 80: 837-870.

Aubert, D., Chen, L., Moon, Y.-H., Martin, D., Castle, L.A., Yang, C.-H., and Sung, Z.R. (2001). EMF1, a novel protein involved in the control of shoot architecture and flowering in Arabidopsis. *Plant Cell* 13: 1865-1875.

Beals, T.P. and Goldberg, R.B. (1997). A novel cell ablation strategy blocks tobacco anther dehiscence. *Plant Cell* 9: 1527-1545.

Blázquez, M.A. (2000). Flower development pathways. *J. Cell Sci.* 113: 3547-3548.

Blázquez, M.A., Soowal, L.N., Lee, I., and Weigel, D. (1997). *LEAFY* expression and flower initiation in *Arabidopsis*. *Development* 124: 3835-3844.

Blázquez, M.A. and Weigel, D. (2000). Integration of floral inductive signals in *Arabidopsis*. *Nature* 404: 889-892.

Brunner, A.M., Dye, S.J., Skinner, J.S., Ma, C., Meilan, R., and Strauss, S.H. (2001). Maturation and flowering in *Populus:* Gene expression changes correlated with reproductive competency and genes for manipulating flowering. *IUFRO/Molecular Biology of Forest Trees,* July 22-27, Stevenson, WA.

Brunner, A.M., Mohamed, R., Meilan, R., Sheppard, L.A., Rottmann, W.H., and Strauss, S.H. (1998). Genetic engineering of sexual sterility in shade trees. *J. Arboriculture* 24: 263-273.

Brunner, A.M., Rottmann, W.H., Sheppard, L.A., Krutovskii, K., DiFazio, S.P., Leonardi, S., and Strauss, S.H. (2000). Structure and expression of duplicate *AGAMOUS* orthologues in poplar. *Plant Mol. Biol.* 44: 619-634.

Chuang, C.-F. and Meyerowitz, E.M. (2000). Specific and heritable genetic interference by double-stranded RNA in *Arabidopsis thaliana*. *Proc. Natl Acad. Sci. USA* 97: 4985-4990.

Clark, J.R. (1983). Age-related changes in trees. *Journal of Arboriculture* 9: 201-205.

Collins, A.J. and Campbell, M.M. (2001). Eucalyptus genes implicated in the control of maturation and flowering. *IUFRO/Molecular Biology of Forest Trees,* July 22-27, Stevenson, WA.

Day, C.D., Galgoci, B.F.C., and Irish, V.F. (1995). Genetic ablation of petal and stamen primordia to elucidate cell interactions during floral development. *Development* 121: 2887-2895.

Decroocq, V., Zhu, X., Kauffman, M., Kyozuka, J., Peacock, W.J., Dennis, E.S., and Llewellyn, D.J. (1999). A *TM3*-like MADS-box gene from *Eucalyptus* expressed in both vegetative and reproductive tissues. *Gene* 228: 155-160.

Dong, Y.-H., Yao, J.-L., Atkinson, R.G., Putterill, J.J., Morris, B.A., and Gardner, R.C. (2000). *MDH1:* An apple homeobox gene belonging to the *BEL1* family. *Plant Mol. Biol.* 42: 623-633.

Dye, S.J., Brunner, A.M., Skinner, J.S., and Strauss, S.H. (2001). Isolation of *TERMINAL FLOWER 1* homologs from *Populus* and expression analyses over a seasonal cycle and continuous maturation gradient. *IUFRO/Molecular Biology of Forest Trees,* July 22-27, Stevenson, WA.

Elo, A.-K., Lemmetyinen, J., Keinonen, K., and Sopanen, T. (2001). Preventing flowering in silver birch by *BpMADS4* antisense. *IUFRO/Molecular Biology of Forest Trees,* July 22-27, Stevenson, WA.

Elo, A.-K., Lemmetyinen, J., Turunen, M.-L., Tikka, L., and Sopanen, T. (2001). Three MADS-box genes similar to *APETALA1* and *FRUITFULL* from silver birch *(Betula pendula). Physiol. Plant.* 112: 95-103.

Ferrándiz, C., Gu, Q., Martienssen, R., and Yanofsky, M.F. (2000). Redundant regulation of meristem identity and plant architecture by *FRUITFULL, APETALA1* and *CAULIFLOWER. Development* 127: 725-734.

Frohlich, M.W. and Parker, D.S. (2000). The mostly male theory of flower evolutionary origins: From genes to fossils. *Syst. Bot.* 25: 155-170.

Fukui, M., Futamura, N., Mukai, Y., Wang, Y., Nagao, A., and Shinohara, K. (2001). Ancestral MADS box genes in sugi, *Cryptomeria japonica* D. Don (Taxodiaceae), homologous to the B function genes in angiosperms. *Plant Cell Physiology* 42: 566-575.

Goldman, M.H.S., Goldberg, R.B., and Mariani, C. (1994). Female sterile tobacco plants are produced by stigma-specific cell ablation. *EMBO J.* 13: 2976-2984.

Hartley, R.W. (1989). Barnase and barstar: Two small proteins to fold and fit together. *TIBS* 14: 450-454.

Hartmann, U., Höhmann, S., Nettesheim, K., Wisman, E., Saedler, H., and Huijser, P. (2000). Molecular cloning of *SVP:* A negative regulator of the floral transition in *Arabidopsis. Plant J.* 21: 351-360.

Hempel, F.D., Weigel, D., Mandel, M.A., Ditta, G.S., Zambryski, P.C., Feldman, L.J., and Yanofsky, M.F. (1997). Floral determination and expression of floral regulatory genes in *Arabidopsis. Development* 124: 3845-3853.

Jack, T. (2001a). Plant development going MADS. *Plant Mol. Biol.* 46: 515-520.

Jack, T. (2001b). Relearning our ABCs: New twists on an old model. *Trends Plant Sci.* 6: 310-316.

Kardailsky, I., Shukla, V.K., Ahn, J.H., Dagenais, N., Christensen, S.K., Nguyen, J.T., Chory, J., Harrison, M.J., and Weigel, D. (1999). Activation tagging of the floral inducer *FT. Science* 286: 1962-1965.

Klopfenstein, N.B., Chun, Y.W., Kim, M.-S., and Ahuja, M.R. (1997). *Micropropagation, Genetic Engineering, and Molecular Biology of Populus. Gen. Tech. Rep. RM-GTR-297.* Fort Collins, CO: U.S. Department of Agriculture, Forest Service, Rocky Mountain Forest and Range Experiment Station.

Koltunow, A.M., Truettner, J., Cox, K.H., Wallroth, M., and Goldberg, R.B. (1990). Different temporal and spatial gene expression patterns occur during anther development. *Plant Cell* 2: 1201-1224.

Koornneef, M., Alonso-Blanco, C., Peeters, A.J.M., and Soppe, W. (1998). Genetic control of flowering time in *Arabidopsis. Annu. Rev. Plant Physiol. Plant Mol. Biol.* 49: 345-370.

Kumar, S. and Fladung, M. (2001a). Controlling transgene integration in plants. *Trends Plant Sci.* 6: 155-159.

Kumar, S. and Fladung, M. (2001b). Gene stability in transgenic aspen *(Populus).* II. Molecular characterization of variable expression of transgene in wild and hybrid aspen. *Planta* 213: 731-740.

Kyozuka, J., Harcourt, R., Peacock, W.J., and Dennis, E.S. (1997). *Eucalyptus* has functional equivalents of the *Arabidopsis AP1* gene. *Plant Mol. Biol.* 35: 573-584.

Landschutze, V., Willmitzer, L., and Muller-Rober, B. (1995). Inhibition of flower formation by antisense repression of mitochondrial citrate synthase in transgenic potato plants leads to a specific disintegration of the ovary tissues of flowers. *EMBO J.* 14: 660-666.

Lee, I., Wolfe, D.S., Nilsson, O., and Weigel, D. (1997). A *LEAFY* co-regulator encoded by *UNUSUAL FLORAL ORGANS. Curr. Biol.* 7: 95-104.

Lemmetyinen, J., Keinonen, K., Elo, A., Lännenpää, M., Pennanen, T., Nowak, A., and Sopanen, T. (2001). Genetic modification of inflorescence development in birch. *IUFRO/Molecular Biology of Forest Trees,* July 22-27, Stevenson, WA.

Lemmetyinen, J., Keinonen-Mettälä, K., Lännenpää, M., von Weissenberg, K., and Sopanen, T. (1998). Activity of the CaMV 35S promoter in various parts of transgenic early flowering birch clones. *Plant Cell Rep.* 18: 243-248.

Lemmetyinen, J., Pennanen, T., Lännenpää, M., Keinonen, K., Mäkelä, H., Hassinen, M., and Sopanen, T. (2001). Birch gene BpMADS1 and its use in prevention of flowering. *IUFRO/Molecular Biology of Forest Trees,* July 22-27, Stevenson, WA.

Lemmetyinen, J., Pennanen, T., Lännenpää, M., and Sopanen, T. (2001). Prevention of flower formation in dicotyledons. *Mol. Breeding* 7: 341-350.

Levy, Y.Y. and Dean, C. (1998). The transition to flowering. *Plant Cell* 10: 1973-1989.

Lin, C. (2000). Photoreceptors and regulation of flowering time. *Plant Physiology* 123: 39-50.

Liu, J., Huang, Y., Ding, B., and Tauer, C.G. (1999). cDNA cloning and expression of a sweetgum gene that shows homology with *Arabidopsis AGAMOUS. Plant Science* 142: 73-82.

Liu, J.J. and Podila, G.P. (1997). Characterization of a MADS box gene (Y09611) from immature female cone of red pine (PGR97-032). *Plant Physiology* 113: 665.

Lohmann, J.U., Hong, R.L., Hobe, M., Busch, M.A., Parcy, F., Simon, R., and Weigel, D. (2001). A molecular link between stem cell regulation and floral patterning in *Arabidopsis. Cell* 105: 793-803.

Longman, K.A. (1976). Some experimental approaches to the problem of phase change in forest trees. *Acta Horticulturae* 56: 81-90.

Mandel, M.A. and Yanofsky, M.F. (1995a). The Arabidopsis *AGL8* MADS box gene is expressed in inflorescence meristems and is negatively regulated by *APETALA1. Plant Cell* 7: 1763-1771.

Mandel, M.A. and Yanofsky, M.F. (1995b). A gene triggering flower formation in *Arabidopsis. Nature* 377: 522-524.

Mariani, C., De Beuckeleer, M., Truettner, J., Leemans, J., and Goldberg, R.B. (1990). Induction of male sterility in plants by a chimaeric ribonuclease gene. *Nature* 347: 737-741.

Meilan, R. and Strauss, S.T. (1997). Poplar genetically engineered for reproductive sterility and accelerated flowering. In N.B. Klopfenstein, Y.W. Chun, M.-S. Kim, and M.R. Ahuja (Eds.), *Micropropagation, Genetic Engineering, and Molecular Biology of Populus* (pp. 212-219). General Technical Report RM-GTR-297. Fort Collins, CO: U.S. Department of Agriculture, Forest Service, Rocky Mountain Forest and Range Experiment Station.

Mellerowicz, E.J., Horgan, K., Walden, A., Coker, A., and Walter, C. (1998). *PRFLL*—A *Pinus radiata* homologue of *FLORICAULA* and *LEAFY* is expressed in buds containing vegetative shoot and undifferentiated male cone primordia. *Planta* 206: 619-629.

Meyer, P. and Saedler, H. (1996). Homology-dependent gene silencing in plants. *Annu. Rev. Plant Physiol. Plant Mol. Biol.* 47: 23-48.

Michaels, S.D. and Amasino, R.M. (2001). Loss of *FLOWERING LOCUS C* activity eliminates the late-flowering phenotype of *FRIGIDA* and autonomous pathway mutations but not responsiveness to vernalization. *Plant Cell* 13: 935-941.

Mizukami, Y., Huang, H., Tudor, M., Hu, Y., and Ma, H. (1996). Functional domains of the floral regulator AGAMOUS: Characterization of the DNA binding domain and analysis of dominant negative mutations. *Plant Cell* 8: 831-845.

Mizukami, Y. and Ma, H. (1992). Ectopic expression of the floral homeotic gene *AGAMOUS* in transgenic *Arabidopsis* plants alters floral organ identity. *Cell* 71: 119-131.

Mizukami, Y. and Ma, H. (1995). Separation of AG function in floral meristem determinacy from that in reproductive organ identity by expressing antisense AG RNA. *Plant Mol. Biol.* 28: 767-784.

Mouradov, A., Glassick, T., Hamdorf, B., Murphy, L., Fowler, B., Marla, S., and Teasdale, R.D. (1998). *NEEDLY*, a *Pinus radiata* ortholog of *FLORICAULA/LEAFY* genes, expressed in both reproductive and vegetative meristems. *Proc. Natl. Acad. Sci. USA* 95: 6537-6542.

Mouradov, A., Glassick, T.V., Hamdorf, B.A., Murphy, L.C., Marla, S.S., Yang, Y., and Teasdale., R.D. (1998). Family of MADS-box genes expressed early in male and female reproductive structures of Monterey pine. *Plant Physiology* 117: 55-61.

Mouradov, A., Hamdorf, B., Teasdale, R.D., Kim, J.T., Winter, K.-U., and Theissen, G. (1999). A *DEF/GLO*-like MADS-box gene from a gymnosperm: *Pinus radiata* contains an ortholog of angiosperm B class floral homeotic genes. *Dev. Genet.* 25: 245-252.

Müller, B.M., Saedler, H., and Zachgo, S. (2001). The MADS-box gene *DEFH28* from *Antirrhinum* is involved in the regulation of floral meristem identity and fruit development. *Plant J.* 28: 169-179.

Nilsson, O. and Weigel, D. (1997). Modulating the timing of flowering. *Curr. Opin. Biotech.* 8: 195-199.

Nilsson, O., Wu, E., Wolfe, D.S., and Weigel, D. (1998). Genetic ablation of flowers in transgenic *Arabidopsis*. *Plant J.* 15: 799-804.

Ohto, M., Onai, K., Furukawa, Y., Aoki, E., Araki, T., and Nakamura, K. (2001). Effects of sugar on vegetative development and floral transition in *Arabidopsis*. *Plant Physiology* 127: 252-261.

Page, T., Macknight, R., Yang, C.-H., and Dean, C. (1999). Genetic interactions of the *Arabidopsis* flowering time gene *FCA*, with genes regulating floral initiation. *Plant J.* 17: 231-239.

Pelaz, S., Gustafson-Brown, C., Kohalmi, S.E., Crosby, W.L., and Yanofsky, M.F. (2001). *APETALA1* and *SEPALLATA3* interact to promote flower development. *Plant J.* 26: 385-394.

Peña, L., Martín-Trillo, M., Juárez, J., Pina, J.A., Navarro, L., and Martínez-Zapater, J.M. (2001). Constitutive expression of *Arabidopsis LEAFY* or *APETALA1* genes in citrus reduces their generation time. *Nat. Biotechnol.* 19: 263-267.

Pnueli, L., Abu-Abeid, M., Zamir, D., Nacken, W., Schwarz-Sommer, Z., and Lifschitz, E. (1991). The MADS box gene family in tomato: Temporal expression during floral development, conserved secondary structures and homology with homeotic genes from *Antirrhinum* and *Arabidopsis. Plant J.* 1: 255-266.

Pnueli, L., Hareven, D., Broday, L., Hurwitz, C., and Lifschitz, E. (1994). The *TM5* MADS box gene mediates organ differentiation in the three inner whorls of tomato flowers. *Plant Cell* 6: 175-186.

Pnueli, L., Hareven, D., Rounsley, S.D., Yanofsky, M.F., and Lifschitz, E. (1994). Isolation of the tomato *AGAMOUS* gene *TAG1* and analysis of its homeotic role in transgenic plants. *Plant Cell* 6: 163-173.

Poethig, R.S. (1990). Phase change and the regulation of shoot morphogenesis in plants. *Science* 250: 923-930.

Ratcliffe, O.J., Amaya, I., Vincent, C.A., Rothstein, S., Carpenter, R., Coen, E.S., and Bradley, D.J. (1998). A common mechanism controls the life cycle and architecture of plants. *Development* 125: 1609-1615.

Ratcliffe, O.J., Nadzan, G.C., Reuber, T.L., and Riechmann, J.L. (2001). Regulation of flowering in *Arabidopsis* by an *FLC* homologue. *Plant Physiology* 126: 122-132.

Richards, D.E., King, K.E., Ait-Ali, T., and Harberd, N.P. (2001). How gibberellin regulates plant growth and development: A molecular genetic analysis of gibberellin signaling. *Annu. Rev. Plant Physiol. Plant Mol. Biol.* 52: 67-88.

Rigola, D., Pè, M.E., Fabrizio, C., Mè, G., and Sari-Gorla, M. (1998). *CaMADS1*, a MADS box gene expressed in the carpel of hazelnut. *Plant Mol. Biol.* 38: 1147-1160.

Rishi, A.S., Nelson, N.D., and Goyal, A. (2001). Genetic modification for improvement of *Populus. Physiol. Mol. Biol. Plants* 7: 7-21.

Rottmann, W.H., Meilan, R., Sheppard, L.A., Brunner, A.M., Skinner, J.S., Ma, C., Cheng, S., Jouanin, L., Pilate, G., and Strauss, S.H. (2000). Diverse effects of overexpression of *LEAFY* and *PTLF*, a poplar *(Populus)* homolog of *LEAFY/ FLORICAULA*, in transgenic poplar and *Arabidopsis. Plant J.* 22: 235-245.

Rutledge, R., Regan, S., Nicolas, O., Fobert, P., Côté, C., Bosnich, W., Kauffeldt, C., Sunohara, G., Séguin, A., and Stewart, D. (1998). Characterization of an *AGAMOUS* homologue from the conifer black spruce *(Picea mariana)* that produces floral homeotic conversions when expressed in *Arabidopsis. Plant J.* 15: 625-634.

Samach, A., Onouchi, H., Gold, S.E., Ditta, G.S., Schwarz-Sommer, Z., Yanofsky, M.F., and Coupland, G. (2000). Distinct roles of CONSTANS target genes in reproductive development of *Arabidopsis. Science* 288: 1613-1616.

Sheldon, C.C., Rouse, D.T., Finnegan, E.J., Peacock, W.J., and Dennis, E.S. (2000). The molecular basis of vernalization: The central role of *FLOWERING LOCUS C (FLC). Proc. Natl. Acad. Sci. USA* 97: 3753-3758.

Sheppard, L.A., Brunner, A.M., Krutovskii, K.V., Rottmann, W.H., Skinner, J.S., Vollmer, S.S., and Strauss, S.H. (2000). A *DEFICIENS* homolog from the dioecious tree black cottonwood is expressed in female and male floral meristems of the two-whorled, unisexual flowers. *Plant Physiology* 124: 627-639.

Shore, P. and Sharrocks, A.D. (1995). The MADS-box family of transcription factors. *Eur. J. Biochem.* 229: 1-13.

Sieburth, L.E. and Meyerowitz, E.M. (1997). Molecular dissection of the *AGAMOUS* control region shows that *cis* elements for spatial regulation are located intragenically. *Plant Cell* 9: 355-365.

Simon, R., Igeño, M.I., and Coupland, G. (1996). Activation of floral meristem identity genes in *Arabidopsis*. *Nature* 384: 59-62.

Simpson, G.G., Gendall, A.R., and Dean, C. (1999). When to switch to flowering. *Annu. Rev. Cell Dev. Biol.* 99: 519-550.

Skinner, J.S., Meilan, R., Brunner, A.M., and Strauss, S.H. (2000). Options for genetic engineering of floral sterility in forest trees. In S.M. Jain and S.C. Minocha (Eds.), *Molecular Biology of Woody Plants* (pp. 135-153). Dordrecht, Netherlands: Kluwer Academic Publishers.

Southerton, S.G., Marshall, H., Mouradov, A., and Teasdale, R.D. (1998). Eucalypt MADS-box genes expressed in developing flowers. *Plant Physiology* 118: 365-372.

Southerton, S.G., Strauss, S.H., Olive, M.R., Harcourt, R.L., Decroocq, V., Zhu, X., Llewellyn, D.J., Peacock, W.J., and Dennis, E.S. (1998). *Eucalyptus* has a functional equivalent of the *Arabidopsis* floral meristem identity gene *LEAFY*. *Plant Mol. Biol.* 37: 897-910.

Southerton, S.G., Uren, T., Yang, Y., Watson, J., Furbank, B., Dennis, L., Peacock, J., and Moran, G. (2001). Flowering control in eucalypts. *IUFRO/Molecular Biology of Forest Trees*, July 22-27, Stevenson, WA.

Strauss, S.H., DiFazio, S.P., and Meilan, R. (2001). Genetically modified poplars in context. *Forest. Chron.* 77: 271-279.

Strauss, S.H., Rottmann, W.H., Brunner, A.M., and Sheppard, L.A. (1995). Genetic engineering of reproductive sterility in forest trees. *Mol. Breeding* 1: 5-26.

Sundås-Larsson, A., Landberg, K., and Meeks-Wagner, D.R. (1998). The *TERMINAL FLOWER2 (TFL2)* gene controls the reproductive transition and meristem identity in *Arabidopsis thaliana*. *Genetics* 149: 597-605.

Sundström, J. (2001). Evolution of genetic mechanisms regulating reproductive development in plants. Characterisation of MADS-box genes active during cone development in Norway spruce. *Acta Universitatis Upsaliensis, Comprehensive Summaries of Uppsala Dissertations from the Faculty of Science and Technology 612* (44 p). Uppsala, Sweden: Uppsala University, Tryck & Medier.

Sundström, J., Carlsbecker, A., Svensson, M.E., Svenson, M., Johanson, U., Theissen, G., and Engström, P. (1999). MADS-box genes active in developing pollen cones of Norway spruce *(Picea abies)* are homologous to the B-class floral homeotic genes in angiosperms. *Dev. Genet.* 25: 253-266.

Sung, S.-K. and An, G. (1997). Molecular cloning and characterization of a MADS-box cDNA clone of the Fuji apple. *Plant Cell Physiology* 38: 484-489.

Sung, S.-K., Yu, G.-H., and An, G. (1999). Characterization of *MdMADS2*, a member of the *SQUAMOSA* subfamily of genes, in apple. *Plant Physiology* 120: 969-978.

Sung, S.-K., Yu, G.-H., Nam, J., Jeong, D.-H., and An, G. (2000). Developmentally regulated expression of two MADS-box genes, *MdMADS3* and *MdMADS4*, in the morphogenesis of flower buds and fruits in apple. *Planta* 210: 519-528.

Tandre, K., Albert, V.A., Sundås, A., and Engström, P. (1995). Conifer homologues to genes that control floral development in angiosperms. *Plant Mol. Biol.* 27: 69-78.

Tandre, K., Svenson, M., Svensson, M.E., and Engström, P. (1998). Conservation of gene structure and activity in the regulation of reproductive organ development of conifers and angiosperms. *Plant J.* 15: 615-623.

Telfer, A. and Poethig, R.S. (1998). *HASTY:* A gene that regulates the timing of shoot maturation in *Arabidopsis thaliana. Development* 125: 1889-1898.

Theissen, G. (2001). Development of floral organ identity: Stories from the MADS house. *Curr. Opin. Plant Biol.* 4: 75-85.

Theissen, G., Becker, A., Di Rosa, A., Kanno, A., Kim, J.T., Münster, T., Winter, K.-U., and Saedler, H. (2000). A short history of MADS-box genes in plants. *Plant Mol. Biol.* 42: 115-149.

Tilly, J.J., Allen, D.W., and Jack, T. (1998). The CArG boxes in the promoter of the *Arabidopsis* floral organ identity gene *APETALA3* mediate diverse regulatory effects. *Development* 125: 1647-1657.

Tzfira, T., Zuker, A., and Altman, A. (1998). Forest-tree biotechnology: Genetic transformation and its application to future forests. *Trends in Biotechnology* 16: 439-446.

Vahala, T., Oxelman, B., and von Arnold, S. (2001). Two *APETALA2*-like genes of *Picea abies* are differentially expressed during development. *J. Exp. Bot.* 52: 1111-1115.

Walden, A.R., Wang, D.Y., Walter, C., and Gardner, R.C. (1998). A large family of TM3 MADS-box cDNAs in *Pinus radiata* includes two members with deletions of the conserved K domain. *Plant Science* 138: 167-176.

Waterhouse, P.M., Wang, M.-B., and Lough, T. (2001). Gene silencing as an adaptive defence against viruses. *Nature* 411: 834-842.

Weigel, D. and Meyerowitz, E.M. (1994). The ABCs of floral homeotic genes. *Cell* 78: 203-209.

Weigel, D. and Nilsson, O. (1995). A developmental switch sufficient for flower initiation in diverse plants. *Nature* 377: 495-500.

Wesley, S.V., Helliwell, C.A., Smith, N.A., Wang, M., Rouse, D.T., Liu, Q., Gooding, P.S., Singh, S.P., Abbott, D., Stoutjesdijk, P.A., et al. (2001). Construct design for efficient, effective and high-throughput gene silencing in plants. *Plant J.* 27: 581-590.

Yanofsky, M.F. (1995). Floral meristems to floral organs: Genes controlling early events in *Arabidopsis* flower development. *Annu. Rev. Plant Physiol. Plant Mol. Biol.* 46: 167-188.

Yao, J.-L., Dong, Y.-H., Kvarnheden, A., and Morris, B. (1999). Seven MADS-box genes in apple are expressed in different parts of the fruit. *J. Amer. Soc. Hort. Sci.* 124: 8-13.

Yao, J.-L., Dong, Y.-H., and Morris, B.A.M. (2001). Parthenocarpic apple fruit production conferred by transposon insertion mutations in a MADS-box transcription factor. *Proc. Natl. Acad. Sci. USA* 98: 1306-1311.

Yoshida, N., Yanai, Y., Chen, L., Kato, Y., Hiratsuka, J., Miwa, T., Sung, Z.R., and Takahashi, S. (2001). EMBRYONIC FLOWER 2, a novel polycomb group protein homolog, mediates shoot development and flowering in *Arabidopsis*. *Plant Cell* 13: 2471-2481.

Zuo, J., Niu, Q.-W., Møller, S.G., and Chua, N.-H. (2001). Chemical-regulated, site-specific DNA excision in transgenic plants. *Nat. Biotechnol.* 19: 157-161.

Chapter 12

Stability of Transgene Expression in Aspen

Sandeep Kumar
Matthias Fladung

INTRODUCTION

The production of transgenic plants is an expanding component of plant biotechnology. The technology has significant potential for the genetic improvement of long-lived forest trees. Genetic transformation protocols are increasingly becoming available for a number of tree species. However, for commercial success of tree transgenics it is important that the expression of transgenes is stable throughout their life cycle. The introduced traits need to be transmitted faithfully through successive generations in a predictable manner before the use of transgenic trees in the forest tree breeding programs. The evaluation of transgene stability is more difficult in trees compared to annual plants because of long reproductive cycles in trees. However, forest trees with extended vegetative periods of several years provide an opportunity to study the somatic stability of a foreign gene which may remain unnoticed in short rotation plants (Kumar and Fladung, 2001b).

A transgene incorporated into the plant genome is integrated randomly and in unpredictable copy numbers. The transgene integrates often in the form of repeats which can suppress the expression of the transgene (Assad et al., 1993; Hobbs et al., 1990, 1993; Kilby et al., 1992; Matzke et al., 1994; Kumar and Fladung, 2000a). Although single-copy transgenes can also be silenced (Matzke and Matzke, 1998; Kooter et al., 1999; Kumar and Fladung, 2001b), the unstable expression is more frequently correlated with the incidence of high transgene copy number (Assad et al., 1993; Kumar and Fladung, 2001b). Complex integration sites containing transgene repeats may either undergo structural instability because of intrachromosomal recombination between multiple copies, sometimes resulting in loss of the transgene (Srivastava et al., 1996; Fladung, 1999), or chemical modification, such as methylation resulting in silencing (Mette et al., 1999; reviewed in Kooter et al., 1999; Vaucheret and Fagard, 2001).

The position of the integration site can also influence the stability of the expression (Matzke and Matzke, 1998). Chromosome locations that promote transgene expression are not well characterized but are thought to be transcriptionally active regions of euchromatin. The heterochromatin state of the chromosome is believed to be inaccessible to transcription factors and often correlates with cytosine hypermethylation and histone hypoacetylation (Ng and Bird, 1999; Meyer, 2000). A consequence of this chromosomal architecture is that the transgenes randomly integrate in the proximity of heterochromatin showing variable expression (Pröls and Meyer, 1992; Van Blokland et al., 1997; Tulin et al., 1998; Iglesias et al., 1997). The transgene expression may also be affected by the regulatory sequences of nearby host genes if it becomes inserted into euchromatin, a transcriptionally active region (Herman et al., 1990; Kerbundit et al., 1991; Koncz et al., 1989; Kumar and Fladung, 2001b).

In this chapter, we focus on the stability of transgene expression in aspen, a model tree species. This includes the molecular analyses of transgenic lines showing inactivation of morphologically visible *rolC* gene. The results obtained were compared to transgenic lines showing stable *rolC* phenotype to determine the molecular basis of stable or unstable transgene expression (Fladung, 1999; Kumar and Fladung, 2000a, 2001b). The molecular analyses included the determination of integrated T-DNA structure and transgene promoter methylation. To study the influence of genomic context on transgene expression, T-DNA flanking genomic DNA sequences were determined both in stable and unstable transgenic lines (Kumar and Fladung, 2001b). Polymerase chain reaction (PCR)-based methods (Kumar and Fladung, 2000b) developed for the early screening of potentially unstable lines for the expression of transgenes and the mechanism of *Agrobacterium* T-DNA (transferred DNA) transfer in aspen (Kumar and Fladung, 2002b) will also be discussed.

ROLC *GENE, A PHENOTYPIC MARKER*
FOR THE STABILITY OF TRANSGENE EXPRESSION

The *rolC* gene of *Agrobacterium rhizogenes* is a dominant pleiotropic gene which can be used as a morphological marker of transgene expression. This system based on visible modification of leaf size and leaf color, together with reduced apical dominance, has already been used in tobacco (Schmülling et al., 1988; Spena et al., 1987) and potato (Fladung and Ballvora, 1992; Fladung and Gieffers, 1993). The *rolC* gene has successfully been transferred to the aspen genome, and similar to the results obtained in tobacco and potato, it was shown that 35S-*rolC* altered the growth and development in aspen as well (Figure 12.1, Fladung et al., 1996, 1997). The morphological variation in rbc-*rolC* transgenic plants, however, was not as conspicuous as 35S-*rolC* transgenic plants. Therefore, 35S-*rolC* model system was used to study and monitor the long-term stable expression of a foreign gene in wild *(Populus tremula)* and hybrid aspen *(P. tremula × P. tremuloides)*. The variations in the morphological features of *rolC* phenotype (Figure 12.1) were later substantiated with the molecular analyses to determine and establish a molecular basis of transgene silencing in wild/ hybrid aspen (Fladung, 1999; Kumar and Fladung, 2001b; Fladung and Kumar, 2002).

*MORPHOLOGICAL ANALYSIS OF 35S-*ROLC *TRANSGENIC ASPEN PLANTS*

The transgenic lines were monitored for the visible *rolC* expression for a period of five to six years during their continuous growth under in vitro, greenhouse, and field conditions. Under in vitro conditions, two transgenic lines obtained from wild aspen showed variable *rolC* phenotype, whereas complete *rolC* inactivation was observed in one line. All in vitro-cultured hybrid aspen-based transgenic lines maintained a consistent *rolC* phenotype. Major alterations in *rolC* expression in most of the transgenic lines were observed after transfer from in vitro conditions to greenhouse or field. Incomplete suppression of the transgene expression was observed in three hybrid-based transgenic lines out of the 15 lines transferred to the greenhouse. Among five wild aspen transgenic lines transferred to the greenhouse, three lines were observed with altered or reverted transgene expression (Kumar and Fladung, 2001b). Two of the field-planted transgenic lines, one each from hybrid and wild aspen, showed variable transgene expression. We did not observe any transgene inactivation in the other three field-grown hybrid aspen lines. Interestingly, all wild aspen-based transgenic lines showed variable *rolC* expression in one stage or another (Kumar and Fladung, 2001b). The Northern blot analysis supported the morphological data for the most of the transgenic lines, showing the absence of *rolC* transcript. However, the *rolC* transcript was observed in some transgenic lines with reverted *rolC* phenotype indicating different forms of transgene silencing in these transgenic lines (Fladung and Kumar, 2002).

FIGURE 12.1. Transgenic aspen (left) harboring *rolC* gene is reduced in size with smaller leaves compared to control plant (right)

TRANSGENE SILENCING

The transgene silencing phenomenon has been distinguished in two general categories. The first class of transgene inactivation represents the homology-dependent gene silencing (HDGS) and occurs when multiple copies of a specific sequence are present in a genome. Depending on the level at which silencing occurs, HDGS has been distinguished into transcriptional gene silencing (TGS) occurring through repression of transcription and posttranscriptional gene silencing (PTGS) via mRNA degradation (reviewed in Vaucheret and Fagard, 2001). Transcriptionally silenced transgenes acquire metastable epigenetic states that are characterized by altered methylation patterns and chromatin structure. Reversible promoter methylation may be used as a useful marker for TGS (Kooter et al., 1999). The second category of transgene silencing is termed position effect, in which the flanking plant DNA and/or chromosomal location negatively influence the expression of the single transgene loci. In a simplistic view, it is "how" and "where" transgene(s) is (are) integrated into the host genome that determines the stability of the expression of transgene(s). Therefore, the factors and mechanisms responsible for the transgene integration and repeat formation might be crucial for the expression of integrated foreign genes.

TRANSGENE REPEAT FORMATION AND UNSTABLE EXPRESSION

The comparison of Southern blot analysis to the morphological observations obtained in aspen revealed that all the transgenic lines containing more than one copy of the transgene were showing variation in the *rolC* phenotype. Further PCR-based characterization of the structure of transgene(s) showed that all multiple-copy transgenic lines contained transgenes in the form of repeats (Kumar and Fladung, 2001b). Some of the single-copy transgenic lines were also found positive to rpPCR (reverse primer PCR; Kumar and Fladung, 2000b), indicating transgene repeats in these lines. The sequencing of the repeat junctions showed that most of these lines contained incomplete repeats, and the part of the T-DNA containing the restriction site for the enzyme used in Southern blotting was deleted. The PCR-based 35S promoter methylation (Kumar and Fladung, 2000b; Fladung and Kumar, 2002a) revealed that the 35S promoter was methylated in all the transgenic lines containing T-DNA repeats and showing variable *rolC* phenotype. These results indicated that rpPCR combined with PCR-based promoter methylation could be a quick and inexpensive method for the initial screening to detect potential unstable transgenic lines for the transgene expression.

The rpPCR utilizes a set of primer pairs, and primers in each pair are oriented in opposite directions in the construct (Figure 12.2; Kumar and Fladung, 2000b) so that no amplification is possible in case of single-copy integration without T-DNA repeat (Figure 12.2B). The amplification products will always be obtained when transgenes are integrated in the form of repeats. The results obtained with different primer pairs explain the position and status (complete or truncated) of the repeat. However, amplification of inverted repeats may sometimes not be feasible when a single primer is used because of palindromic sequences. Therefore, additional pairs of primers have been suggested to address this problem (Figure 12.2C,D; Kumar and Fladung, 2001b). Similarly, promoter methylation can be studied restricting genomic DNA with methylation-sensitive enzymes, for example, *HpaII* or *MspI,* which are differentially sensitive to internal cytosine methylation at the sequence CCGG (Fu et al., 2000; Kumar and Fladung, 2000b). The restricted DNA is then subjected to PCR amplification using primers designed specifically to amplify the promoter region, positive amplification indicating methylation of the promoter at the restriction site. For a positive control, a primer pair is added for the amplification of a small genomic region.

The rpPCR and promoter methylation methods were developed using transgenic lines showing morphological inactivation of *rolC* gene. However, the method has already been extended to screen a large number of independent aspen and hybrid aspen transgenic lines, harboring different gene constructs. Out of randomly selected transgenic lines, roughly 21 percent were found to carry transgene repeats (Kumar and Fladung, 2000a). The molecular analysis of transgenic plants using traditional methods, for example, Southern blotting, is a reliable, but time-consuming and expensive process. In practice, up to ten transgenic lines may have to be analyzed to obtain a single line in which the transgene is expressed in the desired temporal and spatial manner (Wallace et al., 2000). This could be a realistic practice for the annual plants because a few selected transgenic lines for the stable expression of transgene(s) may be used as stock plants for further breeding programs. Unfortunately, long reproductive cycles of trees do not allow making use of a few donor transgenic lines in a forest tree breeding program. Therefore, a large number of stably expressing primary transgenic lines are required for the commercial success of tree transgenics. This, in turn, will require a large-scale screening of transgenic plants. The rpPCR could be a viable method for the primary screening of a large number of transgenic plants. The transgenic lines that did not show transgene repeats may later undergo diagnostic Southern blot analysis to ensure single-copy transgene integration.

(A) Transgene

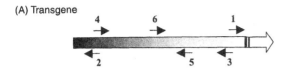

(B) Direct repeat (head to tail)

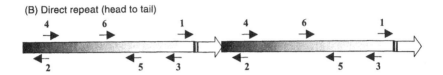

(C) Inverted repeat (tail to tail)

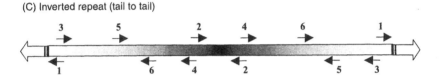

(D) Inverted repeat (head to head)

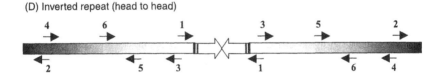

FIGURE 12.2. Schematic diagrams of possible transgene repeat formations in transgenic plants. (A) Transgene with left (shaded) and right (open arrow) borders. The numbers (1-6) represent the locations of different primers used and arrowheads indicate their respective directions (5'-3'). (B) A direct T-DNA complete repeat formation in head (right border)-to-tail (left border) integration. Amplification products of expected sizes using five primer pairs (1+2, 1+5, 1+3, 2+6, and 2+4) will confirm the presence of a complete direct repeat. It is impossible to get PCR amplification using these primers in case of a single-copy integration or when multiple transgene copies are integrated at different chromosomal locations. (C) An inverted transgene repeat formation in tail-to-tail orientation that can be detected using three primer pairs (2+5, 2+3, and 5+3). (D) Similarly, an inverted repeat in head-to-head orientation may be determined by another set of primer pairs (1+6, 1+4, and 6+4). It is possible to obtain amplification from a single primer in case of inverted repeats, leading to multiple PCR products using primer pairs just described.

TRANSGENE INTEGRATION
AND UNSTABLE EXPRESSION

We also observed variable *rolC* expression in three transgenic aspen lines despite single-copy transgene integration. A separate category of transgene silencing concerns position effects, in which genomic position of the integrated transgene affects its expression. This refers to repressive influences exerted on transgenes by flanking plant DNA and/or an unfavorable chromosomal location. This type of silencing presumably reflects the epigenetic state of neighboring host sequences or the relative tolerance of particular chromosomal regions to invasion by alien DNA (Matzke and Matzke, 1998). Position effect may be studied by analyzing genomic environments of differentially expressed transgenes. We studied the genomic context of unstably expressed transgenes in aspen, analyzing flanking genomic regions in three wild aspen-based transgenic lines unstable for the *rolC* expression (Kumar and Fladung, 2001b). In two of these transgenic lines the transgene-flanking genomic regions were found to contain AT-rich nucleotide regions. This was in sharp contrast to the non-AT-rich flanking regions with "normal" nucleotide distribution in three hybrid aspen-based transgenic lines marked stable for the *rolC* expression (Fladung, 1999). The high AT values observed in the wild aspen transgenic lines differ from the compositional distribution reported for the transcribed chromosomal regions in dicots which show AT-content around 54 percent (Salinas et al., 1988). AT-rich and/or tandem repetitive DNA sequences are major DNA components of "constitutive heterochromatin" which is usually found in pericentromeric regions (Hankeln et al., 1996). Euchromatic genes brought into juxtaposition with heterochromatin may exhibit "position effect variegation" (Cryderman et al., 1999).

MECHANISMS OF TRANSGENE INTEGRATION
AND REPEAT FORMATION

Among various methods of plant transformation *Agrobacterium tumefaciens*-based transformation is well established and better studied. *Agrobacterium tumefaciens* is the most widely utilized vector for genetic transformation of plants. It harbors a Ti (tumor inducing) plasmid which encodes most of the major functions required for virulence (Zambryski, 1988; Zambryski et al., 1989). *Agrobacterium* induces neoplastic growth in many plant species by transporting a single-strand version of the T-DNA into the plant genome (Dumas et al., 2001). The T-DNA itself does not encode any gene important for the transfer process. The T-DNA resides on the Ti plasmid and is delimited by 25 base pair (bp)-long imperfect direct repeats at its ends. Although the wild-type T-DNA carries Ti genes, it is possible to place any DNA inserted between the T-DNA borders which will be transferred into the host cell and imported into its nucleus (Ballas and Citovsky, 1997; reviewed in Kumar and Fladung, 2002a).

The mechanisms of transgene integration in annual plants have been characterized based on the nucleotide sequence analysis of the target sites carried out in different *Nicotiana* and *Arabidopsis* transgenic plants lines before and after the integration (Gheysen et al., 1991; Matsumoto et al., 1990; Mayerhofer et al., 1991; reviewed in Tinland and Hohn, 1995; Tinland, 1996). These studies showed that T-DNA integration in plants is a mode of illegitimate recombination, the main feature being that T-DNA integration resulted in the deletion of the target DNA (Mayerhofer et al., 1991). We have analyzed T-DNA/plant junctions in 30 aspen and hybrid aspen transgenic lines (Kumar and Fladung, 2002b). T-DNA integration loci were compared to their respective preinsertion sites for ten transgenic lines. The results suggest that similar to annual plants, T-DNA integration in aspen is also a mode of illegitimate recombination which could be obtained via single strand annealing followed by ligation. However, a second class of T-DNA integration was also found that showed deletions in the right border and/or complex filler formation. A T-DNA integration model has been suggested for these events which is similar to the synthesis-dependent strand-annealing (SDSA) model described to repair double-strand break (DSB) repair in somatic plant cells (Puchta, 1998).

T-DNA repeat formation and subsequent transgene silencing has been described in annual plants (Flavell, 1994; Kooter et al., 1999) which mainly includes *Arabidopsis* (Kilby et al., 1992; Assad et al., 1993), tobacco (Krizkova and Hrouda, 1998; Hobbs et al., 1990, 1993; Matzke et al., 1994; Park et al., 1996), and petunia (Linn et al., 1990; Stam et al., 1997). Several factors are believed to have influence on T-DNA rearrangements, including the *Agrobacterium* strain used, the transformation protocol, the genotype of the plant recipient and tissues thereof, and the selection pressure exerted (Tinland and Hohn, 1995). Knowledge about the influence of plasmid structure on transgene integration and rearrangement is important to optimize plant transformation constructs and methods in the future (Kohli et al., 1999). A number of papers have been published describing the fate of exogenous DNA and suggesting mechanistic models for occurrence of T-DNA repeats in annual plants (Gorbunova and Levy, 1997; Krizkova and Hrouda, 1998; Salomon and Puchta, 1998; De Groot et al., 1994; De Neve et al., 1997; De Buck et al., 1999; Kohli et al., 1999). An attempt has been made to study T-DNA repeat formation in a tree system as well (Kumar and Fladung, 2000a). T-DNA repeat junctions in eleven independent aspen transgenic lines were analyzed. Direct repeats were observed in ten transgenic lines, and in one case an inverted repeat was found with tail-to-tail integration (see Figure 12.2). The transgenic lines analyzed represented two aspen clones and six different transgene constructs from two *Agrobacterium tumefaciens* strains. The data obtained were summarized in a mechanistic model for T-DNA rearrangements in the aspen genome. The model suggests simultaneous integration of two independent T-DNAs into a genomic hotspot leading to transgene repeat formation (Kumar and Fladung, 2000a).

HOW TO OBTAIN STABLY EXPRESSING TRANSGENIC PLANTS

It is obvious from the work reviewed in this chapter on the stability of transgene expression in aspen, including some examples from annual plants, that extensive variation exists in the expression among independent transformants. Whether genomic position and/or HDGS based on multiple transgene copies are involved in the variable expression, unpredictable transgene expression levels can create economic concern. For example, if transgenic trees are used for the plantations that do not express the transgenic trait properly, such as insect or pest resistance or male/female sterility, major economic and/or ecological losses can be incurred. In the basic research studies, the unpredictable expression of a desired introduced trait makes the process lengthy and complex because several individual transformation events must be compared to understand the gene function (Matzke and Matzke, 1998; Fladung, 1999; Vaucheret and Fagard, 2001). Furthermore, variation among different transgenic lines is a major obstacle for the precise comparisons between different gene constructs.

There are two possible approaches to overcome the problem of transgene inactivation in plants. First, extensive screening and testing could be carried out prior to the release of transgenic plants for public use. Between two classes of gene silencing, the silencing based on the presence of multiple transgene copies might be comparatively simple to define and easy to screen using molecular tools including the PCR-based methods described here. Alternatively, transformation protocols may be improved to obtain transgenic plants containing a single-copy transgene in high frequency (Kumar and Fladung, 2001a). However, variation in the transgene expression because of position effect is rather complex and requires testing over multiple generations. Although expensive, this is perhaps a feasible process for annual plants. For long-lived trees, however, this could not be a viable proposition.

The second approach to address the issue of expression variability and gene silencing systematically could be to precisely modify or target defined locations within the genome (reviewed in Vergunst and Hooykaas, 1999; Kumar and Fladung, 2002a). Targeting a single copy of a transgene into a predetermined plant genomic location provides an efficient tool for securing long-term stable expression. This can be achieved by site-specific recombination (SSR) or by homologous recombination (HR). Site-specific recombination can be used to place a recognition target within the genome so that the transgene can be precisely placed into the target. The SSR reaction requires the action of site-specific recombinase and the target sequence that is recognized by the recombinase (reviewed in Kumar and Fladung, 2002a). Transgenic plants containing these target sequences can be created and characterized for the stable expression of the target locus using some marker gene. These transgenic plants will serve as "stock" plants for the future transgenic work integrating new transgenes precisely into the predefined target locus which is characterized for the stable expression free from position effect variegation. However, regardless of the chromosome position, there is still a probability of methylation-associated silencing specific to the transgene (Day et al., 2000).

Homologous recombination involves the exchange of covalent linkages between DNA molecules in the regions of highly similar or identical sequence. The HR differs from the SSR because HR does not require a previously introduced target site and can be used to target a transgene to any chosen locus in the genome. Although a number of studies of gene targeting via homologous recombination in plants have been reported, the process remains challenging and inefficient (Miao and Lam, 1995; Risseeuw et al., 1997; Thykjaer et al., 1997). Some alternative approaches have been suggested to obtain HR-based gene targeting in plants (Kumar and Fladung, 2001a, 2002a) that need to be tested.

FUTURE PERSPECTIVES

Gene transfer technology has an important role to play for the future silvicultural trait development of the tree species. The transgene incorporated into the host genome with the current transgenics procedures is integrated randomly and in unpredictable copy numbers, often in the form of repeats abolishing the expression of the transgene. Therefore, transformation procedures need to be improved to obtain single-copy integration free from methylation, or quick screening methods may be designed to screen multiple integration events. The integration site also has a profound effect on expression of the transgene which is affected by inherent and extrinsic factors that may trigger methylation and reduce stability of the expression. Using site-specific recombination in annual plants, transgenes have been successfully delivered into a specific chromosome position and expressed at a predictable level yielding a higher and more consistent level of transgene expression. Efforts should also be made to test such a recombination system in a tree species in order to break the barriers for the genetic improvement created by long reproductive cycles in these species.

REFERENCES

Assad, F.F., Tucker, K.L., and Signer, E.R. (1993). Epigenetic repeat-induced gene silencing (RIGS) in *Arabidopsis*. *Plant Mol. Biol.* 22: 1067-1085.

Ballas, N. and Citovsky, V. (1997). Nuclear localization signal binding protein from *Arabidopsis* mediates nuclear import of *Agrobacterium* VirD2 protein. *Proc. Natl. Acad. Sci. USA* 94: 10723-10728.

Cryderman, D.E., Morris, E.J., Biessmann, H., Elgin, S.C.R., and Wallrath, L.L. (1999). Silencing of *Drosophila* telomeres: Nuclear organization and chromatin structure play critical roles. *EMBO J.* 18: 3724-3735.

Day, C.D., Lee, E., Kobayashi, J., Holappa, L.D., Albert, H., and Ow, D.W. (2000). Transgene integration into the same chromosome location can produce alleles that express at predictable level, or alleles that are differentially silenced. *Genes and Dev.* 14: 2869-2880.

De Buck, S., Jacobs, A., Van Montagu, M., and Depicker, A. (1999). The DNA sequences of T-DNA junctions suggest that complex DNA loci are formed by recombination process resembling T-DNA integration. *Plant J.* 20: 295-304.

De Groot, M.J., Offringa, R., Groet, J., Does, M.J,, van Hooykaas, P.J., and dan Elzen, P.J. (1994). Non-recombinant background in gene targeting: Illegitimate recombination between *hpt* gene and defective 5" deleted *nptII* gene can restore a km^r phenotype in tobacco. *Plant. Mol. Biol.* 25: 721-733.

De Neve, M., De Buck, S., Jacobs, A., Van Montagu, M., and Depicker, A. (1997). T-DNA integration patterns in co-transformed plant cells suggest that T-DNA repeats originate from co-integration of separate T-DNAs. *Plant J.* 11: 15-29.

Dumas, F., Duckely, M., Pelczar, P., Van Gelder, P., and Hohn, B. (2001). An *Agrobacterium* VirE2 channel for transferred-DNA transport into plant cells. *Proc. Natl. Acad. Sci. USA* 98: 485-490.

Fladung, M. (1999). Gene stability in transgenic aspen *(Populus)*. I. Flanking DNA sequences and T-DNA structure. *Mol. Gen. Genet.* 260: 574-581.

Fladung, M. and Ballvora, A. (1992). Further characterization of *rolC* transgenic tetraploid potato clones, and influence of day length and level of *rolC* expression on yield parameters. *Plant Breed.* 109: 18-27.

Fladung, M. and Gieffers, W. (1993). Resistance reactions of leaves and tubers of *rolC* transgenic tetraploid potato to bacterial and fungal pathogens. Correlation with sugar, starch and chlorophyll content. *Physiol. Mol. Plant. Pathol.* 42: 123-132.

Fladung, M. and Kumar, S. (2002). Gene stability in transgenic aspen-*Populus*. III. T-DNA repeats influence transgene expression differentially among different transgenic lines. *Plant Biology* 4: 329-338.

Fladung, M., Kumar, S., and Ahuja, M.R. (1997). Genetic transformation of *Populus* genotypes with different chimeric gene constructs: Transformation efficiency and molecular analysis. *Trans. Res.* 6: 111-121.

Fladung, M., Muhs, H.-J., and Ahuja, M.R. (1996). Morphological changes observed in transgenic *Populus* carrying the *rolC* gene from *Agrobacterium rhizogenes*. *Silvae Genetica* 45: 349-354.

Flavell, R.B. (1994). Inactivation of gene expression in plants as a consequence of specific sequence duplication. *Proc. Natl. Acad. Sci. USA* 91: 3490-3496.

Fu, X., Kohli, A., Twyman, R.M., and Christou, P. (2000). Alternative silencing effects involve distinct types of non-spreading cystosine methylation at a three gene, single-copy transgenic locus in rice. *Mol. Gen. Genet.* 263: 106-118.

Gheysen, G., Villarroel, R., and Van Montagu, M. (1991). Illegitimate recombination in plants: A model for T-DNA integration. *Genes and Dev.* 5: 287-297.

Gorbunova, V. and Levy, A.A. (1997). Non-homologous DNA end joining in plant cells is associated with deletions and filler DNA insertions. *Nucl. Acids Res.* 25: 4650-4657.

Hankeln, T., Winterpacht, A., and Schmidt, E.R. (1996). Instability of tandem repetitive DNA in "natural" and transgenic organisms. In E.R. Schmidt and T. Hankeln (Eds.), *Transgenic Organisms and Biosafety* (pp. 181-188). Berlin, Heidelberg: Springer-Verlag.

Herman, L., Jacobs, A., Van Montagu, M., and Depicker, A. (1990). Plant chromosome/marker gene fusion assay for study of normal and truncated T-DNA integration events. *Mol. Gen. Genet.* 224: 248-256.

Hobbs, S.L.A., Kpodar, P., and De Long, C.M.O. (1990). The effect of T-DNA copy number, position and methylation on reporter gene expression in tobacco transformants. *Plant. Mol. Biol.* 15: 851-864.

Hobbs, S.L.A., Warkentinm, T.D., and De Long, C.M.O. (1993). Transgene copy number can be positively or negatively associated with transgene expression. *Plant. Mol. Biol.* 21: 17-26.

Iglesias, V.A., Moscone, E.A., Papp, I., Neuhuber, F., Michalowski, S., Phelan, T., Spiker, S., Matzke, M., and Matzke, A.J. (1997). Molecular and cytogenetic analyses of stably and unstably expressed transgene loci in tobacco. *Plant Cell* 9: 1251-1264.

Kertbundit, S., Degreve, H., Deboeck, F., Van Montagu, M., and Hernalsteens, J.P. (1991). *In vivo* random *beta-glucuronidase* gene fusions in *Arabidopsis thaliana*. *Proc. Natl. Acad. Sci. USA* 95: 7203-7208.

Kilby, N.J., Leyser, H.M.O., and Furner, I.J. (1992). Promoter methylation and progressive transgene inactivation in *Arabidopsis*. *Plant. Mol. Biol.* 20: 103-112.

Kohli, A., Griffiths, S., Palacios, N., Twyman, R.M., Vain, P., Laurie, D.A., and Christou, P. (1999). Molecular characterization of transforming plasmid rearrangements in transgenic rice reveals a recombination hotspot in the CaMV 35S promoter and confirms the predominance of microhomology mediated recombination. *Plant J.* 17: 591-601.

Koncz, C., Martini, N., Mayerhofer, R., Koncz, K.Z., Körber, H., Redei, G.P., and Schell, J. (1989). High frequency T-DNA-mediated gene tagging in plants. *Proc. Natl. Acad. Sci. USA* 86: 8467-8471.

Kooter, J.M., Matzke, M.A., and Meyer, P. (1999). Listening to the silent genes: Transgene silencing, gene regulation and pathogen control. *Trends Plant. Sci.* 4: 340-347.

Krizkova, L. and Hrouda, M. (1998). Direct repeats of T-DNA integrated in tobacco chromosome: Characterization of junctions region. *Plant J.* 16: 673-680.

Kumar, S. and Fladung, M. (2000a). Determination of transgene repeat formation and promoter methylation in transgenic plants. *Bio Techniques* 28: 1128-1137.

Kumar, S. and Fladung, M. (2000b). Transgene repeats in aspen: Molecular characterization suggests simultaneous integration of independent T-DNAs into receptive hotspots in the host genome. *Mol. Gen. Genet.* 264: 20-28.

Kumar, S. and Fladung, M. (2001a). Controlling transgene integration in plants. *Trends Plant Sci.* 6: 155-159.

Kumar, S. and Fladung, M. (2001b). Gene stability in transgenic aspen *(Populus)*. II. Molecular characterization of variable expression of transgene in wild and hybrid aspen. *Planta* 213: 731-740.

Kumar, S. and Fladung, M. (2002a). Gene targeting in plants. In S.M. Jain, D.S. Brar, and B.S. Ahloowalia (Eds.), *Molecular Techniques in Crop Improvement* (pp. 481-500). Dordrecht, Netherlands: Kluwer Academic Publishers.

Kumar, S. and Fladung, M. (2002b). Transgene integration in aspen: Structures of integration sites and mechanism of T-DNA integration. *Plant J.* 31: 543-551.

Linn, F., Heidmann, I., Saedler, H., and Meyer, P. (1990). Epigenetic changes in the expression of the maize A1 gene in *Petunia hybrida:* Role of numbers of integrated gene copies and state of methylation. *Mol. Gen. Genet.* 222: 329-336.

Matsumoto, S., Ito, Y., Hosoi, T., Takahashi, Y., and Machida, Y. (1990). Integration of *Agrobacterium* T-DNA into a tobacco chromosome: Possible involvement of DNA homology between T-DNA and plant DNA. *Mol. Gen. Genet.* 224: 309-316.

Matzke, A.J. and Matzke, M.A. (1998). Position effects and epigenetic silencing of plant transgenes. *Curr. Opin. Plant Biol.* 1: 142-148.

Matzke, A.J.M., Neuhuber, F., Park, Y.-D., Ambros, P.F., and Matzke, M.A. (1994). Homology-dependent gene silencing in transgenic plants: Epistatic silencing loci contain multiple copies of methylated transgenes. *Mol. Gen. Genet.* 244: 219-229.

Mayerhofer, R., Koncz-Kalman, Z., Nawrath, C., Bakkeren, G., Crameri, A., Angelis, K., Redei, G.P., Schell, J., Hohn, B., and Koncz, C. (1991). T-DNA integration: A mode of illegitimate recombination in plants. *EMBO J.* 10: 697-704.

Mette, M.F., Winden, J.V.D., Matzke, M.A., and Matzke, A.J.M. (1999). Production of aberrant promoter transcripts contributes to methylation and silencing of unlinked homologous promoters in trans. *EMBO J.* 18: 241-248.

Meyer, P. (2000). Transcriptional transgene silencing and chromatin components. *Plant Mol. Biol.* 43: 221-234.

Miao, Z.H. and Lam, E. (1995). Targeted disruption of the TGA3 locus in *Arabidopsis thaliana*. *Plant J.* 7: 359-365.

Ng, H.-H. and Bird, A.P. (1999). DNA methylation and chromatin modification. *Curr. Opin. Genet. Dev.* 9: 158-163.

Park, Y.-D., Papp, I., Moscone, E.A., Iglesias, V.A., Vaucheret, H., Matzke, A.J.M., and Matzke, M.A. (1996). Gene silencing mediated by promoter homology occurs at the level of transcription and results in meiotically heritable alterations in methylation and gene activity. *Plant J.* 9: 183-194.

Pröls, F. and Meyer, P. (1992). The methylation patterns of chromosomal integration regions influence gene activity of transferred DNA in *Petunia hybrida*. *Plant J.* 2: 465-475.

Puchta, H. (1998). Repair of genomic double-strand breaks in somatic plant cells by one-sided invasion of homologous sequences. *Plant J.* 13: 331-339.

Risseeuw, E., Franke-van Dijk, M.E., and Hooykaas, P.J. (1997). Gene targeting and instability of *Agrobacterium* T-DNA loci in the plant genome. *Plant J.* 11: 717-728.

Salinas, J., Matassi, G., Montera, L.M., and Bernards, G. (1988). Compositional compartmentalization and compositional patterns in nuclear genome of plants. *Nucl. Acids Res.* 16: 4269-4285.

Salomon, S. and Puchta, H. (1998). Capture of genomic and T-DNA sequences during double-strand break repair in somatic plant cells. *EMBO J.* 17: 6086-6095.

Schmülling, T., Schell, J., and Spena, A. (1988). Single genes from *Agrobacterium rhizogenes* influence plant development. *EMBO J.* 7: 2621-2629.

Spena, A., Schmülling, T., Koncz, C., and Schell, J. (1987). Independent and synergistic activity of *rol A, B* and *C* loci in stimulating abnormal growth in plants. *EMBO J.* 6: 3891-3899.

Srivastava, V., Vasil, V., and Vasil, I. K. (1996). Molecular characterization of the fate of transgenes in transformed wheat (*Triticum aestivum* L.). *Theor. Appl. Genet.* 92: 1031-1037.

Stam, M., de Bruin, R., Kenter, S., van der Hoorn, R.A.L., van Blokland, R., Mol, J.N.M., and Kooter, J.M. (1997). Post-transcriptional silencing of chalcone synthase in *Petunia* by inverted transgene repeats. *Plant J.* 12: 63-82.

Thykjaer, T., Finnemann, J., Schauser, L., Christensen, L., Poulsen, C., and Stougaard, J. (1997). Gene targeting approaches using positive-negative selection and large flanking regions. *Plant. Mol. Biol.* 35: 523-530.

Tinland, B. (1996). The integration of T-DNA into plant genomes. *Trends Plant Sci.* 1: 178-184.

Tinland, B. and Hohn, B. (1995). Recombination between prokaryotic and eukaryotic DNA: Integration of *Agrobacterium tumefaciens* T-DNA into the plant genome. *Genet. Eng. (NY)* 17: 209-229.

Tulin, A.V., Naumova, N.M., Aravin, A.A., and Gvozdev, V.A. (1998). Repeated, protein-encoding heterochromatic genes cause inactivation of a juxtaposed euchromatic gene. *FEBS Lett.* 425: 513-516.

Van Blokland, R., Ten Lohuis, M., and Meyer, P. (1997). Condensation of chromatin in transcriptional regions of an inactivated plant transgene: Evidence for an active role of transcription in gene silencing. *Mol. Gen. Genet.* 257: 1-13.

Vaucheret, H. and Fagard, M. (2001). Transcriptional gene silencing in plants: Targets, inducers and regulators. *Trends Genet.* 17: 29-35.

Vergunst, A.C. and Hooykaas, P.J.J. (1999). Recombination in plant genome and its application in biotechnology. *Crit. Rev. Plant Sci.* 18: 1-31.

Wallace, H., Ansell, R., Clark, J., and McWhir, J. (2000). Pre-selection of integration sites imparts repeatable transgene expression. *Nucl. Acids Res.* 28: 1455-1464.

Zambryski, P. (1988). Basic processes underlying *Agrobacterium*-mediated DNA transfer to plant cells. *Annu. Rev. Genet.* 22: 1-30.

Zambryski, P., Tempe, J., and Schell, J. (1989). Transfer and function of T-DNA genes from *Agrobacterium* Ti and Ri plasmids in plants. *Cell* 56: 93-201.

Chapter 13

Asexual Production of Marker-Free Transgenic Aspen Using MAT Vector Systems

Hiroyasu Ebinuma
Koichi Sugita
Etsuko Matsunaga
Saori Endo
Keiko Yamada-Watanabe

INTRODUCTION

Recently, dramatic progress has occurred in the fields of both recombinant DNA and tissue culture, and these have become essential technologies for studies in plant science. They have also been applied to crop improvement, and many transgenic crops with novel characters have been produced. In the United States and Canada, many kinds of transgenic crops have been commercialized and cultivated on a large scale in the field. The commercialization of transgenic trees lags far behind that of transgenic crops, and relevant risk assessments and small-scale field trials are underway to assess their economic value. Forest trees have long generations and rotation cycles because of their perennial nature. Because conventional breeding requires long periods of time for characteristics to be modified, recombinant DNA techniques offer the great advantage of making it possible to shorten the breeding time. However, for the commercialization of transgenic trees, essential technologies (gene isolation, transformation, propagation) must be developed and three major problems must be overcome: (1) Large life-science companies in the European Union and United States have already patented almost all essential technologies. It is difficult to obtain licenses at a reasonable price due to severe competition. (2) The transformation of woody plants is still difficult, since their regeneration frequency is very low compared to that of herbaceous plants. Long periods of tissue culture and selection time are needed to produce transgenic trees. (3) Because transgenic trees have long life cycles, long periods of time and large areas are needed to evaluate their modified characteristics. In particular, a field

309

trial of transgenic trees is indispensable to investigate growth rates and wood characteristics. However, regulatory issues require a lengthy and expensive risk-assessment process at a well-equipped greenhouse before field trials in Japan are allowed. Therefore, for the commercialization of transgenic trees, it is now more important to develop technologies to reduce their environmental impact than to increase their economic value. We have developed a novel transformation method to increase the regeneration frequency of transgenic trees without using antibiotic selection and to reduce their environmental impact by removing a selection marker gene (review in Ebinuma, Sugita, Matsunaga, and Yamakada 1997; Ebinuma, Sugita, Matsunaga, Yamakado, and Komamine 1997; Ebinuma et al. 2000, 2001, 2002; Ebinuma and Komamine 2001). In this chapter, we discuss (1) transformation methods using *Agrobacterium,* (2) problems of antibiotic selection, (3) the principles of the MAT (Multi-Auto-Transformation) vector system, (4) the transformation of hybrid aspen, and (5) gene stacking by repeated transformation.

TRANSFORMATION METHODS
USING AGROBACTERIUM

Agrobacterium tumefaciens is widely used for plant transformation. The epoch-making development in plant genetic engineering resulted from the basic study of plant disease "crown galls." Phytopathogenic *Agrobacterium (Agrobacterium tumefaciens, A. rhizogenes)* infect a wide range of plant species and induce "crown galls" or "hairy roots." During infection, agrobacteria transfer the oncogenes located on the T-DNA of plasmids into plant cells and integrate them into the plant genome. The oncogenes are expressed in plant cells, where they modify the cell's hormonal level and sensitivity. Of the current transformation methods, we have used the *Agrobacterium*-mediated gene transfer system to produce transgenic plants (reviewed in Hooykaas and Schilperoort 1992). The oncogenes are removed from the T-DNA of plasmids, since they cause typical symptoms of crown gall or hairy root disease to transgenic plants. The gene of interest is inserted into the disarmed plasmids instead of the oncogenes and introduced into plant cells using *Agrobacterium*. Figure 13.1 shows the process of *Agrobacterium*-mediated transformation. Usually, the gene of interest is inserted into binary plasmids together with the kanamycin-resistance gene *(nptII)*. The *nptII* gene codes for an enzyme that detoxifies antibiotic kanamycin. Both of these genes are cointroduced into plant cells by infection with *Agrobacterium,* and then infected tissues are cultivated on medium containing kanamycin and plant growth regulators. The transgenic cells can grow

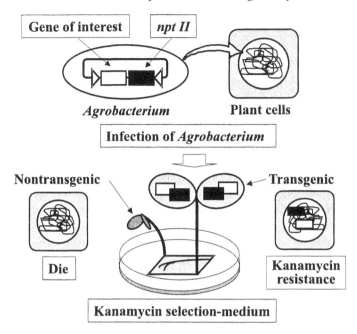

FIGURE 13.1. *Agrobacterium*-mediated transformation. The gene of interest is inserted into a binary plasmid together with the kanamycin-resistance gene *(nptII)*. Both genes are cointroduced into plant cells by infection with *Agrobacterium,* and the infected tissues are then cultivated on medium containing kanamycin and plant growth regulators. The transgenic cells grow and survive, while nontransgenic cells cannot grow and therefore die.

and survive, while nontransgenic cells cannot grow and therefore die. In current transformation methods, selectable marker genes and plant growth regulators are essential for identifying transgenic cells and regenerating transgenic plants.

PROBLEMS OF ANTIBIOTIC SELECTION

Considerable advances in crop improvement have been made through conventional breeding. Usually, breeding programs involve the crossing of parents with desirable traits and the selection of superior progeny. The development of recombinant technologies has resulted in the rapid introduction of novel traits into major crops. Transgenic plants with novel traits have been widely used as breeding materials to produce commercial varieties. However, current transformation methods have three pitfalls regarding their integration into breeding programs that are discussed in this section.

Regeneration of Transgenic Cells

Selective agents (e.g., antibiotics, herbicides) and the corresponding resistance genes (selectable marker genes) are used to identify and separate rare transgenic cells from nontransgenic cells. Transgenic cells can grow and survive on medium containing the antibiotic or herbicide, while nontransgenic cells cannot grow and therefore die. However, dying nontransgenic cells inhibit the supply of nutrients to transgenic cells or excrete toxic compounds. These negative effects decrease the ability of transgenic cells to proliferate and differentiate into transgenic plants.

Environmental Impact

Usually, the selectable marker genes remain in the transgenic plants after transformation. Regardless of the assurances provided in risk-assessment reports, the use of antibiotic-resistance genes has particularly led to public criticism. It is feared that although there might be some pathogenic bacteria in the field, there is also the possibility of bacteria-bacteria spread of antibiotic genes from field bacteria to reservoirs of pathogenic bacteria present in animals and humans due to gene transfer, which could lead to a public health disaster. Regulatory issues also require a lengthy and expensive risk-assessment process for each selectable marker and their products.

Transgene Stacking

Many desirable genes are stacked to improve the complex engineered traits of crops through breeding programs. Sexual crossings or retransformation are used to introduce additional novel traits to transgenic plants. However, when transgenic plants are associated with a selectable marker gene, the same marker gene cannot be used for the selection of double-transformed plants. Compared to the great number of desirable traits, there are very few suitable marker genes. When many transgenes are stacked by crossings, the copy number of selectable marker genes increases. The stacking of highly expressed genes enhances the possibility of homology-dependent gene silencing.

Therefore, it is desirable to develop a selection system using positive markers to reduce these negative effects and a system for removing selectable marker genes so that the same selectable marker gene can be reused for sequential transformation (reviewed in Ebinuma et al. 2001).

PRINCIPLES OF THE MAT
(MULTI-AUTO-TRANSFORMATION) VECTOR SYSTEM

We previously developed a new transformation method to overcome the problems associated with current transformation systems (Ebinuma, Sugita, Matsunaga, Yamakado, and Komamine 1997; Sugita et al. 1999; Sugita, Kasahara, et al. 2000; Sugita, Matsunaga, et al. 2000; Endo et al. 2001, 2002; Matsunaga et al., 2002). We called our system the MAT vector system due to its multifunctional traits (multiple introductions of genes by repeated transformation, autonomous regeneration and selection of transgenic plants, and transformation). The idea behind the MAT vector system is to imitate in vivo transformation with *Agrobacterium*. In the field, agrobacteria can infect a wide range of plant species and induce crown galls or hairy roots, which comprise transgenic cells. They do not need an aseptic facility, tissue culture media, or plant growth regulators for transformation. Recently, great advances have been made in in vitro tissue culture techniques. However, in vitro techniques still have limitations regarding current transformation methods. Therefore, the most promising approach is to develop in vivo transformation methods that plant breeders can use for plant improvement without in vitro tissue culture techniques. An objective of the MAT vector system is to improve in vivo transformation systems with *Agrobacterium* and develop non-tissue culture-based transformation methods.

The amazing capability of *Agrobacterium* for in vivo transformation depends on the function of the oncogenes on the T-DNA of plasmids. During infection, the oncogenes are transferred and integrated into the plant genome. The oncogenes manipulate the hormonal level and sensitivity of transgenic cells and induce their proliferation in vivo (reviewed in Gaudin et al. 1994). Current transformation systems remove the oncogenes from T-DNA and replace them with marker genes or the gene of interest because regenerated transgenic plants exhibit seriously abnormal phenotypes. On the other hand, the MAT vector system modifies the oncogenes and places them on T-DNA together with the gene of interest. Figure 13.2 shows a diagram of the MAT vector plasmid and its mechanism. The MAT vector system is designed to use oncogenes for the proliferation and regeneration of transgenic cells. The oncogenes are combined with the site-specific recombination system (R/*RS*) and removed from transgenic plants after transfor-

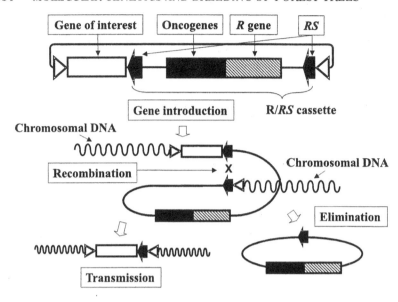

FIGURE 13.2. Mechanism of the MAT vector system. The MAT vector system is designed to use oncogenes for the proliferation and regeneration of transgenic cells. The oncogenes and the recombinase gene *(R)* are flanked by two directly oriented *RS* sites and placed on the T-DNA region of a binary vector plasmid. The gene of interest is inserted into the T-DNA region outside of the R/*RS* cassette. These genes on the T-DNA region are transferred to plant cells and integrated into the genome by infection with *Agrobacterium.* The *R* gene is expressed in transgenic cells, and R recombinase excises the R/*RS* cassette from the plant genome. The resulting transgenic plants have only the gene of interest and one *RS* site in the genome. (*Source:* Adapted from Ebinuma and Komamine 2001.)

mation to recover the normal phenotype. The R/*RS* system is derived from the plasmid pSR1 of *Zygosaccharomyces rouxii* and consists of R recombinase and its recognition site *(RS)* (Matsuzaki et al. 1990; Onouchi et al. 1991). R recombinase mediates recombination between *RS* recognition sites and excises the DNA fragment flanked by the two *RS* sites in the same orientation. The MAT vectors are derivatives of a disarmed binary vector plasmid pBI121 and have oncogenes and the recombinase gene *(R)* flanked by two directly oriented *RS* sites in the T-DNA region (Table 13.1). The gene of interest is inserted into the T-DNA region outside of the R/*RS* cassette and introduced into *Agrobacterium*. During infection, T-DNA is trans-

TABLE 13.1. A list of MAT vector plasmids

Plasmid	Gene of interest	R/RS cassette
pBI121	Nos-*nptII*, 35S-*uidA*	
pIPT5	Nos-*nptII*, 35S-*uidA*, 35S-*ipt*	
pIPT10	Nos-*nptII*, 35S-*uidA*, Native-*ipt*	
pIPT20	Nos-*nptII*, 35S-*uidA*, rbcS-*ipt*	
pIPTIMH	Nos-*nptII*, 35S-*uidA*, Native-*ipt*, *iaaM/H*	
pTL7	*lacZ'*	
pNPI132	Nos-*nptII*, 35S-*uidA*	35S-*R*, 35S-*ipt*
pMAT8	*lacZ'*	GSTII-*R*, Native-*ipt*
pMAT8GUS	35S-*uidA*	GSTII-*R*, Native-*ipt*
pMAT8GFP	Nos-*GFP*	GSTII-*R*, Native-*ipt*
pMATIMH	Nos-*uidA*	GSTII-*R*, Native-*ipt*, *iaaM/H*
pRBI11	Nos-*nptII*, 35S-*uidA*	GSTII-*R*, rbcS-*ipt*

Note: MAT vectors, which are derivatives of the binary plasmid pBI121, have the gene of interest and the R/RS cassette. *NptII* = neomycin phosphotransferase gene; *-uidA* = β-glucuronidase gene; *GFP* = green fluorescent protein gene; *ipt* = isopentenyl transferase gene; *iaaM/H* = tryptophan monooxygenase and indoleacetamide hydrolase genes; *R* = recombinase gene; *RS* = recognition sequence; 35S = CaMV35S promoter; Nos = Nos promoter; GSTII = GST-II-27 promoter; Native = native *ipt* promoter; rbcS = *rbcS* promoter.

ferred to plant cells and integrated into the genome. The *R* gene is expressed in transgenic cells, and R recombinase excises the R/RS cassette from the plant genome. The resulting transgenic plants have only the gene of interest and one RS site in their genome. We refer to these as marker-free transgenic plants.

Agrobacterium tumefaciens infects many kinds of woody plants, and the oncogenes that comprise the *ipt* and *iaaM/H* genes induce crown gall tumors at wound sites (reviewed in Gaudin et al. 1994). The *ipt* gene codes for isopentenyl transferase which catalyzes cytokinin synthesis (Akiyoshi et al. 1984; Barry et al. 1984). The *iaaM* gene codes for tryptophan 2-monooxygenase, and the *iaaH* gene codes for indoleacetamide hydrolase, which catalyze auxin synthesis (IAA) (Tomashow et al. 1984). Cytokinin and auxin are major plant growth regulators that control growth and development in plants. Skoog and Miller (1957) demonstrated that the addition of cytokinin to culture medium containing auxin could induce cell division in cultured tobacco tissues. Later, it was established that cytokinin, in combination with auxin, stimulates plant cell division and determines the direction of plant cell differentiation. A high cytokinin-to-auxin ratio triggers shoot formation but inhibits root induction, whereas a low ratio produces the opposite effect. In current transformation methods, cytokinin and auxin are widely used to control the proliferation and differentiation of transgenic cells. On the other hand, the MAT vector system uses the *ipt* and *iaaM/H* genes to manipulate endogenous hormone levels instead of having to apply them exogenously. This approach has great potential to increase the transformation efficiency of recalcitrant plant species and also to develop non-tissue-culture-based transformation through the autonomous differentiation of transgenic cells in vivo.

TWO-STEP TRANSFORMATION OF TOBACCO PLANTS USING THE GST-MAT VECTOR

The *ipt*-type MAT vector combines the *ipt* gene with the R/*RS* system so that it can be removed from transgenic plants after selection. Figure 13.3 shows a two-step transformation method for generating marker-free transgenic plants. First, transgenic plants are selected as the *ipt*-shooty phenotype, and marker-free transgenic plants that appear from them are then separated. The R/*RS* system removes the *ipt* gene to generate marker-free transgenic shoots. However, the excision of R/*RS* cassettes before shoot formation reduces the regeneration of transgenic shoots. Therefore, we use an inducible promoter to control the expression of the *R* gene and to induce excision after selection of transgenic shoots. Figure 13.4 shows a diagram of the GST-MAT vector pMAT8. The pMAT8 vector has the *ipt* gene with the native promoter and the *R* gene with the GST-II-27 promoter. The GST-II-27 promoter is a chemically inducible promoter of the glutathione-*S*-transferase (GST-II-27) gene from *Zea mays,* and the herbicide antidote "Safener" induces its expression (Holt et al. 1995). The *gusA* gene with the 35S promoter is inserted into the *lacZ* multicloning sites of pMAT8 as a model gene of interest to create a pMAT8GUS vector.

(A) Infection (B) Regeneration (C) *ipt*-shooty (D) Marker-free

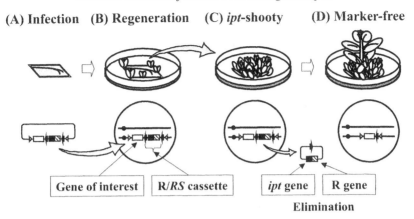

FIGURE 13.3. Two-step transformation method. First, the *ipt* gene regenerates transgenic shoots, and then the R/*RS* system removes the *ipt* and *R* genes to generate marker-free transgenic shoots containing only the gene of interest. (A) Gene transfer by infection with *Agrobacterium;* (B) regeneration of transgenic shoots; (C) formation of the *ipt*-type phenotypes; (D) appearance of marker-free transgenic plants. (*Source:* Adapted from Ebinuma et al. 2001.)

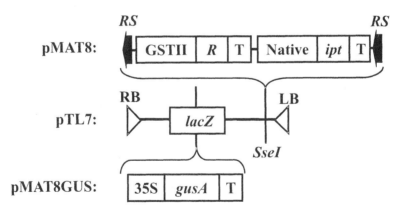

FIGURE 13.4. Diagram of GST-MAT vectors. pTL7 is a derivative of the binary plasmid pBI121 which has *lacZ* multicloning sites and an *SseI* site for insertion of a gene of interest and a MAT cassette, respectively. pMAT8 has an R/*RS* cassette composed of the native *ipt* and GST-II-27-*R* genes. The cassette is inserted into a *SseI* site of pTL7 containing the 35S-*gusA* gene to create pMAT8GUS. *ipt* = isopentenyl transferase gene; *R* = recombinase gene; *RS* = recognition sequence; *gusA* =ß-glucuronidase gene; 35S = CaMV^{35}S promoter; GSTII = GST-II-27 promoter; Native = native *ipt* promoter; T = nopaline synthase terminator; RB and LB = right and left border sequences of a T-DNA.

The production of marker-free transgenic plants using the GST-MAT vector pMAT8GUS involves four main steps (Sugita, Kasahara, et al. 2000; Figure 13.5).

Gene Transfer by Infection

Sixty pieces of leaf segments of tobacco (*Nicotiana tabaccum* cv. SR1) plants were infected with *A. tumefaciens* containing the pMAT8GUS vector and cocultivated on hormone-free MS agar medium containing 50 mg/l acetosyringone for three days.

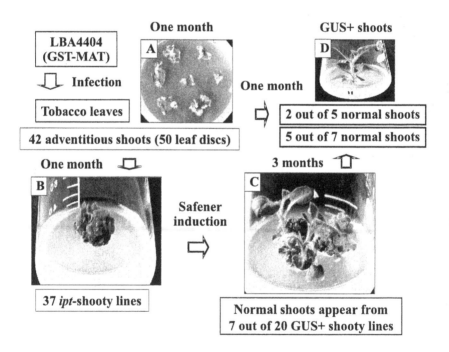

FIGURE 13.5. Transformation of tobacco plants using the GST-MAT vector pMAT8GUS. (A) Regeneration of adventitious shoots from leaf segments; (B) an *ipt*-shooty phenotype lacking apical dominance and rooting ability; (C) appearance of normal shoots exhibiting apical dominance from *ipt*-shooty clones; (D) a marker-free transgenic plant (MF). (*Source:* Adapted from Ebinuma and Komamine 2001.)

Regeneration of Transgenic Plants

The explants were transferred to hormone-free MS agar medium containing 500 mg/l carbenicillin but not kanamycin (nonselective medium). One month after infection, 42 regenerated adventitious buds were separated from the leaf segments and transferred to the same medium. After one month of cultivation, we visually classified these developed buds into two distinct phenotypes: (1) five normal shoots and (2) 37 *ipt*-shooty phenotypes. These abnormal shoots had lost apical dominance and rooting ability due to the overproduction of cytokinin.

Appearance of Marker-Free Plants

Twenty GUS-positive *ipt*-shooty clones were subcultured monthly to the same fresh medium and Safener-induction MS medium containing 30 mg/l Safener. Several normal shoots appeared from seven of 20 *ipt*-shooty clones within three months of induction with Safener. These shoots were transferred to the same medium, grew normally, and rooted.

DNA Analysis of Normal Plants

All 12 normal shoots were subjected to polymerase chain reaction (PCR) analysis. A predicted *ipt* fragment was not amplified in two of five (40 percent) normal shoots that developed directly from the adventitious shoots, and five of seven (70 percent) normal shoots appeared from seven *ipt*-shooty clones, but an excision product was amplified by PCR analysis. Seven of 12 (58 percent) normal shoots were marker-free transgenic plants, and five (42 percent) normal shoots were nontransgenic escapes. These results show that we can both rapidly produce marker-free transgenic plants without the production of *ipt*-shooty intermediates and induce the generation of marker-free transgenic plants using Safener. We investigated the copy number of the *gusA* gene in these seven marker-free transgenic plants by Southern blot analysis and found that six of seven (86 percent) marker-free transgenic tobacco plants had a single *gusA* gene, while the remaining plant had two genes. These results indicate that the GST-MAT vector is useful for producing marker-free transgenic plants containing a single-copy transgene at high frequency.

TRANSFORMATION OF HYBRID ASPEN
USING THE GST-MAT VECTOR

The *ipt*-type MAT vector uses the *ipt* gene to regenerate transgenic shoots and select transgenic plants as the *ipt*-shooty phenotype. However, during transformation, nontransgenic shoots are regenerated together with transgenic shoots, since the overproduction of cytokinin by the *ipt* gene causes it to leak out from transgenic cells and promote the regeneration of nontransgenic cells around the transgenic cells. Therefore, we separated regenerated shoots and subcultured them on hormone-free medium to visually discriminate transgenic shoots from nontransgenic escapes. The endogenous levels of plant hormones and the cell responses to plant growth regulators are very different depending on the plant species, plant tissue, and developmental stage. The state of the plant materials, the choice of a promoter for the *ipt* gene, and the tissue culture conditions greatly affect the regeneration efficiency of transgenic plants and the percentage of transgenic shoots relative to nontransgenic shoots. Therefore, we constructed *ipt* genes with several different promoters to optimize the cytokinin level required for the regeneration of transgenic cells (Matsunaga et al. 2002).

The pIPT5, 10, and 20 vectors are derivatives of the binary vector pBI121 which contain the chimeric *ipt* gene with a 35S promoter, a native *ipt* promoter and an rbcS 3B promoter, respectively. The native *ipt* promoter was isolated from tumor-inducing (Ti) plasmids of *A. tumefaciens* PO22 (Wabiko et al. 1989). It is induced by wounding and is actively expressed in shoots and roots. The rbcS 3B promoter was isolated from tomato by PCR. It is induced by light and is actively expressed only in green tissues (Sugita and Gruissem 1987). We introduced these *ipt* genes into a hybrid aspen clone "Kitakami Hakuyo" and evaluated their efficiency at generating transgenic shoots. The hybrid clone (*Populus sieboldii* × *P. grandidentata*) is vegetatively propagated to maintain the elite genome and is used for paper production.

We infected stem segments of the hybrid aspen clone with *A. tumefaciens* containing the pIPT5, 10, and 20 or pBI121 vector and cocultivated these on hormone-free modified MS agar medium (800 mg/l ammonium nitrate, 2 g/l potassium nitrate) containing 40 mg/l acetosyringone for three days. When we used the pIPT5, 10, and 20 vectors, the explants were transferred to the same hormone-free medium containing 500 mg/l carbenicillin. After two and a half months of cultivation, adventitious buds were regenerated from stem segments and transferred to fresh medium. After one more month of cultivation, we visually classified the developed buds into the normal and *ipt*-shooty phenotypes and subjected them to a β-glucuronidase (GUS) assay. When we used the pBI121 vector, the explants were transferred to modified MS (0.5 mg/l zeatin) medium containing 500 mg/l carbenicillin and

100 mg/l kanamycin. After 4 months of cultivation, the adventitious buds were transferred to hormone-free modified MS medium containing 500 mg/l carbenicillin and 100 mg/l kanamycin. After one month of cultivation, we subjected these developed shoots to kanamycin and a GUS assay and classified them into kanamycin-resistant transgenic shoots and nontransgenic shoots. We obtained four to six times more transgenic shoots (*ipt*-shooty) per stem segment with the pIPT20 vector than with the pIPT5 and ten vectors. Interestingly, the ratio of transgenic shoots to regenerated shoots using the pIPT5 (33/296 [11.1 percent]) and ten (40/292 [13.7 percent]) vectors were much smaller than that using the pIPT20 (247/495 [53.1 percent]) vector. In addition, the ratio of transgenic shoots with GUS activity to transgenic shoots using the pIPT20 (86.8 percent) vector was greater than that using the pIPT5 (62.1 percent) and ten (75.0 percent) vectors. The transformation efficiency using the pIPT20 vector was tenfold greater than that with the pBI121 (3.86/0.35) vector, and the culture time was shortened by nearly half (2.5/4). These results show that the chimeric *ipt* gene with an rbcS 3B promoter preferentially stimulates the regeneration of transgenic shoots in hybrid aspen.

We constructed the GST-MAT vector pRBI11 composed of the *ipt* gene with the rbcS 3B promoter and the *R* recombinase gene with the GST-II-27 promoter (Figure 13.6). The transformation process of hybrid aspen using pRBI11 is as follows (Matsunaga et al. 2002) (Figure 13.7).

Gene Transfer by Infection

Fifty stem segments of hybrid aspen (*Populus sieboldii* × *P. grandidentata*) were infected with *A. tumefaciens* containing the pRBI11 vector and co-cultivated on hormone-free modified MS agar medium (800 mg/l ammonium nitrate, 2 g/l potassium nitrate) containing 40 mg/l acetosyringone for three days.

Regeneration of Transgenic Plants

The explants were transferred to the same hormone-free medium containing 500 mg/l carbenicillin without kanamycin (nonselective medium). After two and a half months of cultivation, the stem segments, together with regenerated adventitious buds, were transferred to the same medium. After one month of cultivation, about half of the regenerated shoots exhibited the *ipt*-shooty phenotype. Without separation and further cultivation, the transgenic shoots with the *ipt* gene could be clearly identified as the *ipt*-shooty phenotype. GUS activity was detected in about 80 percent of the *ipt*-shooty clones regenerated from 23 (46.0 percent) of the 50 stems by GUS assay.

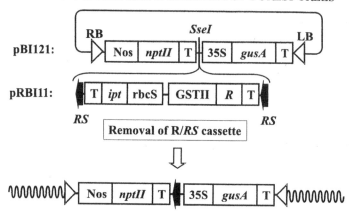

FIGURE 13.6. Diagram of the GST-MAT vector pRBI11. pRBI11 has an R/*RS* cassette composed of the rbcS 3B-*ipt* and GST-II-27-*R* genes. These cassettes are inserted into an *Sse*I site of pBI121. *ipt* = isopentenyl transferase gene; *R* = recombinase gene; *RS* = recognition sequence; *nptII* = neomycin phospho-transferase gene; *gusA* = ß-glucuronidase gene; 35S = CaMV[35]S promoter; Nos = Nos promoter; GSTII = GST-II-27 promoter; rbcS = *rbc*S promoter; T = nopaline synthase terminator; RB and LB = right and left border sequences of a T-DNA.

Appearance of Marker-Free Plants

We previously reported that marker-free transgenic tobacco could be obtained by Safener induction (Sugita, Kasahara, et al. 2000). However, we found that, compared to tobacco plants, it was much more difficult to generate marker-free transgenic hybrid aspen plants because of the low regeneration frequency and high degree of damage with Safener. Furthermore, we observed that the *gusA* gene with the GST-II-27 promoter was induced by wounding and was actively expressed at sites where stem segments of hybrid aspen were cut (data not shown). Therefore, instead of Safener induction, we independently cut 14 GUS-positive *ipt*-shooty clones into small pieces and transferred them to a modified MS medium containing 0.5 mg/l zeatin and 500 mg/l carbenicillin (shoot-inducing medium). Eleven normal shoots appeared from three of 14 (21.4 percent) *ipt*-shooty clones within two months of induction by wounding. These shoots were transferred to 2/3 MS medium containing 0.05 mg/l 3-indoleacetic acid (IBA) (root-inducing medium), in which they grew normally and rooted. Kanamycin resistance and GUS activity of 11 normal shoots were confirmed by kanamycin and GUS assays.

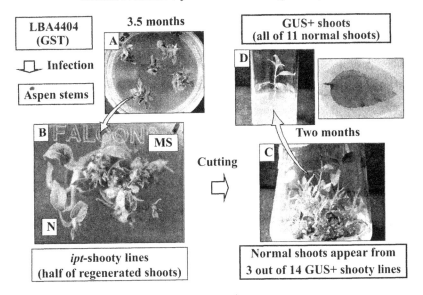

FIGURE 13.7. Transformation of hybrid aspen using the GST-MAT vector (pRBI11). (A) Regeneration of adventitious shoots from stem segments; (B) normal phenotypes (N) and *ipt*-shooty phenotypes (MS) lacking apical dominance and rooting ability; (C) appearance of normal shoots exhibiting apical dominance from *ipt*-shooty clones; (D) a marker-free transgenic plant (MF). (*Source:* Adapted from Ebinuma and Komamine 2001.)

DNA Analysis of Normal Plants

We subjected eight developed shoots from three *ipt*-shooty clones to PCR analysis. In all eight normal plants, a predicted *ipt* fragment was not amplified, but an excision product was amplified by PCR analysis. All eight normal shoots were marker-free transgenic plants, and no nontransgenic escapes were observed. These results indicate that we can produce transgenic aspen plants at a high frequency and induce the generation of marker-free transgenic aspen plants by wounding.

CURRENT TECHNIQUES FOR REMOVING SELECTABLE MARKERS AND THEIR PROBLEMS

About a decade ago, the first transgenic crop, "Flavr Savr tomato," was marketed in the United States. The *nptII* gene has been used as a selectable marker in the production of transgenic tomatoes, which remains in them.

Prior to market introduction, the food and environmental safety of the *nptII* gene and its gene products were assessed by the appropriate regulatory agencies. Many discussions have been conducted on the ethical and social considerations arising from the use of an antibiotic-resistance gene. Although several techniques for removing marker genes from transgenic plants have already been reported, the length of time involved has precluded their use to generate transgenic crops. Subsequently, many different techniques have been developed to remove antibiotic- or herbicide-resistance genes from transgenic plants (reviewed in Ebinuma et al. 2001). These approaches can be divided into two types based on the removal system used.

Cotransformation

It is well known that multiple copies of T-DNAs are transferred into a plant cell and integrated in the plant genome through *Agrobacterium*-mediated transformation. *Agrobacterium* can also cotransform two different T-DNAs into a plant cell. A selectable marker gene and a gene of interest from different T-DNAs are introduced into a plant cell by cotransformation. If the two genes are integrated into unlinked loci, crossing can separate the gene of interest from the selectable marker gene. The marker-free transgenic plants are segregated at the progeny level. Figure 13.8 shows three kinds of cotransformation methods that are described here:

Two-plasmid/two-strain method. Cotransformation is achieved through cocultivation with two *Agrobacterium* strains, which contain a selectable marker gene and a gene of interest on different binary plasmids, respectively. Transformed plants are selected using a selectable marker, and marker-free transgenic plants are then segregated from cotransformed plants by crossing.

Two-plasmid/one-strain method. Cotransformation is achieved through cocultivation with one *Agrobacterium* strain, which contains a selectable marker gene and a gene of interest on different binary plasmids.

Two-T-DNA/one-plasmid method. Cotransformation is achieved through cocultivation with one *Agrobacterium* strain carrying one binary plasmid, which contains a selectable marker gene and a gene of interest on different T-DNAs.

Site-Specific Recombination System

The Cre/*lox* system of bacteriophage P1 is a site-specific recombination system that consists of two components: the recombinase (Cre) and its recognition sites *(loxP)*. A selectable marker gene is flanked by directly adjacent *loxP* sites and linked to a gene of interest. Cre mediates recombination

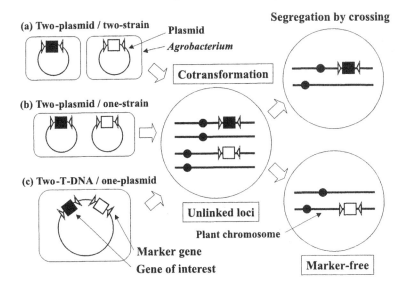

FIGURE 13.8. Cotransformation methods. A selectable marker gene and a gene of interest from different T-DNAs are introduced into a plant cell by cotransformation. If the two genes are integrated into unlinked loci, crossing can separate the gene of interest from the selectable marker gene. Marker-free transgenic plants are segregated at the progeny. (a) Two *Agrobacterium* strains contain a selectable marker gene and a gene of interest on different binary plasmids, respectively; (b) one *Agrobacterium* strain contains a selectable marker gene and a gene of interest on different binary plasmids; (c) one *Agrobacterium* strain carries one binary plasmid, which contains a selectable marker gene and a gene of interest on different T-DNAs. (*Source:* Adapted from Ebinuma et al. 2001.)

events and causes excision of a DNA segment between two directly adjacent *loxP* sites (Odell et al. 1990).

Retransformation or cross-pollination. Transgenic plants containing a gene of interest and the *loxP*-flanked marker gene are produced and the *cre* gene is then introduced by retransformation or cross-pollination. After the *loxP*-flanked marker gene is excised from the genome, crossing can separate the gene of interest from the *cre* gene. The marker-free transgenic plants are segregated at the progeny level (Figure 13.9A).

Induction. A *cre* gene joined to an inducible promoter and a selectable marker gene are flanked by directly adjacent *loxP* sites and linked to a gene of interest. Transgenic plants containing these three genes are produced, and expression of the *cre* gene is then induced to excise the *loxP*-flanked marker gene from the genome (Figure 13.9B). Methods involving cotransformation

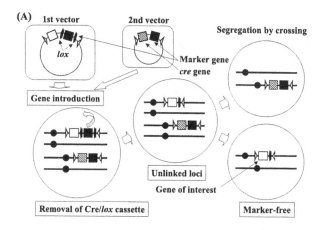

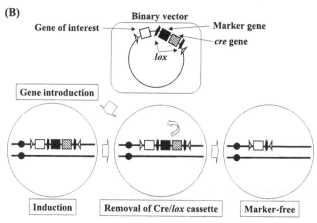

FIGURE 13.9A-B. Site-specific recombination systems. A selectable marker gene is flanked by directly adjacent *loxP* sites and linked to a gene of interest. Cre mediates recombination events and causes excision of a DNA segment between two directly adjacent *loxP* sites. (A) Retransformation or cross-pollination: Transgenic plants containing a gene of interest and the *loxP*-flanked marker gene are produced and the *cre* gene is then introduced by retransformation or cross-pollination. After the *loxP*-flanked marker gene is excised from the genome, crossing can separate the gene of interest from the *cre* gene. The marker-free transgenic plants are segregated at the progeny level. (B) Induction: A *cre* gene joined to an inducible promoter and a selectable marker gene are flanked by directly adjacent *loxP* sites and linked to a gene of interest. Transgenic plants containing these three genes are produced and expression of the *cre* gene is then induced to excise the *loxP*-flanked marker gene from the genome. (*Source:* Adapted from Ebinuma et al. 2001.)

and retransformation or cross-pollination need (site-specific recombination system) crossing for the segregation of marker-free transgenic plants at the progeny level. Therefore, they are difficult to apply in woody plants due to the long period before flowering. They cannot be applied to hybrid clones of woody plants propagated by cutting because crossing disrupts their elite genome. On the other hand, the induction method can be used to remove a selectable marker gene from transgenic plants without crossing. However, it is very difficult to completely remove a marker gene from all of the cells of a transgenic plant and identify marker-free transgenic plants at the R_0 generation. Therefore, this precludes the use of these approaches for the generation of marker-free transgenic woody plants.

ADVANTAGES OF SELECTION
USING THE IPT GENE

The current methods, which can be used to remove a marker gene without crossing, combine an antibiotic- or herbicide-resistance gene with the Cre/*lox* system. The removal system consists of the *loxP*-flanked marker gene and *cre* gene with an inducible promoter. Transgenic plants are selected using the marker gene, and expression of the *cre* gene is then induced to excise them from the genome. However, excision events do not occur uniformly in all of the cells of transgenic plants because expression of the *cre* gene and the recombination ability of *loxP* sites are very different depending on the developmental stage of plant cells. Excision events occur more efficiently in cells of the callus than in those of the embryo or apical meristem (Russell et al. 1992). Therefore, transgenic plants include cells in which excision events do and do not occur during removal. Usually, transgenic cells contain several copies of inserts in the genome. When excision events occur, not all of the copies of marker gene are eliminated from the genome of transgenic cells. Therefore, three kinds of transgenic cells appear and coexist in transgenic plants (Figure 13.10): (1) Transgenic cells have a marker gene in all copies of inserts. (2) Marker-free transgenic cells lose a marker gene in all copies of inserts. (3) Chimeric transgenic cells have a marker gene in several copies of inserts but lose it in other copies. When an antibiotic- or herbicide-resistance gene is used as a selectable marker, it is very difficult to distinguish marker-free transgenic plants from chimeric plants without using a DNA analysis. Once a marker gene is removed from transgenic cells, selective agents cannot be used to identify marker-free transgenic cells, since they are killed in culture medium containing the selective agents. Therefore, transgenic plants are crossed to segregate non-chimeric marker-free transgenic plants at

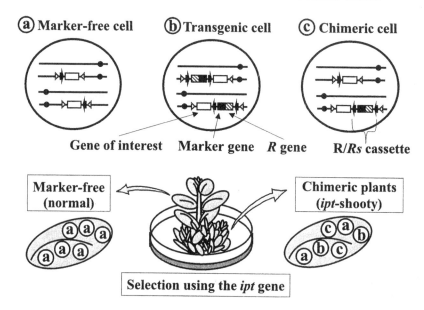

FIGURE 13.10. Advantage of the *ipt* gene. When excision events occur, three kinds of transgenic cells appear and coexist in transgenic plants. (A) Transgenic cells have a marker gene in all copies of inserts. (B) Marker-free transgenic cells lose a marker gene in all copies of inserts. (C) Chimeric transgenic cells have a marker gene in several copies of inserts but lose it in other copies. Once all of the *ipt* genes are removed from transgenic plants exhibiting *ipt*-shooty phenotypes, marker-free transgenic shoots can recover apical dominance and rooting ability, and extend from the *ipt*-shooty clones. (*Source:* Adapted from Ebinuma et al. 2001.)

their progeny, which are identified by DNA analysis. On the other hand, the *ipt*-MAT vectors use the *ipt* gene as a selectable marker. Once all of the *ipt* genes are removed from transgenic plants exhibiting *ipt*-shooty phenotypes, marker-free transgenic shoots can recover apical dominance and rooting ability and extend from *ipt*-shooty clones. Chimeric transgenic plants also appear but exhibit the *ipt*-shooty phenotype because the *ipt* gene remaining in several cells overproduces cytokinin. Therefore, transgenic plants that exhibit a normal phenotype are identified as marker-free transgenic plants that only contain cells in which the *ipt* gene is eliminated from all copies of inserts. The *ipt* gene is very useful for the selection of marker-free transgenic plants. The *ipt*-MAT vectors need neither DNA analysis nor sexual crossing to identify nonchimeric marker-free transgenic plants.

TRANSGENE STACKING USING
THE IPT-*MAT VECTOR*

Presently, many transgenic trees have been developed for commercial purposes. The first generation of transgenic trees has single genes that confer resistance to pests or tolerance to herbicides. Recently, many genes that control pathways of second metabolism have been isolated and identified. The next step has been taken with transgenic crops, in that several genes can be introduced to manipulate complex agronomic traits. Transgenic crops containing a single gene are widely used as breeding materials to stack multiple traits by sexual crossing. In contrast, forest trees require long periods of time for sexual crossings due to their long life cycles. Therefore, it is desirable to develop a gene stacking method through retransformation. However, retransformation is difficult using the current methods because the marker gene remaining in transgenic plants precludes the use of the same marker gene for selection of the double-transformed plants. Because there are only a limited number of suitable marker genes, a system for removing marker genes without crossing is a prerequisite for stacking multiple genes by retransformation in woody plants.

We have demonstrated that the *ipt*-MAT vector could efficiently generate marker-free transgenic plants. However, since the *ipt*-MAT vector uses the R/*RS* system to remove a selectable marker gene, one recognition site *(RS)* remains in the genome of marker-free transgenic plants. In yeast, the R/*RS* system can mediate recombination between two *RS* sites, which are present about 180 kb apart on one chromosome or on two nonhomologous chromosomes (Matsuzaki et al. 1990). Recombination leads to chromosomal excision, inversion, or translocation. When many genes of interest are introduced through retransformation using *ipt*-MAT vectors, recombination between the remaining *RS* sites and the introduced *RS* sites might cause chromosomal rearrangement. Therefore, we retransformed transgenic plants using the *ipt*-MAT vector and examined their chromosomal DNA by Southern blot analysis.

We used a transgenic tobacco line (132BMO6) produced by the *ipt*-MAT vector pNPI132 for retransformation. The 132BMO6 line contains one copy of insert that includes the *nptII* and *gusA* genes and one *RS* site. We inserted the chimeric *GFP* gene with a nos promoter into the *lacZ* multi-cloning sites of the GST-MAT vector pMAT8 as a model gene of interest to construct pMAT8GFP plasmid (Figure 13.11). The transformation of tobacco using pMAT8GFP can be summarized as follows (Sugita, Matsunaga, et al. 2000).

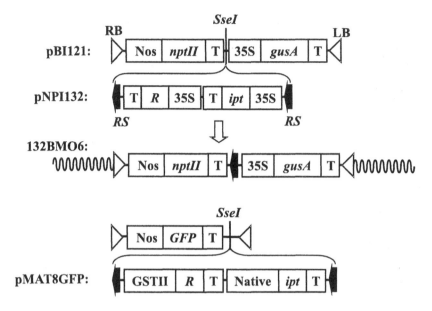

FIGURE 13.11. Diagram of the MAT vectors used for gene stacking. pNPI132 has an R/*RS* cassette composed of the 35S-*ipt* and 35S-*R* genes. These cassettes are inserted into an *SseI* site of pBI121. The 132BMO6 line contains one copy of an insert that includes the *nptII* and *gusA* genes and one *RS* site. pMAT8GFP has a chimeric *GFP* gene with a nos promoter which is inserted into the *lacZ* multicloning sites of the GST-MAT vector pMAT8. *ipt* = isopentenyl transferase gene; *R* = recombinase gene; *RS* = recognition sequence; *gusA* = ß-glucuronidase gene; *GFP* = green fluorescent protein gene; 35S = CaMV^{35}S promoter; Nos = Nos promoter; GSTII = GST-II-27 promoter; Native = native *ipt* promoter; T = nopaline synthase terminator; RB and LB = right and left border sequences of a T-DNA. (*Source:* Adapted from Ebinuma et al. 2002.)

Gene Transfer by Infection

Twenty pieces of leaf segments of transgenic tobacco (132BM06) plants were infected with *A. tumefaciens* containing the pMAT8GFP vector and cocultivated on hormone-free MS agar medium containing 50 mg/l acetosyringone for three days.

Regeneration of Transgenic Plants

The explants were transferred to hormone-free MS agar medium containing 500 mg/l carbenicillin but not kanamycin (nonselective medium).

One month after infection, 36 regenerated adventitious buds were separated from the leaf segments and transferred to the same medium. After one month of cultivation, we visually identified 34 (94 percent) *ipt*-shooty phenotypes and subjected them to PCR analysis.

Appearance of ipt-*Free Plants*

Twenty (59 percent) *ipt*-shooty explants, in which the predicted excision products were amplified by PCR, were subcultured to fresh Safener-induction MS medium containing 30 mg/l Safener (R29148) every month. Several normal shoots appeared from seven of 20 (35 percent) *ipt*-shooty clones within three months of Safener induction. These shoots were transferred to the same medium, grew normally, and rooted.

DNA Analysis of Normal Plants

Seven normal shoots that appeared from independent *ipt*-shooty clones were subjected to PCR analysis. Predicted *nptII* and *gusA* fragments and an excision product of pNPI132 were amplified in all seven normal plants. In five of seven normal plants, a predicted *GFP* fragment and an excision product of pMAT8GFP were amplified, as was a predicted *ipt* fragment in one of five GFP-positive normal plants. These results indicated that four *ipt*-free transgenic plants in which multiple transgenes were stacked had been generated from 20 excision-positive *ipt*-shooty lines (20 percent). We investigated the integrated *nptII* and *gusA* genes of four *ipt*-free transgenic plants by Southern blot analysis. Genomic DNA of the untransformed 132BMO6 line and four *ipt*-free transgenic plants were digested with two restriction enzymes and hybridized with the *gusA* coding regions. No rearrangement of the integrated genes was detected (Figure 13.12). These results showed that the GST-MAT vector did not cause DNA rearrangement between the first and second transformations. Because the GST-MAT vector tends to generate *ipt*-free transgenic plants containing a single transgene, this vector might be effective for avoiding undesirable DNA rearrangements.

COMBINATION OF THE IPT
AND IAAM/H GENES

In current transformation methods, two kinds of plant growth regulators (cytokinin and auxin) are added to the tissue culture medium for the regeneration of transgenic plants. Although the endogenous hormone levels are

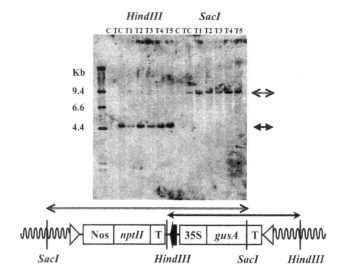

FIGURE 13.12. Southern analysis of transgene-stacked plants. Genomic DNA was digested with two restriction enzymes (*HindIII* or *SacI*) and hybridized with the *gusA* coding regions. Lanes C = nontransgenic control; TC = original marker-free transgenic tobacco plants; T1-T5 = independent transgene-stacked marker-free transgenic tobacco plants.

very different among plant species as well as among plant tissues, it has been well established that adventitious shoots regenerate from a plant tissue culture with an increase in the amount of cytokinin relative to that of auxin in the culture medium. The *ipt*-type MAT vector uses only the *ipt* gene for the regeneration of transgenic plants instead of plant growth regulators (Endo et al. 2001). Because the control of both cytokinin and auxin is needed to optimize the hormone levels in plant tissue and regenerating transgenic shoots in many plant species, we combined the *ipt* genes with the *iaaM/H* genes to manipulate both the auxin and cytokinin levels. The *iaaM/H* genes code for a tryptophan monooxygenase and an indoleace-tamide hydrolase, which catalyze auxin synthesis (Thomashow et al. 1984). The *iaaM/H* genes were isolated from *A. tumefaciens* PO22 (Wabiko et al. 1989), which induces a large tumor on the trunks of hybrid aspen.

pIPTIMH are derivatives of the binary vector pBI121 which contain the native *ipt* and *iaaM/H* genes. We transformed tobacco (*Nicotiana tabaccum* cv. SR1) and poplar (*Populus tomentosa*) plants with *Agrobacterium* containing pIPT10 and pIPTIMH to compare the regenerative abilities of the *ipt* gene and the *ipt* gene combined with the *iaaM/H* genes (Endo et al. 2002). We observed that both the proliferation of calli and the differentiation of shoots were induced faster from tobacco leaf discs and poplar stem segments infected with pIPTIMH than from those infected with pIPT10. Transgenic plants with pIPTIMH also exhibited *ipt*-shooty phenotypes which lacked apical dominance and rooting ability and were easily distinguished from nontransgenic shoots. These results show that the *iaaM/H* genes promote both the proliferation of calli and the differentiation of shoots in combination with the native *ipt* gene, and the *ipt* gene combined with the *iaaM/H* gene can be used as a selectable marker to identify transgenic plants.

We constructed the GST-MAT vector pMATIMH composed of the native *ipt* and *iaaM/H* genes and the *R* gene with the GST-II-27 promoter (Figure 13.13). The production of marker-free transgenic plants using pMATIMH involves four main steps (Endo et al. 2002).

Gene Transfer by Infection

Twenty pieces of leaf segments of tobacco (*Nicotiana tabaccum* cv. SR1 plants) were infected with *A. tumefaciens* containing the pMATIMH vector and cocultivated on hormone-free MS agar medium containing 50 mg/l acetosyringone for three days.

Regeneration of Transgenic Plants

The explants were transferred to hormone-free MS agar medium containing 500 mg/l carbenicillin but not kanamycin (nonselective medium).

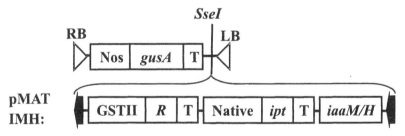

FIGURE 13.13. Diagram of the GST-MAT vector pMATIMH. pMATIMH has an R/*RS* cassette composed of the native *ipt, iaaM/H,* and GST-II-27-*R* genes. The cassette is inserted into an *SseI* site of pTL7 containing the Nos-*gusA* gene to create pMATIMHGUS. *ipt* = isopentenyl transferase gene; *iaaM/H* = tryptophan monooxygenase and indoleacetamide hydrolase genes; *R* = recombinase gene; *RS* = recognition sequence; *gusA* = ß-glucuronidase gene; Nos = Nos promoter; GSTII = GST-II-27 promoter; Native = native *ipt* promoter; T = nopaline synthase terminator; RB and LB = right and left border sequences of a T-DNA. (*Source:* Adapted from Sugita, Matsunaga, et al. 2000.)

One month after infection, 20 regenerated adventitious buds were separated from the leaf segments and transferred to the same medium. After one month of cultivation, we visually identified 15 *ipt*-shooty phenotypes that lacked apical dominance and rooting ability.

Appearance of Marker-Free Plants

Twelve GUS-positive *ipt*-shooty clones were transferred to fresh Safener-induction MS medium containing 30 mg/l Safener (R29148) every two weeks. Several normal shoots appeared from five of 12 (41.7 percent) *ipt*-shooty clones within three months of Safener induction. These shoots were transferred to the same medium, grew normally, and rooted.

DNA Analysis of Normal Plants

Normal shoots that appeared from *ipt*-shooty clones were subjected to PCR analysis. A predicted *gusA* fragment and an excision product were amplified but not a predicted *ipt* fragment. We examined the copy number of normal shoots and *ipt*-shooty clones by Southern analysis. All of the normal shoots contained only a single copy of the transgene. These results indicate that the *ipt*-type MAT vector in combination with *iaaM/H* genes can increase both the regeneration efficiency of transgenic plants and the generation efficiency of marker-free transgenic plants.

DECREASE IN ENVIRONMENTAL IMPACT
USING STERILE TRANSGENIC TREES

Many kinds of transgenic crops (maize, soybean, rapeseed, cotton, etc.) have been cultivated on a large scale in Canada and the United States. The safety of the introduced genes was assessed by the appropriate regulatory agencies. However, despite scientific assurance, recent public concerns about their safety have caused considerable delays in the market performance of transgenic crops. Although transgenic trees do not need to be assessed for food safety, the use of antibiotic-resistance genes as selectable markers also raises a concern that these genes may be transferred to pathogenic bacteria. A method for removing marker genes should be very useful for obtaining both regulatory and public approval for transgenic trees due to a reduction in their perceived environmental impact. Furthermore, many strategies have been explored for the genetic containment of transgenic trees, since the escape of transgenes via sexual production is a serious issue in environmental assessments due to their potential for outcrossing with native forest trees and to their potential invasiveness as weeds. A sterile technology, which prevents pollen and seed production, is a very promising way to decrease the spread of transgenes. Strauss and colleagues (1995) reported the use of sterile tree clones as recipients for transformation. These include a triploid clone of hybrid poplar and a transgenic clone in which flowering is genetically manipulated. Several genes for the control of flowering have already been isolated from pine and poplar trees (see Chapter 12; Weigel and Nilsson 1995; Mellerowicz et al. 1998; Mouradov et al. 1998; Wang et al. 1997). These include meristem identity genes, homeotic genes, and male or female cone-specific genes. These genes have been modified and introduced into pine and poplar trees to reduce or abolish the formation of reproductive tissues. This sterile technology may also be advantageous with regard to increased wood yield. Particularly in pines, large amounts of energy and nitrogen are used to produce reproductive organs. Sterility might change the proportion of this investment and increase the biomass production of tree plantations. This approach has great potential to overcome environmental concerns but presents several drawbacks for the commercialization of transgenic trees. Because sterile transgenic clones cannot propagate themselves by means of seeds, cutting, or tissue culture, technologies are needed for the mass production of seedlings. Retransformation is also essential to improve their characteristics since valuable genes cannot be stacked by sexual crossings. The MAT vector system is desirable for using sterile clones as recipients in transformation and for stacking multiple genes through retransformation.

CONCLUSION

Wild *Agrobacterium* has acquired the ability to infect a wide range of plant species through a long history of evolution. Plants have also developed defense mechanisms to fight against their infection. Among current methods using *Agrobacterium,* the efficiency of transformation is greatly influenced by the state of the plant material and the *Agrobacterium* strain. The major factors that limit transformation events by *Agrobacterium* are (1) the attachment of bacteria to plant cells, (2) the interaction of bacteria with plant cells, (3) the transfer of T-DNA from bacteria to plant cells, (4) the acceptance of T-DNA into plant cells, and (5) the proliferation of transformed plant cells. The oncogenes of *Agrobacterium* play an important role in breaching the defense mechanisms of plants and promoting the interaction of bacteria with plant cells and the proliferation of infected tissues. We have improved several kinds of oncogenes and put them together to construct new MAT vectors. The objective of these new MAT vectors is to imitate in vivo transformation systems of wild *Agrobacterium* strains and develop an efficient transformation system for recalcitrant plant species.

Over the next decade, rapid progress should be made in recombinant DNA technologies and many kinds of transgenic crops and trees should be generated. For the commercialization of transgenic crops and trees, it will become more important to reduce their environmental impact than to increase their economic value. The MAT vector system, which can remove marker genes without sexual crossings and stack valuable genes in sterile clones, is a promising technology for obtaining both regulatory and public approval and to accelerate the commercialization of transgenic trees. We plan to stack several valuable genes, the safety of which has been assessed in crops, into eucalyptus using the MAT vector system. Our ultimate goal is to produce high stress-resistant trees with high biomass yields and to make tree plantations in unutilized lands for a stable supply of paper materials.

REFERENCES

Akiyoshi, D.E., Klee, H., Amasino, R.M., Nester, E.W., and Gordon, M.P. (1984). T-DNA of *Agrobacterium tumefaciens* encodes an enzyme of cytokinin biosynthesis. *Proc Natl Acad Sci USA* 81: 5994-5998.

Barry, G.F., Rogers, S.G., Fraley, R.T., and Brand, L. (1984). Identification of a cloned cytokinin biosynthetic gene. *Proc Natl Acad Sci USA* 81: 4776-4780.

Ebinuma, H. and Komamine, A. (2001). MAT (Multi-Auto-Transformation) vector system. The oncogenes of *Agrobacterium* as positive markers for regeneration and selection of marker-free transgenic plants. *In Vitro Cell Dev Biol-Plant* 37: 114-119.

Ebinuma, H., Sugita, K., Matsunaga, E., Endo, S., and Kasahara, T. (2000). Selection of marker-free transgenic plants using the oncogenes *(ipt, rol A, B, C)* of *Agrobacterium* as selectable markers. In S.M. Jain and S.C. Minocha (Eds.), *Molecular biology of woody plants*, Volume 2 (pp. 25-46). Dordrecht, Netherlands: Kluwer Academic Publishers.

Ebinuma, H., Sugita, K., Matsunaga, E., Endo, S., and Yamada, K. (2002). GST-MAT vector for the efficient and practical removal of marker genes from transgenic plants. In J.F. Jackson, H.F. Linskens, and R.B. Inman (Eds.), *Molecular methods of plant analysis*, Volume 22, *Testing for genetic manipulation* (pp. 95-117). Heidelberg, Germany: Springer-Verlag GmbH and Co.

Ebinuma, H., Sugita, K., Matsunaga, E., Endo, S., Yamada, K., and Komamine, A. (2001). Systems for the removal of a selection marker and their combination with a positive marker. *Plant Cell Rep* 20: 383-392.

Ebinuma, H., Sugita, K., Matsunaga, E., and Yamakado, M. (1997). Selection of marker-free transgenic plants using the isopentenyl transferase gene as a selectable marker. *Proc Natl Acad Sci USA* 94: 2117-2121.

Ebinuma, H., Sugita, K., Matsunaga, E., Yamakado, M., and Komamine, A. (1997). Principle of MAT vector. *Plant Biotechnol* 14: 133-139.

Endo, S., Kasahara, T., Sugita, K., and Ebinuma, H. (2002). A new GST-MAT vector containing both the *ipt* gene and the *iaaM/H* genes can produce marker-free transgenic plants with high frequency. *Plant Cell Rep* 20: 923-928.

Endo, S., Kasahara, T., Sugita, K., Matsunaga, E., and Ebinuma, H. (2001). The isopentenyl transferase gene is effective as a selectable marker gene for plant transformation in tobacco (*Nicotiana tabacum* cv. Petite Havana SR1). *Plant Cell Rep* 20: 60-66.

Gaudin, V., Vrain, T., and Jouanin, L. (1994). Bacterial genes modifying hormonal balances in plants. *Plant Physiol Biochem* 32: 11-29.

Holt, D.C., Lay, V.J., Clarke, E.D., Dinsmore, A., Jepson, I., Bright, S.W.J., and Greenland, A.J. (1995). Characterization of the Safener-induced glutathione *S*-transferase isoform II from maize. *Planta* 196: 295-302.

Hooykaas, P.J.J. and Schilperoort, R.A. (1992). *Agrobacterium* and plant genetic engineering. *Plant Mol Biol* 19: 15-38.

Matsunaga, E., Sugita, K., and Ebinuma, H. (2002). An asexual production of selectable marker-free transgenic woody plants, vegetatively propagated species. *Mol Breed*. 10: 95-106.

Matsuzaki, H., Nakajima, R., Nishiyama, J., Araki, H., and Oshima. Y. (1990). Chromosome engineering in *Saccharomyces cerevisiae* by using a site-specific recombination system of a yeast plasmid. *J. Bact* 172: 610-618.

Mellerowicz, E.J., Horgan, K., Walden, A., Coker, A., and Walter C. (1998). PRFLL - a *Pinus radiata* homologue of FLORICAULA and LEAFY is expressed in buds containing vegetative shoot and undifferentiated male cone primordial. *Planta* 206: 619-629.

Mouradov, A., Glassick, T., Hamdorf, B., Murphy, L., Fowler, B., Marla, S., and Teasdale, R.D. (1998). NEEDLY, a *Pinus radiata* ortholog of FLORICAULA/ LEAFY genes, expressed in both reproductive and vegetative meristems. *Proc Natl Acad Sci USA* 95: 6537-6542.

Odell, J., Caimi, P., Sauer, B., and Russell, S. (1990). Site-directed recombination in the genome of transgenic tobacco. *Mol Gen Genet* 223: 369-378.

Onouchi, H., Yokoi, K., Machida, C., Matsuzaki, H., Oshima, Y., Matsuoka, K., Nakamura, K., and Machida, Y. (1991). Operation of an efficient site-specific recombination system of *Zygosaccharomyces rouxii* in tobacco cells. *Nucl Acids Res* 19: 6373-6378.

Russell, S.H., Hoopes, J.L., and Odell, J.L. (1992). Directed excision of a transgene from the plant genome. *Mol Gen Genet* 234: 49-59.

Skoog, F. and Miller, C.O. (1957). Chemical regulation of growth and organ formation in plant tissues cultured in vitro. *Symp Soc Exp Biol* 11: 118-130.

Strauss, S.H., Rottmann, W.H., Brunner, A.M., and Sheppard, L.A. (1995). Genetic engineering of reproductive sterility in forest trees. *Mol Breed* 1: 5-26.

Sugita, K., Kasahara, T., Matsunaga, E., and Ebinuma, H. (2000). A transformation vector for the production of marker-free transgenic plants containing a single copy transgene at high frequency. *Plant J* 22: 461-469.

Sugita, K., Matsunaga, E., and Ebinuma, H. (1999). Effective selection system for generating marker-free transgenic plants independent of sexual crossing. *Plant Cell Rep* 18: 941-947.

Sugita, K., Matsunaga, E., Kasahara, T., and Ebinuma, H. (2000). Transgene stacking in plants in the absence of sexual crossing. *Mol Breed* 6: 529-536.

Sugita, M. and Gruissem, W. (1987). Developmental, organ-specific, and light-dependent expression of the tomato ribulose-1,5-bisphosphate carboxylase small subunit gene family. *Proc Natl Acad Sci USA* 84: 7104-7108.

Thomashow, L.S., Reeves, S., and Thomashow, M.F. (1984), Crown gall oncogenesis: Evidence that a T-DNA gene from the *Agrobacterium* Ti plasmid pTiA6 encodes an enzyme that catalyses synthesis of indoleacetic acid. *Proc Natl Acad Sci USA* 81: 5071-5075.

Wabiko, H., Kagaya, M., Kodama, I., Masuda, K., Kodama, Y., Yamamoto, H., Shibano, Y., and Sano, H. (1989). Isolation and characterization of diverse nopaline type Ti plasmids of *Agrobacterium tumefaciens* from Japan. *Arch Microbiol* 152: 119-124.

Wang, D.Y., Bradshaw, R.E., Walter, C., Connett, M.B., and Fountain, D.W. (1997). Structural characterization of *Pinus radiata* MADS box DNA sequences isolated by PCR cloning. *New Zeal J For Sci* 27(1): 3-10.

Weigel, D. and Nilsson, O. (1995). A developmental switch sufficient for flower initiation in diverse plants. *Nature* 377: 495-500.

PART IV:
GENOME MAPPING
IN FOREST TREES

Chapter 14

High-Density Linkage Maps in Conifer Species and Their Potential Application

Enrique Ritter
Santiago Espinel
Jean-Michel Favre
Matthias Fladung

INTRODUCTION

Molecular markers have provided a major contribution to the genetic knowledge of many cultivated plant species. In addition to their basic importance for genetic and evolutionary studies, molecular markers are useful to construct linkage maps and to localize monogenic and polygenic traits, which allow the efficient introgression and selection of individuals with specific characteristics. Such markers are particularly useful in forest species with long generative cycles, since they would enable early selection tests in breeding programs. Moreover, molecular markers also play a crucial role in the isolation and cloning of plant genes by map-based cloning.

The development of extremely dense genetic linkage maps is generally the first step to comprehensively describing the genome of an organism. Such dense genetic linkage maps have been available in humans (> 8,000 markers; Broman et al., 1998) and laboratory rat (> 7,000 markers; Steen et al., 1999), and more recently in plants, such as cereal rice (> 2,000 markers; Harushima et al., 1998). The most spectacular example is the ultra-high-density (UHD) map of potato which contains nearly 10,000 markers (<http://www.dpw.wau.nl/uhd/>). The development of amplified fragment-length polymorphism (AFLP) technology generating large amounts of markers has increased the number of dense linkage maps, and now other crops such as tomato, maize, and barley also have maps consisting of over 1,000 markers.

Several linkage maps have also been produced in forest species using different marker types (restriction fragment-length polymorphism [RFLP], random amplification of polymorphic DNA [RAPD], AFLP, simple sequence repeats [SSR], isozymes, proteins). Due to their economic importance, conifers represent promising target species for molecular marker analyses. In conifers, linkage maps are available for *Pinus taeda* (Remington et al., 1999; Devey et al., 1999, 1994; Groover et al., 1994; Sewell et al., 1999), *Pinus pinaster* (Costa et al., 2000; Plomion et al., 1997), *Pinus radiata* (Devey et al., 1996), *Pinus edulis* (Travis et al., 1998), and for species belonging to other genera such as *Larix* (Arcade et al., 2000) or *Picea* (Binelli and Bucci, 1994; Paglia et al., 1998; Gosselin et al., 2002). These maps have a restricted number of 400 to 500 markers. Recently a high-density map of *P. pinaster* based on 620 AFLP markers has been constructed (Chagne et al., 2002). Within an ongoing European research and development project (QLK5-1999-01159), another ultra-high-density linkage map of *Pinus pinaster* descending from the cross 0024 Landes × C803 Corsica is under construction. This map will contain several thousands of AFLP markers, numerous landmark SSRs and expressed sequence tags (ESTs), and several quantitative trait loci (QTL) for important characters related to growth characteristics, wood quality, and resistance to biotic and abiotic stresses. Details of this new UHD linkage map of *P. pinaster* have been published recently (Ritter et al., 2002). The actual map contains more than 1,500 markers (over 120 markers per chromosome) and covers a genome length of around 2,000 cM. The Web page <http://www.neiker.net/UHDfor> contains the regularly updated status of this linkage map and further results of interest. Here, we focus on the particular aspects of high-density map construction in conifer species and their possible application. Some of these aspects may apply generally also to other crop species, while others consider specific characteristics of conifers.

BASIC ASPECTS FOR CONSTRUCTING HIGH-DENSITY MAPS IN CONIFER SPECIES

Plant Material

Setting up suitable plant material for linkage mapping is more complicated in woody species than in most herbaceaous species. In particular, conifers and many angiosperm tree species have to be given special attention. Generally, seeds from these species require more time to mature for germination (Ruiz de la Torre and Ceballos, 1979). Furthermore, besides mapping usually a progeny is also used for QTL analyses, which requires more space for planting and time for evaluation of characters. In order to optimize available resources, breeders tend to reduce plant material for selection as soon as possible. Because conifers are generally outbreeders, setting up classical F2 progenies derived from inbred lines is nearly impossible due to the large generation cycles. Therefore, different population structures can be found for linkage mapping such as pedigrees of several generations (Sewell et al., 1999; Devey et al., 1999), open-pollinated families combined with megagametophyte analysis (Remington et al., 1999; Hayashi et al., 2001), full-sib progenies (Arcade et al., 2000; Hurme and Savolainen, 1999), or self-pollinated progenies (Costa et al., 2000). However, these designs are not always the most informative, depending on the resulting marker configuration for the analyses (Ritter and Salamini, 1996).

Marker Types and Polymorphism

Linkage maps in forest species, and particularly in *Pinus,* are based on different marker types such as RFLPs in *Pinus taeda* (Devey et al., 1994; Groover et al., 1994; Sewell et al., 1999) and RAPDs in *Pinus pinaster* (Costa et al., 2000) or *Pinus radiata* (Devey et al., 1996). Also, AFLP maps are available for *Pinus pinaster* (Costa et al., 2000; Chagne et al., 2002) or *Pinus edulis* (Travis et al., 1998). Other marker types such as proteins and isozymes (Plomion et al., 1997; Costa et al., 2000; Devey et al., 1994), as well as EST (Cato et al., 2001) and SSR markers (Devey et al., 1999), have been integrated in these maps, which may be useful for aligning different maps.

Marker types have particular properties that influence their efficiency for map construction and require different amounts of resources. RFLP markers are based on hybridizations between genomic or cDNA probes and digested genomic DNA. In general, this marker type is codominant. One advantage of this technique is that it allows, depending on the hybridization conditions, the mapping of probes from even distant species based on partial sequence homologies and ensures in this way a high degree of transferability. However, the RFLP technique itself is laborious and requires relatively large amounts of DNA. Therefore, this technique has serious shortcomings for routine purposes, which hinders its application in commercial breeding programs.

In contrast, markers generated by polymerase chain reaction (PCR) require less input in terms of manpower and DNA quantities and are, therefore, generally preferred for linkage mapping. Initially, RAPD were most popular to rapidly obtain linkage maps. However, RAPD markers raise problems in term of reproducibility, and possible occurrence of comigrating fragments of different composition is well known (summarized in Black, 1993).

AFLP markers, however, can be generated in a high number, are highly reproducible, and are generally dominant. They are useful to rapidly increase the density of linkage maps and are generally used as standard markers for this purpose, as we did for constructing the high-density map of *P. pinaster.*

Other PCR marker types such as microsatellites or SSRs and ESTs are useful for map alignment. The SSR markers are codominant, highly polymorphic, and represent powerful tools for different genetic analyses (see Chapter 15; Wang et al., 1994). SSR markers have been described in a variety of plant species. They have also been developed in different conifer species and were used in several mapping experiments (Echt et al., 1996; Fisher et al., 1996, 1998; Mariette, Chagne, Decroocq, et al., 2001; Pfeiffer et al., 1997; Smith and Devey, 1994; Soranzo et al., 1998; Van de Ven and McNicol, 1996). They seem to be conserved among species and to a certain degree within families as well (Shepherd et al., 2002). The same also holds true for EST markers that have been applied in several mapping experiments with forest species (Harry et al., 1998; Temesgen et al., 2001; Brown et al., 2001; Iwata et al., 2001).

The resulting amount of segregating polymorphisms in a mapping experiment depends on the selected marker types and on the plant material under study. For example, due to the very large genome size of conifer species (Gerber and Rodolphe, 1994; Wakamiya et al., 1993), especially when using AT-rich primer combinations (PCs), numerous comigrating but different amplification products were generated which could not be resolved properly in sequencing gels. Therefore, using PCs with lower AT content and/or increasing the number of selective nucleotides in the primers to four potentially reduced the number of amplification products and revealed more segregating bands of better quality (Ritter et al., unpublished).

Despite the different parental ecotypes constituting our mapping population, an unexpectedly low degree of segregating polymorphism of AFLP as well as SSR and EST markers was observed in the progeny. On average, seven to eight segregating bands were obtained per PC ranging between one and 25 bands. Several PCs did not show any segregating polymorphism. This fact is surprising, considering the well-marked differentiation between the original provenances of the parents (Mariette, Chagne, Lezier, et al., 2001) and the similar level of genetic diversity encountered in *P. pinaster* and other *Pinus* species (Smith and Devey, 1994; Echt et al., 1996). However, the analysis of only mapping populations allows us to distinguish between segregating and nonsegregating polymorphisms. The larger number of nonsegregating but polymorphic fragments indicated an elevated degree of homozygosity in the progeny analyzed. This increased homozygosity is probably due to a low degree of biodiversity that exists at the specific sites (i.e., trees may be quite different between sites but very similar within a site).

Statistical Analyses

The designs used in conifer species for map construction require specific methods for linkage analyses. Half-sib families can be treated as backcross data for parent specific fragments which are absent in all complementary parents of the design. For pedigree, data-specific methods are provided by the MAPMAKER program (CEPH data type; Lander et al., 1987). For F1 progenies derived from heterozygous parents, as is the case in our *P. pinaster* high-density map, methods provided by Ritter and colleagues (1990) and Ritter and Salamini (1996) are available. Basically, the procedure consists of constructing independent linkage maps based on fragments specific to either parent using appropriate LOD (likelihood of odds) thresholds. Subsequently, fragments common to both parents are mapped into linkage groups as anchor points, codominant allelic fragments are identified, and then linkage groups are integrated as described in Ritter and Salamini (1996). The integrated linkage map of our *P. pinaster* population was produced following this method, based on 70 anchor points.

Some specific characteristics are applicable to the construction of high-density maps. The computational efforts and complexities increase considerably with extending marker numbers. Determination of optimal orders of many tightly linked markers is more complex. Framework maps based on selected precision markers and a posteriori integration of closely linked markers into the corresponding intervals may solve this problem. The novel, pattern-based "bin" concept also represents a possible solution (Vision et al., 2000; <http://www.dpw.wau.nl/uhd/>). Interestingly, this concept may be used for visualization of the list of bins indicating all markers within each bin. On the other hand, scoring errors are gaining increasing importance for high-density map construction. Scoring errors automatically produce recombination events and consequently inflate marker distances in the map. Furthermore, they complicate the determination of the correct marker order. Methods must be applied that analyze double or multiple crossing-over events in the same genotype and allow removal of the most unlikely recombination events.

Map Alignments

In order to maximize the usefulness of a particular linkage map, it should be aligned with other existing maps. In this way information on markers and QTL in different genetic backgrounds can be cross-referenced and compared. Map alignments are based on sequence homologies within and between species, and markers such as RFLPs, SSRs, or ESTs can be used for this purpose. Large efforts are made through international cooperation to align different maps in forest species using microsatellite and EST markers (<http://dendrome.ucdavis.edu/Synteny/>). In the literature only a few examples of map alignment, such as between *Pinus taeda* and *P. elliottii* (Brown et al., 2001) or between *P. taeda* and *P. radiata* (Devey et al., 1999) can be found.

In our study, we also evaluated numerous SSR and EST markers. Several of them could be used to associate and align linkage groups from different maps. However, the low level of polymorphism observed in our population hampered the mapping of many of these markers and subsequently the possibilities for full alignment between linkage maps. Because this goal is crucial for the usefulness of our map, additional SSR and EST primers are being evaluated in order to achieve complete alignments. Furthermore, we observed that SSR and EST markers sometimes revealed different numbers of loci depending on the mapping population. The interspecies transferability, in particular of SSR markers, seems to be low.

Alignments between maps have also been achieved in potato with comigrating AFLP markers involving different *Solanum* species (Rouppe van der Voort et al., 1997). It is important to prove this in conifer species, too, by comparing parental profiles and map locations of comigrating fragments from common AFLP primer combinations which have been used in different mapping experiments. Our first results indicate that the degree of possible comigration is very reduced in conifer species. This type of analysis requires the inclusion of parental samples from other populations in the gels in order to determine comigrating bands with precision. Frequently, base pair (bp) information of AFLP fragments from other populations is missing, or, if available, identical fragments are difficult to identify due to certain variation of bp values based on migration patterns.

Genome Coverage

Increased marker density is generally associated with increased genome coverage. Several formulas exist to calculate genome coverage (Chakravarti et al., 1991) or to calculate the proportion c of a genome within d cM of a marker (Fishman et al., 2001). Conifers are characterized by a large amount of repetitive DNA (Wakamiya et al., 1993) in which recombination events are difficult to detect. These repeats seem to be organized in contiguous genomic regions, since we find several gaps in our map despite the high marker density. Similar gaps can also be observed in the high-density map published in the same species by Chagne and colleagues (2002). These findings also explain why more linkage groups than number of chromosomes of the corresponding species are found in other conifer maps with reduced marker densities, as observed for example in *Picea abies* (Binelli and Bucci, 1994), *Pinus elliottii* (Nelson et al., 1993), and *P. thunberhii* (Hayashi et al., 2001).

APPLICATION OF HIGH-DENSITY MAPS
IN FOREST SPECIES

A high-density map is an ideal platform for elaborating an integrated genetic, functional, and physical map of crop genomes with several exploitation possibilities that are discussed in this section.

Qualitative and Quantitative Trait Analyses

Progenies designed for linkage maps are generally used to integrate qualitative characters determined by single genes as well as quantitative characters determined by polygenic (inter)actions (Thoday, 1961). In principle, qualitative characters can be treated for linkage mapping in principle analogous to segregating markers (i.e., presence or absence of specific character expressions [for example, resistant or susceptible]). For QTL analyses, several methods are available (Lander and Botstein, 1989; Knapp et al., 1990; Martínez and Curnow, 1992; Jansen and Stam, 1994; Zeng, 1994).

QTLs for important characters, such as growth characteristics and wood-specific gravity, were obtained in *Pinus taeda* (Sewell et al., 2002; Kaya et al., 1999; Groover et al., 1994), in hybrids of *Pinus elliottii* × *Pinus caribaea* (Dale and Teasdale, 1995), and in *Pinus pinaster* (Plomion and Durel, 1996) and *Picea abies* (Skov and Wellendorf, 1998). In our high-density map, several characters related to growth characteristics and wood properties have also been determined (<http://www.neiker.net/UHDfor>).

With respect to QTL analyses, the availability of a high-density map offers particular advantages for efficient *marker-assisted selection* (MAS). Frequently, insufficient linkage disequilibrium exists between the marker and the gene controlling the trait. Therefore, markers beyond a certain genetic distance may not be diagnostic for the trait, making their value questionable. Dense linkage maps facilitate the development of markers, which are physically very tightly linked to the trait of interest and allow diagnostic markers linked to genes (e.g., EST or SSR) to be identified.

Reference Maps for Alignment of Other Maps
and for Comparative Genome and QTL Analyses

Comparative genome analysis in many crops has been, to date, mostly focused on the development of different genetic linkage maps based on different and generally unrelated crosses. These maps have been subsequently used to define chromosomal locations for a variety of simple and complex (or quantitative) traits, with varying levels of success and precision. In contrast, high-density linkage maps that are cross-referenced by multiple types of reference markers (EST, SSR, candidate genes) offer good possibility for alignments and can result in a satisfactory "averaged" universal map which truly reflects the organization of the genome.

The degree of homology within and between species with regard to linkage order and sequence divergence makes our high-density map informative for other conifer and forest species. This will allow alignment of other maps and comparative genome analysis. Several reduced "satellite" maps have been constructed in a variety of *Pinus* species and related gymnosperms which are being aligned with our high-density map (<http://www.neiker.net/UHDfor>).

In order to facilitate switching between different genetic backgrounds and species, it would be useful to make parental DNA samples mutually available and to develop a reference collection of comigrating markers, which can be revealed by using a standardized set of primer combinations of SSR, EST, or other types of markers. Such a reference collection would allow us to (1) rapidly construct a framework map with good genome coverage in a new mapping population, (2) facilitate the dissection of complex traits into single genetic components and the identification of the genomic regions associated to these genes (QTL analyses), and (3) provide many putative, indirect markers from the high-density map for wide application. A high-density map can serve in this way as a reference map for other present and future mapping experiments of forest species.

Physical Mapping of "Strategically Important" Regions

We consider our high-density map to be the first step toward physical mapping of a conifer genome. Physical maps have been constructed in humans, animals, and many other organisms. They are useful for genome-wide gene discovery, EST mapping, and comparative genomics. Unique clones of a bacterial artificial chromosome (BAC) library can be conveniently assigned to mapped segregating AFLP fragments or other markers, using pools of BAC clones through hybridization or by PCR (Childs et al., 2001). In plants such BAC libraries have been established in *Arabidopsis* (Mozo et al., 1999), sorghum (Klein et al., 2000), rice (Tao et al., 2001), and many other species. In conifers, to our knowledge, a BAC library is available only for *Picea* (M. Morgante, personal communication).

Physical maps will soon become central to all future molecular genetic studies in conifer species. The characterization of model genomes such as human, yeast, and *Arabidopsis* has followed a progressive evolution from classical genetic studies to the development of increasingly dense genetic linkage maps (with concomitant trait mapping), large-scale EST programs, physical map development, gene anchoring on physical maps, and, finally, whole-genome sequencing. Considering the large genome size of conifers, it seems unrealistic from an economic point of view to propose the whole-genome sequencing of a conifer genome at this stage. However, the technology and resources exist to bring a conifer genome to the penultimate stage in the previously mentioned evolution within a relatively short period of time. High-density linkage maps increase the average number of markers per recombination unit. This helps to build contigs of overlapping BAC clones with the following advantages.

Routine isolation of any major gene known only by trait. Considerable input is currently required to carry out fine-scale mapping, BAC library construction, and chromosome walking on a case by case basis. A physical map would remove/ease many of these steps.

Efficient mapping of a large number of genetic markers (cDNAs, ESTs). Meiotic mapping is generally low resolution and requires detection of polymorphism between parental alleles. No prerequisite for polymorphism exists for physical mapping, making it suitable for high-throughput approaches.

Sequencing of strategic regions. Studies in a variety of plant genomes have demonstrated that a large percentage of resistance genes in any plant species are restricted to a relatively small number of genomic regions (Bakker et al., 2000). BAC contigs of such strategic regions facilitate the study of genome organization and function through reduced sequencing efforts.

Gene/Function Mapping to Identify and Exploit Genes Controlling Important Characters

High-density maps are most useful for integrating markers encoding concrete genes and advance in this way toward a functional map of a genome. Mapping "strategically important" genes/functions provides new resources to develop markers for marker-assisted selection, preferably located in genes controlling the traits of interest.

In addition to the EST studies described, the markers used for mapping in conifer species are, to date, mostly neutral, targeting anonymous, genomic DNA. Moreover, markers obtained in QTL analysis are determined through a statistical procedure. They are generally associated (more or less linked) only with the traits of interest. Therefore, they are of limited use for identifying the responsible gene itself, or for searching for more efficient allelic variants of the gene in germplasm collections. Moreover, depending on the genetic background, the allelic configuration among marker and QTL alleles may vary due to free recombination between these loci (for example, from coupling to repulsion, homozygosity of the marker). Consequently, the allelic states at both loci and the allele numbers must be evaluated again for each new cross, in order to use these markers for MAS. To overcome these difficulties, attempts have been recently made to directly identify the genes influencing character of interest. Different promising techniques have been developed and successfully applied. They include EST and candidate gene mapping, physical mapping, cDNA-AFLP, and cDNA microarrays. Allelic variants of these genes can be used in MAS in different genetic backgrounds or for genetic transformation.

EST and Candidate Gene Markers

EST markers typically represent candidate gene markers. In conifers, several projects exist to develop EST markers. One such example is Loblolly Pine Genomics (<http://pinetree.ccgb.umn.edu/>). Numerous ESTs and cDNAs representing candidate genes involved in a variety of processes are currently being mapped by different groups (Temesgen et al., 2001; Brown et al., 2001). In this context, the availibility of a BAC library would also be useful to rapidly capture the full-length transcripts.

cDNA Markers

The cDNA-AFLP method allows detection of differentially expressed transcripts using PCR (Bachem et al., 1996). It allows identification and analysis of genes involved in or controlling various biological processes ranging from development to responses to environmental cues. DNA microarrays also provide a convenient tool for analysis of genome-wide expression, but detailed genome sequence information, or at least a large cDNA collection, is necessary for this purpose. If these requisites are not met, cDNA-AFLP provides a more appropriate method. Furthermore, this method can complement the array technology and constitutes a useful tool for gene discovery (Breyne and Zabeau, 2001). Enrichment procedures can be embedded for target cDNAs using PCR and hybridization steps as described by Diatchenko and colleagues (1996) or Ivashuta and colleagues (1999). cDNA-AFLP has been applied successfully for identifying and isolating differentially expressed genes in different organisms including animals and plants.

Fukuda and colleagues (1999) cloned several genes that are dominantly expressed in highly metastatic tumor lines of rats with this approach. The method was used to identify pathogenicity factors in the potato cyst nematode *Globodera rostochiensis* (Qin et al., 2000). cDNA-AFLPs were applied to identify and isolate specifically expressed genes after *Peronospora* infection in *Arabidopsis* (Van der Biezen et al., 2000) and after *Cladosporium* inoculation in tobacco (Durrant et al., 2000). This method was also used to identify cell-cycle modulated genes in tobacco (Breyne and Zabeau, 2001) and to identify and clone genes involved in potato tuber development (Bachem et al., 1996; Bachem et al., 2001). However, only a limited cDNA-AFLP study has been reported in conifers so far (Whetten et al., 2001).

cDNA Microarrays

The microarray technique allows analysis of the expression of many cDNAs at a time by hybridizing cDNAs spotted on microchips with appropriate substrates. Recent studies applying this technique to analysis of cell-wall biosynthesis and wood formation are available in pine and aspen (Whetten et al., 2001; Hertzberg et al., 2002).

Allele Mining in Germplasm

Mapped gene markers or even closely linked neutral markers associated with important characters can be exploited by analyzing their variability in related germplasm. These analyses could provide possible associations between allelic variants and specific trait expression leading to the identification of more efficient allelic variants of a particular gene. Allele-specific PCR markers could thus be developed and more comfortably applied for screening procedures in marker-assisted breeding.

WWW-Accessible Database Containing an Integrated Map

It is convenient to unify all existing genetically mapped traits, DNA markers, gene markers, and BAC clones in a unique "consensus map." Placing all mapped traits relative to one another on a single integrated genetic and physical map will advance the potential for candidate-gene approaches for the dissection of these traits and provide a solid framework for mapping the same traits in different genetic backgrounds. The volume of data generated in such a project is expected to be vast. To fully enable community exploitation, the combined, interpreted data must be presented in a manner accessible to any end user. To maximize dissemination and to stimulate the exploitation of the resources generated, a Web-accessible database capable of supporting simple and complex queries should be developed and maintained in a format compatible with "stand alone" use and integration into appropriate international databases.

CONCLUSIONS AND PERSPECTIVES

High-density molecular marker maps are now available for humans, different animal and plant species, and recently for the conifer species *Pinus pinaster*, as well. Setting up suitable progenies for mapping and character evaluation is generally more complex in trees than in herbaceous species. The resulting polymorphisms in the mapping population will depend on the particular plant material under study and on the selected marker types. Special statistical procedures are necessary to obtain accurate linkage maps from the vast amount of marker data. A high-density map is an ideal platform for elaborating an integrated genetic, functional, and physical map with different potential applications. Dense linkage maps can be used to develop markers for qualitative and quantitative characters which can be used for marker-assisted selection. High-density maps that are cross-referenced by different codominant marker types allow alignment to other linkage maps and completion of comparative genome and QTL analyses in the same or related species. Physical maps can then be constructed efficiently by hybridizing mapped AFLP markers with pools of BAC clones or by performing PCR with these pools. Using EST, candidate genes, cDNA AFLP, or cDNA microarrays, gene/function mapping can be performed to identify and exploit genes controlling important characters. Furthermore, allele mining may allow detection of more efficient allelic variants in related germplasm.

REFERENCES

Arcade, A., Anselin, F., Faivre-Rampant, P., Lesage, M.C., Laurans, F., Paques, L.E., and Prat, D. (2000). Application of AFLP, RAPD and ISSR markers to genetic mapping of European larch and Japanese larch. *Theor. Appl. Genet* 100: 299-307.

Bachem, C.W.B., Horvath, B., Trindade, L., Claassens, M., Jordi, W., and Visser, R.G.F. (2001). A potato tuber-expressed mRNA with homology to steroid dehydrogenases affects gibberellin levels and plant development. *Plant J.* 25: 595-604.

Bachem, C.W.B., van der Hoeven, R.S., de Bruijn, S.M., Vreugdenhil, D., Zabeau, M., and Visser, R.G.F. (1996). Visualization of differential gene expression using a novel method of RNA fingerprinting based on AFLP: Analysis of gene expression during potato tuber development. *Plant J.* 9: 745-753.

Bakker, J., Stiekema, W., and Klein-Lankhorst, R. (2000). Homologues of a single resistance gene cluster in potato confer resistance to distinct pathogens: A virus and a nematode. *Plant J.* 23: 1-11.

Binelli, G. and Bucci, G. (1994). A genetic linkage map of *Picea abies* Karts, based on RAPD markers, as a tool in population genetics. *Theor. Appl. Genet* 88: 283-288.

Black, W.C. (1993). PCR with arbitrary primers: Approach with care. *Insect Mol. Biol.* 2: 1-6.

Breyne, P. and Zabeau, M. (2001). Genome-wide expression of plant cell cycle modulated genes. *Curr. Opin. Plant Biol.* 4: 136-142.

Broman, K.W., Murray, J.C., Sheffield, V.C., White, R.L., and Weber, J.L. (1998). Comprehensive human genetic maps: Individual and sex-specific variation in recombination. *Am. J. Hum. Genet.* 63: 861-869.

Brown, G.R., Kadel, E.E. III, Bassoni, D.L., Kiehne, K.L., Temesgen, B., van Buijtenen, J.P., Sewell, M.M., Marshall, K.A., and Neale, D.B. (2001). Anchored reference loci in loblolly pine (*Pinus taeda* L.) for integrating pine genomics. *Genetics* 159: 799-809.

Cato, S.A., Gardner, R.C., Kent, J., and Richardson, T.E. (2001). A rapid PCR-based method for genetically mapping ESTs. *Theor. Appl. Genet* 102: 296-306.

Chagne, D., Lalanne, C., Madur, D., Kumar, S., Frigerio, J.-M., Krier, C., Decroocq, S., Savoure, A., Bou-Dagher-Karrat, M., Bertocchi, E., et al. (2002). A high density genetic map of maritime pine based on AFLPs. *Ann. For. Sci.* 59: 627-636.

Chakravarti, A., Lasher, L.K., and Reefer, J.E. (1991). A maximum likelihood method for estimating genome length using genetic linkage data. *Genetics* 128: 175-182.

Childs, K.L., Klein, R.R., Klein, P.E., Morishige, D.T., and Mullet, J.E. (2001). Mapping genes on an integrated sorghum genetic and physical map using cDNA selection technology. *Plant J.* 27: 243-255.

Costa, P., Pot, D., Dubos, C., Frigerio, J.-M., Pionneau, C., Bodénès, C., Bertocchi, E., Cervera, M., Remington, D.L., and Plomion, C. (2000). A genetic map of maritime pine based on AFLP, RAPD and protein markers. *Theor. Appl. Genet* 100: 39-48.

Dale, G. and Teasdale, B. (1995). Analysis of growth, form and branching traits in an F2 population of the *Pinus elliottii* × *Pinus caribaea* interspecific hybrid using RAPD markers. In *Proceedings of the 25th Southern Forest Tree Improvement Conference* (pp. 242-253), New Orleans, LA.

Devey, M.E., Bell, J.C., Smith, D.N., Neale, D.B., and Moran, G.F. (1996). A genetic map for *Pinus radiata* based on RFLP, RAPD and microsatellite markers. *Theor. Appl. Genet.* 92: 673-679.

Devey, M.E., Fiddler, T.A., Liu, B.-H., Knapp, S.J., and Neale, B.D. (1994). An RFLP linkage map for loblolly pine based on a three-generation outbred pedigree. *Theor. Appl. Genet.* 88: 273-278.

Devey, M.E., Sewell, M.M., Uren, T.L., and Neale, D.B. (1999). Comparative mapping in loblolly and radiata pine using RFLP and microsatellite markers. *Theor. Appl. Genet.* 99: 656-662.

Diatchenko, L., Lau, Y.F.C., Campbell, A., Chenchick, A., and Sverdlov, E.D. (1996). Suppression subractive hybridization: A method for generating differentially regulated or tissue-specific cDNA probes. *P. Natl. Acad. Sci. USA* 93: 6025-6030.

Durrant, W.E., Rowland, O., Piedras, P., Hammond-Kosack, K.E., and Jones, J.D.G. (2000). cDNA-AFLP reveals a striking overlap in race-specific resistance and wound response gene expression profiles. *Plant Cell* 12: 963-977.

Echt, C.S., May-Marquardt, P., Hseih, M., and Zahorchak, R. (1996). Characterization of microsatellite markers in eastern white pine. *Genome* 39: 1102-1108.

Fisher, P.J., Gardner, R.C., and Richardson, T.E. (1996). Single locus microsatellites isolated using 5' anchored PCR. *Nucleic Acids Research* 24: 4369-4371.

Fisher, P.J., Richardson, T.E., and Gardner, R.C. (1998). Characteristics of single- and multi-copy microsatellites from *Pinus radiata. Theor. Appl. Genet.* 96: 969-979.

Fishman, L., Kelly, A.J., Morgan, E., and Willis, J.H. (2001). A genetic map of the *Mimulus guttatus* species complex reveals transmission ratio distortion due to heterospecific interactions. *Genetics* 159: 1701-1716.

Fukuda, T., Kido, A., Kajino, K., Tsustsumi, M., Miyauchi, Y., Tsujiuchi, T., Konishi, Y., and Hino, O. (1999). Cloning of differentially expressed genes in highly and low metastatic rat osteosarcomas by a modified cDNA-AFLP method. *Biochem. Bioph. Res. Co.* 261: 35-40.

Gerber, S. and Rodolphe, F. (1994). An estimation of the genome length of maritime pine (*Pinus pinaster* Ait). *Theor. Appl. Genet.* 88: 289-292.

Gosselin, I., Zhou, Y., Bousquet, J., and Isabel, N. (2002). Megagametophyte-derived linkage maps of white spruce *(Picea glauca)* based on RAPD, SCAR and ESTP markers. *Theor. Appl. Genet.* 104: 987-997.

Groover, A., Devey, M., Fiddler, T., Lee, J., Megraw, R., Mitchel-Olds, T., Sherman, B., Vujcic, S., Williams, C., and Neale, D. (1994). Identification of quantitative trait loci influencing wood specific gravity in an outbred pedigree of loblolly pine. *Genetics* 138: 1293-1300.

Harry, D.E., Temesgen, B., and Neale, D.B. (1998). Codominant PCR-based markers for *Pinus taeda* developed from mapped cDNA clones. *Theor. Appl. Genet.* 97: 327-336.

Harushima, Y., Yano, M., Shomura, A., Sato, M., Shimano, T., Kuboki, Y., Yamamoto, T., Lin, S.Y., Antonio, B.A., Parco, A., et al. (1998). A high-density rice genetic linkage map with 2275 markers using a single F2 population. *Genetics* 148: 479-494.

Hayashi, E., Kondo, T., Terada, K., Kuramoto, N., Goto, Y., Okamura, M., and Kawasaki, H. (2001). Linkage map of Japanese black pine based on AFLP and RAPD markers including markers linked to resistance against the pine needle gall midge. *Theor. Appl. Genet.* 102: 871-875.

Hertzberg, M., Aspeborg, H., Schrader, J., Andersson, A., Erlansson, R., Blomqvist, K., Bhalero, R., Uhlen, M., Teeri, T.T., Lundeberg, J., et al. (2002). A transcriptional roadmap to wood formation. *Proc. Natl. Acad. Sci. USA* 98: 14732-14737.

Hurme, P. and Savolainen, O. (1999). Comparison of homology and linkage of random amplified polymorphic DNA (RAPD) markers between individual trees of Scots pine (*Pinus sylvetris* L.). *Mol. Ecol.* 8: 15-22.

Ivashuta, S., Inai, R., Uchiyama, K., and Gau, M. (1999). The coupling of differential display and AFLP approaches for nonradioactive mRNA fingerprinting. *Mol. Biotechnol.* 12: 137-141.

Iwata, H., Ujino-Ihara, T., Yoshimura, K., Nagasaka, K., Mukai, Y., and Tsumura, Y. (2001). Leaved amplified polymorphic markers in sugi, *Cryptomeria japonica* D. Don, and their locations on linkage map. *Theor. Appl. Genet.* 103: 881-895.

Jansen, R.C. and Stam, P. (1994). High resolution of quantitative traits into multiple loci via interval mapping. *Genetics* 136: 1447-1455.

Kaya, Z., Sewell, M.M., and Neale, D.B. (1999). Identification of quantitative trait loci influencing annual height- and diameter-increment growth in loblolly pine (*Pinus taeda* L.). *Theor. Appl. Genet.* 98: 586-592.

Klein, P.E., Klein, R.R., Cartinhour, S.W., Ulanch, P.E., Dong, J., Obert, J.A., Morishige, D.T., Schlueter, S.D., Childs, K.L., Ale, M., and Mullet, J.E. (2000). A high-throughput AFLP-based method for constructing integrated genetic and physical maps: Progress toward a sorghum genome map. *Genome Research* 10: 789-807.

Knapp, S.J., Bridges, W.C., and Birkes, D. (1990). Mapping quantitative trait loci using molecular marker linkage maps. *Theor. Appl. Genet.* 79: 583-592.

Lander, E.S. and Botstein, D. (1989). Mapping mendelian factors underlying quantitative traits using RFLP linkage maps. *Genetics* 121: 185-199.

Lander, E.S., Green, P., Abrahamson, J., Barlow, A., and Daly, M. (1987). MAP-MAKER: An interactive computer package for constructing primary genetic linkage maps of experimental and natural populations. *Genomics* 1: 174-181.

Mariette, S., Chagne, D., Decroocq, S., Vendramin, G.G., Lalanne, C., Madur, D., and Plomion, C. (2001). Microsatellite markers for *Pinus pinaster* Ait. *Ann. For. Sci.* 58: 203-206.

Mariette, S., Chagne, D., Lezier, C., Pastuszka, P., Raffin, A., Plomion, C., and Kremer, A. (2001). Genetic diversity within and among *Pinus pinaster* populations: Comparison between AFLP and microsatellite markers. *Heredity* 86: 469-479.

Martínez, O. and Curnow, R.N. (1992). Estimating the locations and the sizes of the effects of quantitative trait loci using flanking markers. *Theor. Appl. Genet.* 85: 480-488.

Mozo, T., Dewar, K., Dunn, P., Ecker, J., Fischer, S., Kloska, S., Lehrach, H., Marra, M., Martienssen, R., Meier-Ewert, S., and Altmann, T. (1999). A complete BAC-based physical map of the *Arabidopsis thaliana* genome. *Nat. Genetics* 22: 271-275.

Nelson, C.D., Nance, W.L., and Doudrick, R.L. (1993). A partial genetic linkage map of Slash pine (*Pinus elliotti* Englem var. *elliottii*) based on random amplified polymorphic DNA's. *Theor. Appl. Genet.* 87: 145-151.

Paglia, G.P., Olivieri, A.M., and Morgante, M. (1998). Towards second-generation STS (sequence-tagged sites) linkage maps in conifers: A genetic map of Norway spruce (*Picea abies* K.). *Mol. Gen. Genet.* 258: 466-478.

Pfeiffer, A., Olivieri, A.M., and Morgante, M. (1997). Identification and characterization of microsatellites in Norway spruce (*Picea abies* K.). *Genome* 40: 411-419.

Plomion, C., Costa, P., and Bahrman, N. (1997). Genetic analysis of needle proteins in Maritime pine. 1. Mapping dominant and codominant protein markers assayed on diploid tissue, in a haploid-based genetic map. *Silvae Genetica* 46: 161-165.

Plomion, C. and Durel, C.E. (1996). Estimation of the average effects of specific alleles detected by the pseudo test cross QTL mapping strategy. *Genet. Sel. Evol.* 28: 223-235.

Qin, L., Overmars, H., Helder, J., Popeijus, H., van der Voort, J.R., Groenink, W., van Koert, P., Schots, A., Bakker, J., and Smant, G. (2000). An efficient cDNA-AFLP based strategy for the identification of putative pathogenicity factors from the potato cyst nematode *Globodera rostochiensis*. *Mol. Plant Microb. In.* 13: 830-836.

Remington, D.L., Whetten, R.W., Liu, B.H., and O'Malley, D.M. (1999). Construction of an AFLP genetic map with nearly complete genome coverage in *Pinus taeda*. *Theor. Appl. Genet.* 98: 1279-1292.

Ritter, E., Aragones, A., Markussen, T., Acheré, V., Espinel, S., Fladung, M., Wrobel, S., Faivre-Rampant, P., Jeandroz, S., and Favre, J.-M. (2002). Towards construction of an ultra high density linkage map for *Pinus pinaster*. *Ann. For. Sci.* 59: 637-643.

Ritter, E., Gebhardt, C., and Salamini, F. (1990). Estimation of recombination frequencies and construction of RFLP linkage maps in plants from crosses between heterozygous parents. *Genetics* 224: 645-654.

Ritter, E. and Salamini, F. (1996). The calculation of recombination frequencies in crosses of allogamous plant species with application to linkage mapping. *Genetic Research* 67: 55-65.

Rouppe van der Voort, J.N.A.M., van Zandvoort, P., van Eck, H.J., Folkertsma, R.T., Hutten, R.C.B., Draaistra, J., Gommers, F.J., Jacobsen, E., Helder, J., and Bakker, J. (1997). Use of allele specificity of comigrating AFLP markers to align genetic maps from different potato genotypes. *Mol. Gen. Genet.* 255: 438-447.

Ruiz de la Torre, J. and Ceballos, L. (1979). *Arboles y Arbustos*. Escuela Técnica Superior de Ingenieros de Montes. Sección de Publicaciones. Madrid. [*Trees and Shrubs*. Technical High School of Mountain Engineers. Publication Section. Madrid, Spain.]

Sewell, M.M., Davis, M.F., Tuscan, G.A., Wheeler, N.C., Elam, C.C., Bassoni, D.L., and Neale, D.B. (2002). Identification of QTLs influencing wood property traits in loblolly pine (*Pinus taeda* L.). II. Chemical wood properties. *Theor. Appl. Genet.* 104: 214-222.

Sewell, M.M., Sherman, B.K., and Neale, D.B. (1999). A consensus map for loblolly pine (*Pinus taeda* L.). I. Construction and integration of individual linkage maps from two outbred three-generation pedigrees. *Genetics* 152: 321-330.

Shepherd, M., Cross, M., Maguire, T.L., Dieters, M.J., Williams, C.G., and Henry, R.J. (2002). Transpecific microsatellites for hard pines. *Theor. Appl. Genet.* 104: 819-827.

Skov, E. and Wellendorf, W. (1998). A partial linkage map of *Picea abies* clone V6470 based on recombination of RAPD-marker in haploid megagametophytes. *Silvea Genetica* 47: 273-282.

Smith, D.N. and Devey, M.E. (1994). Occurrence and inheritance of microsatellites in *Pinus radiata. Genome* 37: 977-983.

Soranzo, N., Provan, J., and Powell, W. (1998). Characterization of microsatellite loci in *Pinus sylvestris* L. *Mol. Ecol.* 7: 1260-1261.

Steen, A.E., Kwitek-Black, R.G., Glenn, C., Gullings-Handley, J., Van Etten, W., Atkinson, O.S., Appel, D., Twigger, S., Muir, M., Mull, T., et al. (1999). A high-density integrated genetic linkage and radiation hybrid map of the laboratory rat. *Genome Research* 9: AP1-8.

Tao, Q., Chang, Y.L., Wang, J., Chen, H., Islam-Faridi, M.N., Scheuring, C., Wang, B., Stelly, D.M., and Zhang, H.B. (2001). Bacterial artificial chromosome-based physical map of the rice genome constructed by restriction fingerprint analysis. *Genetics* 158: 1711-1724.

Temesgen, B., Brown, G., Harry, D.E., Kinlaw, C.S., Sewell, M.M., and Neale, D.B. (2001). Genetic mapping of expressed sequence tag polymorphism (ESTP) markers in loblolly pine (*Pinus taeda* L.). *Theor. Appl. Genet.* 102: 664-675.

Thoday, J.M. (1961). Location of polygenes. *Nature* 191: 368-370.

Travis, S.E., Ritland, K., Whitman, T.G., and Keim, P. (1998). A genetic linkage map of Pinyon pine *(Pinus edulis)* based on amplified fragment length polymorphisms. *Theor. Appl. Genet.* 97: 871-880.

Van de Ven, W.T.G. and McNicol, R.J. (1996). Microsatellites as DNA markers in sitka spruce. *Theor. Appl. Genet.* 93: 613-617.

Van der Biezen, E.A., Juwana, H., Parker, J.E., and Jones, J.D.G. (2000). CDNA-AFLP display for the isolation of *Peronospora parasitica* genes expressed during infection in *Arabidopsis thaliana. Mol. Plant Microb. In.* 13: 895-898.

Vision, T.J., Brown, D.G., Shmoys, D.B., Durrett, R.T., and Tanksley, S.D. (2000). Selective mapping: A strategy for optimizing the construction of high-density linkage maps. *Genetics* 155: 407-420.

Wakamiya, I., Newton, R.J., Johnston, J.S., and Price, H.J. (1993). Genome size and environmental factors in the genus *Pinus. Am. J. Bot.* 80: 1235-1241.

Wang, Z., Weber, J. L., Zhong, G., and Tanksley, S. D. (1994). Survey of plant short tandem DNA repeats. *Theor. Appl. Genet.* 88: 1-6.

Whetten, R., Sun, Y.H., Zhang, Y., and Sederoff, R. (2001). Functional genomics and cell wall biosynthesis in loblolly pine. *Plant Mol. Biol.* 47: 275-291.

Zeng, Z.-B. (1994). Precision mapping of quantitative trait loci. *Genetics* 136: 1457-1468.

Chapter 15

Microsatellites in Forest Tree Species: Characteristics, Identification, and Applications

Giovanni G. Vendramin
Ivan Scotti
Birgit Ziegenhagen

INTRODUCTION

Population geneticists often strive to apply the ultimate, most variable, most widely distributed, easier-to-score molecular markers. This attitude is well explained by the need to extract as much information as possible from every experiment, which generally involves large samples and plenty of lab work. Once the data are scored, the quest for the meaning of the results begins. Marker data need to be analyzed in depth, and this task is easier when markers are better suited to "tell tales" of their own. Therefore, markers are sought from which as many kinds of inference as possible can be drawn: population heterozygosity, mating system, population structure, population differentiation, and population history. Microsatellites, with their peculiar structure, seem to be ideally designed to meet population geneticists' desires because they are highly polymorphic, codominant, widespread across the genome of any species, and suitable for reconstructing population history. Thanks to all these features, they have become the marker system of choice, and a rich statistical literature to treat microsatellite data has bloomed in recent years, so that the capacity to extrapolate information from these DNA fragments is probably rivaled only by the analysis of sequence haplotypes. For all these reasons, attention is devoted to microsatellites in all branches of genetics, including forestry.

MICROSATELLITES IN THE NUCLEAR GENOME:
DISTRIBUTION AND VARIABILITY

Microsatellites (simple sequence repeats = SSRs) are a class of repeated DNA that has been found within the genome of all eucaryote groups searched for the presence of SSRs. Their variability and widespread distribution make them one of the most efficient marker systems in population genetics. Microsatellites are repeats of short (2 to 6 base pair [bp]) sequences, the composition of which may vary (Tautz 1989). The frequency of repeat units in the genome depends in general on their composition and length, as do their variability, average number of repeats, and mutation rate.

Two potential mechanisms can explain the high mutation rates of microsatellites. The first is recombination between DNA molecules by unequal crossing-over or by gene conversion (Smith 1976; Jeffreys et al. 1994). The second mechanism involves slipped-strand mispairing during DNA replication (Levinson and Gutman 1987). Studies using yeast and *Escherichia coli* as model organisms have shown that replication slippage seems to be the main mechanism generating length mutations in microsatellites (Levinson and Gutman 1987; Henderson and Petes 1992). The length of the microsatellite repeats may have an effect on the mutation rate such that longer repeats are more polymorphic than shorter ones (Weber 1990; Chakraborty et al. 1997; Sia et al. 1997; Primmer et al. 1998; Ellegren 2000). This is probably due to the higher stability of misaligned configurations in longer repeat arrays. The second parameter that influences microsatellite stability is the purity of the repeat. Interrupted microsatellite repeats (due to insertion or base substitution) appear to have lower mutation rates than perfect repeats. This might be due to a lower probability of slipped intermediates formation in the presence of sequence interruptions (Petes et al. 1997).

Most microsatellites arrays are shorter than a few tens of repeats units. This strongly suggests that there must be size constraints restricting the expansion of repeat arrays. However, there is no direct evidence for selective constraints acting on allele length at microsatellite loci, although several mechanisms have been suggested (Samadi et al. 1998; Taylor et al. 1999). In general, mutational processes of microsatellites seem to be very complex. It is very likely that these processes are heterogeneous with differences among loci and alleles (Ellegren 2000).

The usefulness of SSRs also depends on their dispersion in the genome, as simple sequence repeats have been found in association with most of its fractions. Repeats have been found in translated and untranslated regions of genes (introns as well as 5'- and 3'-untranslated regions), as well as in the nonexpressed, noncoding portion of the genome and in association with the repeated fraction of it. The following section deals with the data that have been collected on the abundance and genomic distribution of micro-satellites in the genome of conifer species.

The Frequency of Microsatellite Classes in Conifers

The relative abundance of di-, tri-, and tetranucleotide microsatellites has been surveyed in *Pinus taeda* and *Pinus strobus* by Echt and May-Marquardt (1997). Their paper shows that dinucleotide repeats are found at frequencies between one every 220 kbp (kilobase pairs) (AC repeats in *Pinus strobus*) and one every 520 kbp (AC stretches in *Pinus taeda*); trinucleotide repeats (namely AAT stretches) can be as frequent as dinucleotides or almost absent from the genome; tetranucleotides are at least as rare as one every 1,500 kbp. The estimation of the frequency of microsatellites has also been reported for other forest tree species. Kostia and colleagues (1995) provide an estimate between one repeat every 100 kbp and one every 500 kbp for dinucleotide SSRs in *Pinus sylvestris,* while Lagercrantz and colleagues (1992) report a frequency of one GT repeat per 180 kbp and one CT repeat every 150 kbp in Norway spruce. In the same species, a more recent estimate (Pfeiffer et al. 1997) brings these figures to one GT repeat every 406 kbp and one CT repeat every 194 kbp. A third measure for Norway spruce is taken on the subset of genomic DNA represented by expressed sequences (Scotti et al. 2000), in which the same repeats were found once every 6,700 kbp (GT repeats) and once every 4,000 kbp (CT repeats). Thus, the latter two estimates are in accordance on the relative frequency of these two classes of repeats, with CTs more frequent than GTs; the same holds true in *Pinus strobus* (Echt and May-Marquardt 1997). By comparison, in well-studied species such as *Arabidopsis thaliana,* rice, soybean, maize, and wheat (Morgante et al. 2002), CT repeats are always more frequent than GT repeats, thus resembling the situation in conifers. The study by Morgante and colleagues (2002) reports much higher frequencies for all classes of repeats in these five species. This may be due to differences in the method used to obtain the estimates (inference from colony hybridizations in the studies on conifers; direct counts from sequenced genomic fragments in Morgante and colleagues [2002]), and, therefore, it cannot be excluded that the frequency of SSRs in conifers is much higher than reported.

SSRs in the High-Copy-Number DNA

SSRs are known to be dispersed in the genome, although evidence indicates that in plant genomes they tend to be overrepresented in low-copy-number regions (Morgante et al. 2002) and that they are not entirely randomly distributed in conifer chromosomes (Elsik and Williams 2001). Nevertheless, they also occur in the highly repeated fraction of genomic sequences, and this hampers their use as molecular markers. Proof of the high amount of SSR sequences found within what has been called "junk" DNA, which is DNA present in a high number of copies and with no (so far) known function, is given in Pfeiffer and colleagues (1997). The same feature is implicitly assumed in other papers, in which the amplification patterns of some SSR markers are described as multilocus or multiband (Scotti et al. 2002). Although this is a drawback from the point of view of marker development and scoring, the presence of SSRs in all fractions of genomic DNA grants that all parts of the chromosomes can be explored using these markers, for example, in mapping programs. An interesting outcome would be that large islands of repeated ("junk") DNA exist in the genome of conifers.

SSRs in Low-Copy-Number Regions

On the other hand, the presence of microsatellites in low-copy-number regions has also been demonstrated. Straightforward experiments such as dot-blot analyses (Pfeiffer et al. 1997) show that at least part of the SSRs that can be isolated from a genomic library are unique or low-copy-number clones. This determines the strategies for marker development. A different approach by Elsik and Williams (2001) has led to the isolation of SSRs enriched in single-locus clones. It has to be noted that both methods do not need a priori selection of clones with structural or functional characteristics (e.g., hypomethylated DNA; cDNA clones) and therefore makes no assumption on the relationships between copy number and other features of chromosome fragments. Elsik and Williams (2001) show that some SSR motifs, such as AC, AG, and ATC, are preferentially found in the slowly reassociating component of the genome, and this is in accordance with results obtained in other species (Morgante et al. 2002). The general conclusion that can be drawn from these experiments and from others (Schmidt et al. 2000) is that microsatellite repeats, while representing themselves a class of widely distributed repeated DNA, tend to be associated to the nonrepetitive fraction of the genome.

Microsatellites and Genes: Association of Microsatellites and Expressed Sequenced Tags (ESTs)

A common approach for the recovery of SSR markers is the search for SSR-containing clones in sequence databases. However, no extensive genomic program has been undertaken for any conifer, which is a serious hindrance in the development of new methods in conifer genetics—especially in comparison with the power reached by high-throughput techniques (e.g., single-nucleotide polymorphism [SNP] detection and analysis) in other species. EST collection and sequencing efforts have been started, however, for at least three species (*Pinus taeda, P. pinaster,* and *Picea abies*), and thus it is likely that SSRs will be found in these libraries. The presence of SSRs in conifer cDNA sequences has been proven by Scotti and colleagues (2000). This is not surprising, considering that several kinds of SSR sequences can be found in public EST databases. The variability at these microsatellite loci has also been shown in the same study. Dinucleotide microsatellites can be generally found in the 5'- and 3'-untranslated regions (UTR), and, as they may have a role in gene expression (Morgante et al. 2002), it is not known whether they can be considered fully neutral markers. However, arguments in favor of the essential neutrality of SSRs, with constraints, even within the coding regions, have been raised (Young et al. 2000). Therefore, we can expect that the chance of recovery of (polymorphic) SSRs from databases will increase along with the increase of the number of the available EST clones.

A Survey of the Composition of Microsatellite Repeats in Conifers

We have reviewed the literature describing the development of SSR markers in conifers in order to determine the amount of structural complexity of their sequences. According to the current classification of microsatellites (Weber 1990), from a total of 240 markers we have found 185 perfect repeats (77 percent), of which 141 are simple perfect and 44 compound perfect, and 55 imperfect repeats (23 percent), of which 17 are simple imperfect and 38 compound imperfect. Compared to this, in maize Taramino and Tingey (1996) found 31 perfect repeats out of 34 (91 percent); in *Sorghum,* Taramino and colleagues (1997) found 12 perfect repeat out of 13 (92 percent); Steinkellner, Fluch, and colleagues (1997) found 15 perfect repeats out of 17 (88 percent) in red oak, while in bur oak, Dow and colleagues (1995) reported 46 percent compound microsatellites; and Huang and colleagues (1998) found 38 perfect repeats out of 40 (95 percent) in the kiwi tree. Thus, conifers appear to have a lower fraction of perfect repeats compared to other species, and this means that complexity in the sequence of microsatellite alleles tends to be higher. This feature is related to the "ageing" of microsatellite stretches, as the tendency to the expansion of the repeats is counterbalanced by point mutations within the stretch (Kruglyak et al. 1998). The consequences of the presence of this structural complexity are further discussed in the next section.

FROM GENOME TO POPULATIONS: THE ISOLATION OF MICROSATELLITES AND THE DEVELOPMENT OF MOLECULAR MARKERS

Different methods have been used for the isolation of microsatellites from the genome of conifers. These species have complex and large genomes, in which the fraction represented by repeated elements is high; as already discussed, SSR stretches are likely to fall into this portion of the chromosomes. Conifer geneticists have therefore faced the need for methods that allow an efficient recovery of microsatellite markers from DNA libraries. After the first attempts with nonenriched libraries that resulted in a low percentage of positive clones and of single-locus markers (e.g., 19 percent of single-locus SSRs in Pfeiffer et al. 1997), two requirements were addressed: (1) the enrichment of libraries for positive, microsatellite-containing clones and (2) the enrichment of positive clones for single-copy fragments.

Methods for the Preparation of Microsatellite-Enriched Genomic Libraries

The former problem has been typically solved in two ways: one according to the method of Tenzer and colleagues (1999), including selection of clones by biotinylated primers and streptavidin-coated magnetic beads, followed by amplification with adapter-specific primers; and the other according to Edwards and colleagues (1996) that allows the isolation of several types of microsatellite repeats at once. The approach proposed by Edwards and colleagues (1996) revealed its efficiency also in a Mediterranean conifer species, *Pinus halepensis,* in which more than 90 percent of the clones of the enriched library contained a mono- (16 percent), a di- (77 percent), or a trinucleotide (7 percent) repeat (Keys et al. 2000).

Methods for the Selection of Low-Copy-Number Clones

The methods for the selection of low-copy-number clones can be classified according to the relative order of the selection and the cloning phase. Some methods involve the selection of the low-copy fraction(s) of the genome prior to the cloning phase. This category includes methods such as the search of microsatellites in cDNA libraries (Scotti et al. 2000) and the method described by Elsik and colleagues (2000). A third method, based on the selection of hypomethylated DNA, may be introduced in the future, as it promises to enrich for both low-copy-number sequences and microsatellite stretches (Morgante et al. 2002). Postcloning enrichment for low-copy-number, microsatellite-positive clones via dot-blot selection has been used by Scotti and colleagues (2002) following the method established in Pfeiffer and colleagues (1997), which turned out to be the most efficient system for the development of single-copy markers so far, with 66 percent of primer pairs producing a clear single-locus pattern on gels.

Finding the Optimal Conditions for Polymerase Chain Reaction (PCR) Amplification of SSRs

Conifer SSR markers tend to be sensitive to amplification conditions. This is probably due to the size of the genome and to its composition, and even low-copy-number clones rarely occur as a single copy throughout the genome. Unpublished data by Scotti show that Southern hybridizations of probes deriving from the hypomethylated fraction of the genome typically hybridize to several bands from independent loci, showing that even DNA fragments that typically occur in low- to single-copy are represented several times in the genome. This may explain why primers designed for SSR amplification tend to amplify more bands than expected. Therefore, care must be taken in the fine-tuning of PCR conditions, and downscaling of reaction volumes and of the concentration of template DNA and *Taq* polymerase seem to affect the quality of banding patterns (Scotti et al. 2002). Quality of the DNA seems to have minor effects on the quality of PCR products, since a variety of DNA extraction methods are reported in papers describing microsatellite markers in conifers. Different types of polymerase are reported as well throughout the literature, although it is well known that different polymerases display different degrees of specificity. Therefore, it is advisable to keep the same enzyme, once the protocol for a set of markers has been established.

MICROSATELLITES IN POPULATIONS: SSRs AS A TOOL FOR THE SCREENING OF DIVERSITY, DIFFERENTIATION, AND EVOLUTION

Theoretical Models of Microsatellite Mutations

To estimate population differentiation measures and genetic distances from microsatellite data, theoretical models for the evolutionary processes of microsatellites are needed. The development of statistics accurately reflecting genetic structuring requires an accurate understanding of the mutation model underlying microsatellite evolution (Balloux and Lugon-Moulin 2002). Two main and extreme theoretical models have been considered for microsatellites. In the infinite allele model (IAM, Kimura and Crow 1964) mutation can involve any number of tandem repeats and always results in a new allele status not previously existing in the population. As a consequence, this model does not allow for homoplasy. However, the slipped-strand mispairing is currently accepted as the main mechanism for microsatellite length variation. This mechanism mostly causes small changes in the repeat numbers such that alleles of similar lengths should be more closely related to one another than alleles of completely different sizes. Alleles may also mutate toward allele states that are already present in the population. The stepwise mutation model (SMM) better describes this kind of evolutionary process. Under a SMM, each mutation creates a novel allele either by adding or deleting a single repeat, with a probability in both directions. Consequently, more different alleles in terms of sizes are also more evolutionarily distant than alleles having similar sizes; therefore, SMM has a memory. Rienzo and colleagues (1994) proposed an alternative approach named two-phase model (TPM), in which a limited portion of mutations involve several repeats. The K-allele model (KAM) could also be considered for microsatellites. Under this model, there are K possible allelic states, and any allele has a constant probability of mutating toward any K-1 allelic status (Crow and Kimura 1970).

Different kinds of repeat-number variance estimators based on the stepwise mutation model have been developed for genetic distances and population differentiation (Goldstein et al. 1995; Shriver et al. 1995; Slatkin 1995; Balloux and Lugon-Moulin 2002). These estimators are based on many assumptions that take into account factors related to evolution of microsatellites (e.g., size constraints, multistep mutations, directional changes in allele size). Mutation rates have not been included in these models. Mutation rates may not only vary among repeat types (di-, tri-, and tetranucleotide), base composition of the repeat (Bachtrog et al. 2000), and microsatellite type (perfect, compound, or interrupted), but also among taxonomic groups. Moreover, an important role may be played by the nature of the flanking sequences, by the position of a microsatellite in the chromosome, and by the length of the alleles (the longer alleles being more prone to mutation than the shorter alleles, Schlotterer et al. 1998). Therefore, the frequency of interrupted and complex repeats tends to be high in conifers. This suggests that the interpretation of data analyses should be made very cautiously in case the available models do not include the appropriate assumptions.

Several studies have been performed during the past decade to understand the evolution of microsatellite regions. These studies of evolutionary processes have shown that (1) the mutation process is upwardly biased; (2) the mutation of the repeat units depends on the allele size and purity; and (3) the maximum possible size of microsatellite is constrained. Theoretical mutation models such as SMM and TPM may accurately represent the evolutionary processes of microsatellites when closely related populations are considered. However, over long evolutionary distances the mutation process seems to be more complex: frequencies of nonstepwise mutation events range between 4 percent and 7 percent depending on the different taxonomic groups (Ellegren 2000; for more details about the appropriate models and estimates of genetic differentiation parameters using microsatellite data see Balloux and Lugon-Moulin 2002).

Short- and Long-Term Evolution

The evolution and persistence of microsatellite loci has been studied both within species and between different species. The most common approach to study the evolution of microsatellites was to sequence alleles within populations and loci among species. In addition, the potential effect of the repeat areas on the evolutionary rate in the immediately adjacent flanking sequences was also tested.

Microsatellite evolution within populations was studied by sequencing microsatellite alleles from *Pinus radiata* populations by Karhu (2001), with the main aims being (1) to confirm that detected fragments were really alleles from one locus and not from closely related loci and (2) to find the cause of the odd numbers of base pairs differences and unexpected large gaps between allele lengths in some populations. The results showed that the amplification products of each primer pair were alleles from a single locus, although some indels and/or base substitutions were found that may cause size homoplasy. Size homoplasy has also been reported among microsatellite alleles from the same population or same species (Viard et al. 1998; Colson and Goldstein 1999; Makova et al. 2000). The study of Karhu (2001) also demonstrated that the repeat areas of microsatellite loci do not necessarily mutate purely in a stepwise fashion due to the indels or base substitutions in the repeat areas; consequently, the assumptions of statistical methods based on variances in the number of repeats is not always appropriate.

Microsatellite persistence and evolutionary change was studied among different pine species (*P. sylvestris, P. resinosa, P. radiata, P. strobus,* and *P. lambertiana*) belonging to different subgenera (*Pinus* and *Strobus*) that diverged over 100 million years ago (MYA) (Karhu, 2001), with the main objectives to verify the possibility to cross-amplify and to find changes in the structure of the repeat areas as well as in the flanking sequences. In *P. strobus,* 32 percent of the (subgenus *Strobus*) primer pairs resulted in specific amplification in *P. sylvestris* (subgenus *Pinus*). The probability of successful cross-species amplification of microsatellites depends on the relatedness between species. For instance, within species belonging to the family of the Hirundinidae, 90 percent of all the studied marker-species combinations worked (Primmer et al. 1996). Echt and colleagues (1999) showed that it is possible to share primers among members of the subgenus *Pinus*. The amplification rate was 29 percent when the primer pairs were tested among subgenus *Strobus* and subgenus *Pinus*. Homology of the amplification products was assessed by comparing flanking sequences to the known phylogeny of these species.

Sequence comparisons among the five previously mentioned *Pinus* species revealed that microsatellite repeat sequences had persisted in all species despite the very different population sizes, in accordance with the prediction of Stephan and Kim (1998). Some rapid expansion of the repeat area of the microsatellite was observed in *Pinus strobus*. Recently, Kutil and Willimas (2001) found that using trinucleotide microsatellites significantly improved trans-specific microsatellite recovery among hard and soft pine species. Thereafter, two *P. taeda* microsatellites had conserved regions and repeat motifs in all seven analyzed hard pines. Microsatellite loci transferred to *Pinus elliottii* and *P. caribaea* equally from *P. radiata* and *P. taeda* (about 58 percent of the tested primer pairs amplified) (Shepherd et al. 2002), with similar transfer rates. Changes in the repeat structure or flanking regions were minor and consistent with current taxonomic and phylogenetic relationships. Moreover, homologous microsatellite loci display similar levels of polymorphism in all the analyzed species (Shepherd et al. 2002).

In angiosperm species Steinkellner, Lexer, and colleagues (1997), for instance, identified conserved (GA_n) microsatellites among various *Quercus* species. Even among genera it was to a certain extent possible to cross-amplify SSR loci with primers originating from *Quercus petraea*. However, with increasing evolutionary distance the success of cross-amplification and obtaining homologous and polymorphic SSR loci decreased (Steinkellner, Lexer, et al. 1997).

Potential Problems Associated with Microsatellites

Microsatellites also have some drawbacks as markers. The first problem is a putative reduction or complete loss of amplification of some alleles due to base substitutions or indels within the priming site (null alleles). A heterozygote carrying one *null* allele cannot be distinguished on gel from a homozygote for the only DNA fragment which can be scored in the same plant. This can lead to an underestimation of heterozygosity, compared to the expected heterozygosity under the Hardy-Weinberg equilibrium. Segregation analysis in full-sib families helps to identify null alleles. Consequently, the respective loci can be excluded from multilocus analysis. Inheritance and segregation analysis, therefore, are prerequisites for validating SSR variants as markers in population genetics (Gillet 1999). Another problem is associated to the *Taq* polymerase which may generate slippage during PCR and therefore generate problems in microsatellite size determination by means of sequencing (Liepelt et al. 2001). Furthermore, allelic "drop-out" is a phenomenon that may occur when the template DNA is of extremely low quantity. In this case, PCR was shown to amplify only the shorter of two alleles (Taberlet and Luikart 1999). This phenomenon can also lead to an overestimation of homozygosity.

Hypervariability of markers may not always represent an advantage. Microsatellite variation is based on length variation in base pairs of the amplified fragments. Fragments of the same length (identical in state) may not be derived from the same ancestral sequence (and therefore may not be identical by descent), thus introducing the possibility of size homoplasy. Under the IAM sequence homoplasy is relevant, while parameters estimated on the basis of SMM and TPM are less influenced by this phenomenon. On the other hand, size homoplasy can confound estimates of population parameters based on all the previously mentioned models. The degree of homoplasy increases with the mutation rate and the time of divergence (Estoup and Cornuet 1999). In addition, the selective size constraints that reduce the possible allelic states increase size homoplasy (Nauta and Weising 1996). Size homoplasy results in an underestimation of population subdivision and genetic divergence between population and species (Viard et al. 1998; Taylor et al. 1999). Size homoplasy is taken into account by several distance measures based on the SMM (Goldstein et al. 1995; Slatkin 1995; Rousset 1996; Feldman et al. 1997).

Applications

Microsatellites have become the preferred markers in many studies because of their high variability, ease and reliability of scoring, and codominant inheritance. Originally microsatellite markers were used for genetic mapping and as a diagnostic tool to detect human disease (Murray et al. 1992). In forest trees they have been used as codominant anchor points in genome mapping (e.g., oak: Barreneche et al. 1998; Norway spruce: Paglia et al. 1998; maritime pine: Ritter et al., 2002). At the individual level, SSRs are furthermore reported to have potential to detect somaclonal variation (Rahman and Rajora 2001) and also to identify single individuals or clones (Lefort et al. 2000; Gomez et al. 2001). Most frequently, microsatellites are used in population and ecological studies. Microsatellites are excellent markers for studying gene flow, effective population size, migration and dispersal processes, and parentage and relatedness (e.g., Dow and Ashley 1996; Streiff et al. 1999; Godoy and Jordano 2001). The usually great numbers of effective alleles per locus allow parentage analysis with extremely high exclusion percentages. For example, using only six SSR loci Streiff and colleagues (1999) reached an exclusion percentage of about 99.99 in oak. Microsatellites that commonly exhibit a high degree of heterozygosity can also be used to study the effects and level of inbreeding.

Nuclear microsatellite analysis, in combination with the development and application of a new statistical test, allowed the identification of the origin of some *Pinus pinaster* populations (Derory et al. 2002). The test can be theoretically applied to all other forest species, provided that the microsatellite information is available and that the distribution curves of haplotype frequencies do not overlap.

A combination of microsatellites with other markers aimed at a differential display of the distribution of genetic diversity among and within populations assuming different evolutionary drivers beyond the variability of the different markers. Coanalyzing allozymes and microsatellites in oak populations, Degen and colleagues (1999), for example, found a smaller interlocus variation for microsatellites than for allozymes and concluded that microsatellite variation may reflect less interaction of population processes. Comparing the distribution of diversity using allozymes and various DNA markers including microsatellites with that of an adaptive morphological trait in Scots pine populations, Karhu and colleagues (1996), however, did not find any correlation. The conclusion is that microsatellites tremendously contributed to an increasing knowledge about stochastic aspects but so far not about population processes governed by selection or adaptation. New strategies are needed that may aim to identify and analyze microsatellites linked to adaptively relevant regions of the genome. For this purpose, combined use of SSRs and SNPs would be promising. At the species level, Muir and colleagues (2000) demonstrated a surprising result with microsatellites indicating the species status of commonly hybridizing *Quercus robur* and *Q. petraea*.

WHAT IS DIFFERENT ABOUT MICROSATELLITES OF THE ORGANELLE GENOMES?

Powell and colleagues (1995) were the first to describe a variable stretch of nucleotide repeats occurring in the chloroplast genome of conifers. Since then, chloroplast SSRs (cpSSRs) have developed widespread markers in population genetics or have occasionally been used in phylogenetic reconstructions. Compared to chloroplast SSRs, microsatellites of the mitochondrial genome (mtSSRs) have been reported only once, namely in *Pinus* in which they were found to differ among various *Pinus* species (Soranzo et al. 1999). To date, no study demonstrates any intraspecific variability of mtSSRs, which could be useful in studies on intraspecific genetic diversity within *Pinus* species or within any other conifer species. Thus, in this chapter we focus on the description of chloroplast microsatellites.

Targeting cpSSR Loci Using Universal Primers

The discovery of a polymorphic SSR stretch in the chloroplast DNA of *Pinus* species (Powell et al. 1995) enhanced the development of a set of universal primers amplifying 20 cpSSR loci in *Pinus* (Vendramin et al. 1996). The sequences of the 20 PCR-amplified loci revealed a characteristic structural trait: they consist of variable numbers of repeat units that are exclusively mononucleotides (Vendramin et al. 1996).

Chloroplast microsatellites have also been identified in angiosperms. The availability of the complete sequences of the chloroplast genomes of some species (e.g., *Arabidopsis thaliana, Nicotiana tabacum, Oryza sativa, Spinacia oleracea, Zea mays*) as well as of numerous partial chloroplast sequences allowed the identification of some hundreds of microsatellite motifs ($n > 10$) (Provan et al. 1999, 2001). Among these, about 96 percent were represented by repetition of a mononucleotide (A/T), about 1.2 percent of G/C, while only 2.8 percent consisted in repetition of di- (AT and TC) and trinucleotide (ATT) stretches. On the basis of the sequences of *Nicotiana tabacum,* Weising and Gardner (1999) designed a set of ten "universal" primers for the amplification of mononucleotide repeats.

The cpSSRs were found in intergenic spacer regions as well as in nontranscribed regions of chloroplast genes. They occurred as pure, composed, or interrupted repeats (Vendramin et al. 1996). The conserved arrangement and sequences of chloroplast genes as well as an obviously conserved position of microsatellites within the chloroplast genome facilitated a cross-amplification of cpSSR loci throughout different taxa. As demonstrated in several studies, a number of polymorphic cpSSR loci could be amplified not only in various *Pinus* species (Powell et al. 1995; Vendramin et al. 1996; Cato and Richardson 1996; Bucci et al. 1998; Echt et al. 1998; Morgante et al. 1998; Provan et al. 1998; Ribeiro et al. 2001; Richardson et al. 2002) but also in other genera of the Pinaceae (European *Abies:* Vendramin and Ziegenhagen 1997, Parducci et al. 2001; American *Abies:* Clark et al. 2000; *Picea:* Vendramin et al. 2000, *Cedrus:* Vendramin et al., unpublished data) and to a certain extent even in other conifer families including Cupressaceae, Taxodiaceae, and Taxaceae (Vendramin and Ziegenhagen, unpublished). The same is true for the angiosperm "universal" primers (Weising and Gardner 1999). The chloroplast microsatellite primers designed for *Nicotiana tabacum* frequently cross-amplified, not only in related species but also in species belonging to different taxonomic units and phylogenetically very distant groups. The analysis of six chloroplast microsatellites (ccmp2, ccmp3, ccmp4, ccmp6, ccmp7, ccmp10; Weising and Gardner 1999) in about 30 species belonging to the families of Aceraceae, Aquifoliaceae, Araliaceae, Betulaceae, Ericaceae, Fagaceae, Oleacea, Rosaceae, Salicaceae, and Ulmaceae produced amplified fragments of the expected sizes in all species (Vendramin et al., unpublished). Sequence data revealed that (1) the microsatellite stretches are generally present in the amplified fragments of all species; (2) some microsatellites are short (less than 8 bp) and interrupted; (3) the most polymorphic microsatellites are those with perfect stretches, and the variation is generally due to differences in the number of repeats; and (4) the microsatellites monomorphic within species have generally interrupted microsatellites with short A/T and/or G/C stretches (Vendramin et al., unpublished). Therefore, a clear relationship between level of polymorphism, molecular organization (simple and noninterrupted stretches), and number of repeats within the stretch seems to exist. A strong dependency of the polymorphisms of cp SSRs on their size was clearly shown by Deguilloux and colleagues (in preparation). The longer tracts are more variable than the shorter ones. A very high and significant correlation ($R^2 = 0.72$) was observed between allele number and maximum repeat number (Deguilloux et al., in preparation). Imperfect microsatellite contains nucleotide repeats that are interrupted by one or more nonrepeat nucleotides. Structural interruptions decrease the number of tandem repeats and may stabilize microsatellites loci, rendering them less prone to slippage mutations (Van Treuren et al. 1997). Ccmp10, the chloroplast microsatellite displaying a high level of variation within and among populations of *Fagus sylvatica, Fraxinus excelsior, Fraxinus angustifolia, Fraxinus ornus, Tilia cordata, Quercus petraea, Q. pubescens, Ilex aquifolium, Alnus glutinosa, Hedera helix, Corylus avellana, Betula pendula, Salix alba* (Vendramin et al., unpublished), and *Zelkova sicula* (Fineschi et al. 2002) is represented by a long mononucleotide repeat (A/T). The sequence of the ccmp10 fragment of the only species that showed no variation *(Calluna vulgaris)* revealed the presence of an imperfect $[(T)_7C(T)_6]$ stretch (Vendramin et al., unpublished).

The sequences of different haplotypes detected within species also confirmed that generally size variation is due to a different number of repeats within the microsatellite stretches (e.g., *Fraxinus* and *Fagus,* Vendramin

et al., unpublished; *Zelkova sicula* and *Z. abelicea,* Fineschi et al. 2002).
Similarly, conifer sequence data also confirmed the presence of the mononucleotide stretch as well that the variation is due to differences in the number of repeats of the stretch (*Picea abies:* Vendramin et al. 2000; *Pinus
pinaster*: Ribeiro et al. 2001; *Pinus halepensis* and *P. brutia*: Vendramin
et al., unpublished data; *Pinus cembra, P. sibirica,* and *P. pumila:* Gugerli
et al. 2001). Only in *Pinus pinaster* (Ribeiro et al. 2001) and *Pinus sylvestris*
(Provan et al. 1998) an indel of 5 bp and 9 bp, respectively, in the flanking
regions at one cpSSR region was observed. Hence, the development of
cpSSR markers in any newly studied conifer species is a comparably com-
fortable and low-cost procedure. This is an advantage compared to nuclear
microsatellites in which a trans-specific amplification is mainly restricted to
closely related species of the same genus.

Transmission of Chloroplast Microsatellites to the Filial Generation

Chloroplasts, and consequently chloroplast DNA, are predominantly paternally inherited in conifers (Owens and Morris 1990, 1991; Wagner
1992). Accordingly, the haploid genome of chloroplast is not subjected to
sexual recombination, which allows researchers to trace pure paternal lineages. Cato and Richardson (1996) were the first to describe a uniparentally
paternal transmission of cpSSRs to the filial generation in the conifer species *Pinus radiata*. Later on the paternal transmission of chloroplast microsatellites was also confirmed in *Abies alba* and *Picea abies* (Vendramin and
Ziegenhagen 1997; Sperisen et al. 1999). As a result of the inheritance analysis in *P. radiata* and *A. alba,* 1 percent of the analyzed seed offspring were
found to be heteroplasmic at the cpSSR marker loci, revealing both the paternal and the maternal size variant (Cato and Richardson 1996; Vendramin and
Ziegenhagen 1997). This was explained by the phenomenon of "maternal
leakage" occurring in the due course of the fertilization process. This effect
of maternal leakage was first postulated from the results of cytological investigations in the Pinaceae species *Pseudotsuga menziesii* (Owens and
Morris 1990, 1991). The occurrence of a low percentage of heteroplasmy is
supposed not to severely affect studies on the distribution of genetic diversity or on contemporal gene flow, as heteroplasmic individuals can easily be
detected and excluded from the analyses. However, the question arises
whether maternal leakage is confounding a long-term tracking of paternal
lineages.

The paternal transmission of cpSSRs is a prerequisite for tracing the efficiency of pollen-mediated gene flow in conifers. In addition, chloroplast

microsatellites can also be used for tracing seed dispersal from a mother tree when dispersed seeds are trapped and the primary haploid endosperm is analyzed at chloroplast microsatellite loci. As demonstrated by Vendramin and Ziegenhagen (1997), the endosperm of seeds from a controlled cross in *Abies alba* carried the maternal size variants. This gives clear genetic evidence that the maternal plastids, including plastid microsatellites, are not eliminated from the endosperm tissue as is the rule in egg cells or pre-embryos of conifers.

When haploid uniparentally inherited markers are considered in the monoecious conifers, the effective population is supposed to be half the size of that when nuclear markers are used (Birky et al. 1989). Thus, variation at pollen-mediated chloroplast microsatellite loci is believed to be indicative for genetic drift or for isolation in combination with bottleneck effects.

Chloroplast microsatellites are generally maternally inherited in angiosperms and therefore transmitted through seeds. Maternal inheritance has been tested and confirmed in *Fagus sylvatica* analyzing an intraspecific full-sib family (Vendramin et al., unpublished data) and in *Fraxinus* analyzing an interspecific full-sib family (*Fraxinus excelsior* × *Fraxinus angustifolia*) (Morand-Prieur et al. 2002). The maternal inheritance of the chloroplast genome of the broad-leaved species (in conifers, the plastids are generally paternally inherited) has important consequences on the distribution of haplotypic diversity within and among populations compared to conifers.

Mutation Rate and Interlocus Variability

Compared to nuclear microsatellites, the mutation rate of chloroplast microsatellites is supposed to be much lower. It was estimated to vary between 3.2×10^{-5} to 7.9×10^{-5} in the chloroplast genome of *Pinus torreyana* (Provan et al. 1999), which is 10- to 100-fold lower than the estimate for nuclear SSRs (Weber and Wong 1993). From rangewide studies of conifer species analyzing numerous populations it became evident that the variability of homologous cpSSR loci considerably differed between different species. This resulted in largely varying numbers of multilocus haplotypes as well as in the presence of private haplotypes in numerous analyzed populations or provenances, respectively. For example, the analysis of two cpSSR loci in 17 European populations of *A. alba* resulted in 90 different haplotypes (Vendramin et al. 1999), whereas 41 haplotypes were found when 97 populations of *Picea abies* were analyzed at three cpSSR loci (Vendramin et al. 2000). One of the analyzed loci in *A. alba* was also included in the *P. abies* analysis. A closer look at the highly polymorphic locus Pt 30204 in *A. alba* and North American *Abies* species revealed a composed locus with two variable SSR stretches, one being an interrupted repeat. This locus that contributed to an extremely high number of tow-locus haplotypes was argued to represent a "mutational hot spot" in the genus *Abies* (Vendramin et al. 1999; Clark et al. 2000). Such a phenomenon has to be considered as a result of so far unknown mutation processes also affecting flanking regions of the microsatellite. It is not likely that a stepwise mutation process is driving the evolution of complex loci, while a stepwise mutation process may be assumed for pure SSR stretches. Consequently, population genetic parameters operating under the assumption of any of the mutational models should be used with care, especially when complex loci including interrupted repeats are contributing to the multilocus haplotypic variation.

Applications

These characteristics of organelle microsatellites open up a wide range of applications in analyzing genetic diversity of conifers at different spatio-temporal scales, although they impose certain limitations as well. Advantages and disadvantages of the application of organelle microsatellites following different organismic organization are described in the following sections.

Individual Level

At the individual level, nuclear microsatellites have proved to be the marker of choice for individual identification and parentage analysis. In conifer species in which nuclear SSRs are not yet available, the potential usefulness of chloroplast microsatellites for individual identification or paternity analysis has been demonstrated. In three *Abies* species, Parducci and colleagues (2001) attributed 74 percent of the total gene diversity to the within-population variation. Thus, they confirmed the potential usefulness of cpSSR markers for individual identification in *Abies* which had been suggested before by Ziegenhagen and colleagues (1998). According to these investigations in *Abies alba,* certain individual members of a natural regeneration could be unambiguously ruled out to be pollen offspring of the directly neighbored adult tree. The latter study took advantage of a comparably high degree of cpSSR polymorphism available in the study area. However, in *Abies alba* such a situation is not to be expected at all sites of its natural range (Vendramin et al. 1999). This would diminish the utility of cpSSR markers as a general marker for paternity analysis or for tracking the pollen donors in seed orchards or seedlots. Moreover, the exclusion probability is supposed not to approach 99 percent, a value that is commonly achieved by means of nuclear microsatellite markers (e.g., Streiff et al. 1999). Nevertheless, there is an advantage of the haploid cpSSR marker in cases in which the application of diploid nuclear SSR markers is suffering from "unordered genotypes," for example, when none of the putative parents is known a priori. Hence, in a recent study on *Abies alba* endosperm, the haplotype of the mother could directly be identified, and by combining a cpSSR with a multilocus allozyme study, all mother trees involved could be identified from their respective endosperm (Cremer et al. in preparation).

Population Level

Numerous studies have been devoted to the distribution of chloroplast microsatellite diversity among and within populations. The interpretations mainly address effects of stochastic gene flow via pollen or seeds and thus rely on the assumption of neutrality of the cpSSR markers. To date there is not any indication for a selective relevance of these markers, for example, by hitchhiking effects within the genome. As in wind-pollinated species, pollen is the main agent of gene flow (Latta and Mitton 1997; Liepelt et al. 2003); the power of chloroplast DNA markers for differentiating among populations of the wind-pollinated conifer species is generally weak. This also holds true while comparing to the results obtained from maternally inherited markers, particularly when gene flow via gravity-dispersed seeds is traced. For example, G_{ST} in *Abies alba* using cpSSR marker was as low as 13 percent (Vendramin et al. 1999), and the apportionment of gene diversity between populations of *Picea abies* was even lower (R_{ST} = 10 percent; Vendramin et al. 2000). In contrast, the gene diversity distributed among the same *Picea abies* populations estimated using maternally inherited mitochondrial DNA markers was as high as 40 percent (Sperisen et al. 2001). Though harboring some problems of putative homoplasy as well as difficulties in distinguishing between historical and contemporary gene flow, cpSSR markers could be used in rangewide studies to shed light on certain aspects of population history in conifer species (Echt et al. 1998; Morgante et al. 1998; Vendramin et al. 1999; Parducci et al. 2001; Ribeiro et al. 2001; Richardson et al. 2002). A considerable degree of polymorphism in combination with a uniparentally paternal mode of inheritance is advantageous for identifying combined effects of past bottlenecks with an isolation of populations (Morgante et al. 1998; Vendramin et al. 1999). Obviously, the usefulness of cpSSR markers in conifers is based on their sensitivity against a reduction of the reproductively effective population size.

Chloroplast microsatellite polymorphism can also be used for the certification at population level. A detailed study about genetic diversity in *Pinus pinaster* using chloroplast microsatellites showed a homogeneous distribution of the polymorphism within French and Iberian populations and a significant differentiation between the two groups of populations (Ribeiro et al. 2002). This information, combined with a specifically designed statistical test, allowed the origin of some French stands of unknown origin to be determined. The results obtained using chloroplast microsatellites and terpene were concordant, but the cpSSR based test gave faster and more accurate results (Ribeiro et al. 2002).

As described, the apportionment of the chloroplast microsatellite haplotypic diversity within and among populations in angiosperm species was different from that of conifers, the chloroplast genome being maternally inherited in these organisms. The greatest part of the diversity resides among populations. The analysis of about 450 and 200 populations of *Fagus sylvatica* (beech) and *Fraxinus excelsior* (ash), respectively, using six chloroplast microsatellites revealed G_{ST} values higher than 85 percent (Vendramin et al., unpublished data). The distribution of the haplotypes reflects the migration history of the different species, and a significant congruence between genetic and pollen data was observed for beech and ash (Vendramin et al., in preparation) as well as for *Corylus avellana* (Palme and Vendramin 2002). The life history traits of the different species, and in particular the mode of seed dispersal, seem to play an extremely important role in shaping the distribution of diversity estimated using chloroplast microsatellites. This has more effect for species characterized by small and cottony seeds, easily transportable by wind, with much lower G_{ST} values (about 10 percent for *Populus tremula,* Salvini et al., 2001) than for angiosperm species with heavy seeds transported by animals and/or gravity (G_{ST} higher than 80 percent in *Fagus sylvatica* and *Fraxinus excelsior:* Vendramin et al., in preparation; *Quercus petraea:* Grivet 2002).

Among-Species Level

Several studies revealed the potential usefulness of cpSSRs for differentiating among closely related *Pinus* and *Abies* species (Bucci et al. 1998; Clark et al. 2000; Parducci et al. 2001; Teufel, personal communication). In *Abies,* Clark and colleagues (2000) and Parducci and colleagues (2001) used statistics to analyze the distribution of gene diversity among species. In *Pinus,* species-specific size variants of cpSSR loci were found to unambiguously differentiate among species (Bucci et al. 1998; Teufel, personal communication). Using cpSSR marker loci, Bucci and colleagues (1998) found strong evidence for a paternal introgression of *Pinus halepensis* haplotypes into *P. brutia.* Using species-specific cpSSR markers, former crosses between two pine species were revisited in order to verify the "pure-species" nature of the involved parents (Wachowiak, personal communication). In future studies, the use of paternally inherited species-specific markers in combination with maternally inherited species-specific markers will enable a thorough analysis of gene flow, introgression, or hybridization. In broad-leaved trees, chloroplast microsatellites also revealed a very high discriminate power; for example, *Fraxinus excelsior* and *Fagus sylvatica* do not share haplotypes with the closely related species *Fraxinus ornus* and *Fagus orientalis,* respectively (Vendramin et al., in preparation).

CONCLUSIONS

Microsatellites are powerful markers for genetics of forest trees. In the present work, both categories, the biparentally transmitted nuclear and the uniparentally transmitted organelle microsatellites, were reviewed. Once they become available as markers for a broad range of forest tree species, a powerful toolbox will be at hand to make complementary use of the two marker categories. Both marker categories together will provide stochastic markers for answering questions of genetic diversity at all spatio-temporal scales. Markers with different features, mutation rates, and structure can be applied to the evaluation of processes taking place on different time and space scales. This flexibility, combined with the methods becoming available (e.g., SNP detection), will enhance the efficiency of SSRs as witnesses of biological processes in natural populations, as well as tools for the genetic dissection for characters and for breeding. Alone or in combination, nuclear and chloroplast microsatellites represent very useful tools for the identification and certification of the origin of genetic material in many forest tree species, thus showing an important potential practical use in combatting illegal logging.

These markers will be extremely valuable for future studies related to genetic diversity and to saturate genetic linkage maps. For example, linkage of microsatellites with genomic regions of increasing interest might be a future topic. Linkage of microsatellite loci, for example, with the gametophytic self-incompatibility locus in *Prunus* species (Schueler et al. 2003) will allow tracing of sex-biased mating processes. Microsatellites in the neighborhood of fast-evolving genes will enhance the isolation of genomic regions relevant for adaptive processes (Schmid and Tautz 2000). The development of microsatellites depends on the amount of sequence information made available for each species. It is expected that genome sequencing projects, complete or partial, will be established in the future for an increasing number of species. Microsatellites, which are easy to identify and widely distributed in all classes of sequences, once discovered, can be exploited for wide applications.

REFERENCES

Bachtrog, D., Agis, M., Imhof, M., and Schlotterer, C. (2000). Microsatellite variability differs between dinucleotide repeat motifs-evidence from *Drosophila melanogaster*. *Mol. Biol. Evo.* 17: 1277-1285.

Balloux, F. and Lugon-Moulin, N. (2002). The estimation of population differentiation with microsatellite markers. *Mol. Ecol.* 11: 155-166.

Barreneche, T., Bodenes, C., Lexer, C., Trontin, J.-F., Fluch, S., Streiff, R., Plomion, C., Roussel, C., Steinkellner, H., Burg, K., et al. (1998). A genetic linkage map of *Quercus robur* L. (pedunculate oak) based on RAPD, SCAR, microsatellite, minisatellite, isozyme and 5S rDNA markers. *Theor. Appl. Genet.* 97: 1090-1103.

Birky, C.W. Jr., Fuerst, P., and Maruyama, T. (1989). Organelle gene diversity under migration, mutation and drift: Equilibrium, effects of heteroplasmic cells, and comparison to nuclear genes. *Genetics* 121: 613-627.

Bucci, G., Anzidei, M., Madaghiele, A., and Vendramin, G.G. (1998). Detection of haplotypic variation and natural hybridisation in *halepensis*—Complex pine species using chloroplast SSR markers. *Mol. Ecol.* 7: 1633-1643.

Cato, S.A. and Richardson, T.E. (1996). Inter- and intraspecific polymorphism at chloroplast SSR loci and the inheritance of plastids in *Pinus radiata* D. Don. *Theor. Appl. Genet.* 93: 587-592.

Chakraborty, B., Kimmel, M., Stivers, D.N., Davison, J., and Deka, R. (1997). Relative mutation rates at di-, tri-, and tetranucleotide microsatellite loci. *Proc. Natl. Acad. Sci. USA* 94: 1041-1046.

Clark, C.M., Wentworth, T.R., and O'Malley, D.M. (2000). Genetic discontinuity revealed by chloroplast microsatellites in eastern North American *Abies* (Pinaceae). *Am. J. Bot.* 87: 774-782.

Colson, I. and Goldstein, D.B. (1999). Evidence for complex mutations at microsatellite loci in *Drosophila*. *Genetics* 152: 617-627.

Crow, J.F. and Kimura, M. (1970). *An Introduction to Population Genetics Theory*. New York, Evanston, and London: Harper and Row.

Degen, B., Streiff, R., and Ziegenhagen, B. (1999). Comparative study of genetic variation and differentiation of two pedunculate oak *(Quercus robur)* stands using microsatellite and allozyme loci. *Heredity* 83: 597-603.

Derory, J., Mariette, J., Gonzalés-Martìnez, S., Chagnè, D., Madur, D., Gerber, S., Brach, J., Persyn, F., Riberio, M.M., and Plomion, C. (2002). What can nuclear microsatellites tell us about maritime pine genetic resources conservation and provenance certification strategies? *Ann. Sci. For.* 59: 699-708.

Dow, B.D. and Ashley, M.V. (1996). Microsatellite analysis of seed dispersal and parentage of saplings in the bur oak, *Quercus macrocarpa. Mol. Ecol.* 5: 615-627.

Dow, B.D., Ashley, M.V., and Howe, H.F. (1995). Characterization of highly variable (GA/TC)n microsatellites in the bur oak, *Quercus macrocarpa. Theor. Appl. Genet.* 91: 137-141.

Echt, C.S., De Verno, L.L., Anzidei, M., and Vendramin, G.G. (1998). Chloroplast microsatellites reveal population genetic diversity in red pine, *Pinus resinosa* Ait. *Mol. Ecol.* 7: 307-317.

Echt, C.S. and May-Marquard, P. (1997). Survey of microsatellite DNA in pine. *Genome* 40: 9-17.

Echt, C.S., Vendramin, G.G., Nelson, C.D., and Marquardt, P. (1999). Microsatellite DNA as shared genetic markers among conifer species. *Can. J. For. Res.* 29: 365-371.

Edwards, K.J., Barker, J.H.A., Daly, A., Jones, C., and Karp, A. (1996). Microsatellite libraries enriched for several microsatellite sequences in plants. *BioTechniques* 20: 758-760.

Ellegren, H. (2000). Heterogeneous mutation processes in human microsatellite DNA sequences. *Nat. Genetics* 24: 400-402.

Elsik, C.G., Minihan, V.T., Hall, S.E., Scarpa, A.M., and Williams, C.G. (2000). Low-copy microsatellite markers for *Pinus taeda* L. *Genome* 43: 550-555.

Elsik, C.G. and Williams, C.G. (2001). Families of clustered microsatellites in a conifer genome. *Mol. Genet. Genomics* 265: 535-542.

Estoup, A. and Cornuet, J.M. (1999). Microsatellite evolution: Inferences from population data. In D.B. Goldstein and C. Schlotterer (Eds.), *Microsatellite Evolution and Applications* (pp. 49-65). New York: Oxford University Press.

Feldman, M., Bergman, A., Pollock, D.D., and Goldstein, D.B. (1997). Microsatellite genetic distances with range constraints: Analytic description and problems of estimation. *Genetics* 145: 207-216.

Fineschi, S., Anzidei, M., Cafasso, D., Cozzolino, S., Garfì, G., Pastorelli, R., Salvini, D., Taurchini, D., and Vendramin, G.G. (2002). Molecular markers reveal a strong genetic differentiation between two European relic tree species: *Zelkowa abelicea* (Lam.) Boissier and *Z. Sicula* Di Pasquale, Garfì and Quezel (Ulmaceae). *Conserv. Genet.* 3: 145-154.

Gillet, E.M. (Ed.) (1999). Which marker for which purpose? *Final compendium of the Research Project "Development, Optimization and Validation of Molecular Tools for Assessment of Biodiversity in Forest Trees" in the European Union DGXII Biotechnology FW IV Research Programme "Molecular Tools for Biodiversity."* URL: <http://www.webdoc.gwdg.de/ebook/y/1999/whichmarker/cmarker.htm>.

Godoy, J.A. and Jordano, P. (2001). Seed dispersal by animals: Exact identification of source trees with endocarp DNA mircosatellites. *Mol. Ecol.* 10: 2275-2283.

Goldstein, D.B., Rui-Linares, A., Cavalli-Sforza, L.L., and Feldman, M.W. (1995). An evaluation of genetic distances for use with microsatellite loci. *Genetics* 139: 463-471.

Gomez, A., Pintos, B., Aguiriano, E., Manzanera, J.A., and Bueno, M.A. (2001). SSR markers for *Quercus suber* tree identification and embryo analysis. *J. Heredity* 92: 292-295.

Grivet, D. (2002). Phylogèographie et èvolution molèculaire compare d'arbres forestiers à l'aide des marqueurs chloroplastiques. [Chloroplast marker for molecular studies of phylogeography and evolution in forest trees.] These de Doctuer, Universitè Henri Poincarè, Nanci-I, France.

Gugerli, F., Senn, J., Anzidei, M., Madaghiele, A., Buchler, U., Sperisen, C., and Vendramin, G.G. (2001). Chloroplast microsatellites and mitochondrial nad1 intron 2 sequences indicate congruent phylogenetic relationships of Swiss stone pine *(Pinus cembra),* Siberian stone pine *(P. sibirica)* and Siberian dwarf pine *(P. pumila). Mol. Ecol.* 10: 1489-1497.

Henderson, S.T. and Petes, T.D. (1992). Instability of simple sequences DNA in *Saccharomyces cerevisiae. Mol. Cell Biol.* 12: 2749-2757.

Huang, W.G., Cipriani, G., Morgante, M., and Testolin, R. (1998). Microsatellite DNA in *Actinidia chinensis:* Isolation, characterisation, and homology in related species. *Theor. Appl. Genet.* 97: 1269-1278.

Jeffreys, A.J., Tamaki, K., MacLeod, A., Monckton, D.G., Neil, D.L., and Armour, J.A.L. (1994). Complex gene conversion events in germline mutation at human minisatellites. *Nat. Genet.* 6: 136-145.

Karhu, A. (2001). Evolution and applications of pine microsatellites. Acta Universitatis Ouluensis. Academin Dissertation, University of Oulu, Oulu University Press, ISBN 951-42-5923-8.

Karhu, A., Hurme, P., Karjalainen, M., Karvonen, P., Karkkainen, K., Neale, D., and Savolainen, O. (1996). Do molecular markers reflect patterns of differentiation in adaptive traits of conifers? *Theor. Appl. Genet.* 93: 215-221.

Keys, R.N., Autino, A., Edwards, K.J., Fady, B., Pichot, C., and Vendramin, G.G. (2000). Characterisation of nuclear microsatellites in *Pinus halepensis* Mill. and their inheritance in *P. halepensis* and *Pinus brutia* Ten. *Mol. Ecol.* 9: 2157-2159.

Kimura, M. and Crow, J.F. (1964). The number of alleles that can be maintained in a finite population. *Genetics* 49: 725-738.

Kostia, S., Varvio, S.L., Vakkari, P., and Pulikkinen, P. (1995). Microsatellite sequences in a conifer, *Pinus sylvestris. Genome* 38: 1244-1248.

Kruglyak, S., Durrett, R.T., Schug, M.D., and Aquadro, C.F. (1998). Equilibrium distributions of microsatellite repeat length resulting from a balance between slippage events and point mutations. *Proc. Natl. Acad. Sci. USA* 95: 10774-10778.

Kutil, B.L. and Willimas, C.G. (2001). Triplet-repeat microsatellites shared among hard and soft pines. *Am. Genetic Assoc.* 92: 327-332.

Lagercrantz, U., Ellegren, H., and Anderson, L. (1992). The abundance of various polymorphic microsatellite motif differs between plants and vertebrates. *Nucleic Acids Research* 21: 1111-1115.

Latta, R.G. and Mitton, J.B. (1997). A comparison of population differentiation across four classes of gene markers in limber pine *(Pinus flexibilis* James). *Genetics* 146: 1153-1163.

Lefort, F., Douglas, G.C., and Thompson, D. (2000). Microsatellite DNA profiling of phenotypically selected clones of Irish oak *(Quercus* spp.) and ash *(Fraxinus excelsior* L.). *Silvae Genetica* 49: 21-28.

Levinson, G. and Gutman, G.A. (1987). Slipped-strand mispairing: A major mechanism for DNA sequence evolution. *Mol. Biol. Evol.* 4:203-221.

Liepelt, S., Bialozyt, R., and Ziegenhagen, B. (2003). Wind-dispersed pollen mediates postglacial gene flow among refugia. *Proc. Natl. Acad. Sci. USA* 9: 14590-14594.

Liepelt, S., Kuhlenkamp, V., Anzidei, M., Vendramin, G.G., and Ziegenhagen, B. (2001). Pitfalls in determining size homoplasy of microsatellite loci. *Mol. Ecol. Notes* 1: 332-335.

Makova, K.D., Nekrutenko, A., and Baker, R.J. (2000). Evolution of microsatellite alleles in four species of mice (Genus *Apodemus*). *J. Mol. Evol.* 51: 166-172.

Morand-Prieur, M.E., Vedel, F., Raquin, C., Brachet, D., Sihachakr, D., and Frascaria-Lacoste, N. (2002). Maternal inheritance of a chloroplast microsatellite marker in controlled hybrids between *Fraxinus excelsior* and *Fraxinus angustifolia*. *Mol. Ecol.* 11: 613-618.

Morgante, M., Felice, N., and Vendramin, G.G. (1998). The analysis of hypervariable chloroplast microsatellites in *Pinus halepensis* reveals a dramatic genetic bottleneck. In A. Karp, P.G. Isaac, and D.S. Ingram (Eds.), *Molecular Tools for Screening Biodiversity: Plants and Animals* (pp. 407-412). London: Chapman and Hall.

Morgante, M., Hanafey, M., and Powell, W. (2002). Microsatellites are preferentially associated with nonrepetitive DNA in plant genomes. *Nat. Genetics* 30: 194-200.

Muir, G., Fleming, C.C., and Schlotterer, C. (2000). Species status of hybridizing oaks. *Nature* 405: 1016.

Murray, J.C., Bennet, S.R., Kwitek, A.E., Small, K.W., Schinzel, A., Alward, W.L., Weber, J.L., Bell, G.I., and Buetow, K.H. (1992). Lonkage of Reiger syndrome to the region of the epidermal growth factor gene on chromosome 4. *Nat. Genetics* 143: 1021-1032.

Nauta, M.J. and Weising, F.J. (1996). Constraints on allele size at microsatellite loci: Implications for genetic differentiation. *Genetics* 143: 1021-1032.

Owens, J.N. and Morris, S.J. (1990). Cytological basis for cytoplasmic inheritance in *Pseudotsuga menziesii*. I. Pollen tube and archegonial development. *Am. J. Bot.* 77: 433-445.

Owens, J.N. and Morris, S.J. (1991). Cytological basis for cytoplasmic inheritance in *Pseudotsuga menziesii*. II. Fertilization and proembryo development. *Am. J. Bot.* 78: 1515-1527.

Paglia, G.P., Olivieri, A.M., and Morgante, M. (1998). Towards second-generation STS (sequence-tagged sites) linkage maps in conifers: A genetic map of Norway spruce (*Picea abies* K.) *Mol. Gen. Genet.* 258: 466-478.

Palme, A. and Vendramin, G.G. (2002). Chloroplast DNA variation, postglacial recolonisation and hybridisation in hazel, *Corylus avellana*. *Mol. Ecol.* 11: 1769-1780.

Parducci, L., Szmidt, A.F., Madaghiele, A., Anzidei, M., and Vendramin, G.G. (2001). Genetic variation at chloroplast microsatellites (cpSSRs) in *Abies nebrodensis* (Lojac.) Mattei and three neighbouring *Abies* species. *Theor. Appl. Genet.* 102: 733-740.

Petes, T.D., Greenwell, P.W., and Dominska, M. (1997). Stabilization of microsatellite sequences by variant repeats in the yeast *Saccharomices cerevisiae*. *Genetics* 146: 491-498.

Pfeiffer, A., Olivieri, A.M., and Morgante, M. (1997). Identification and characterization of microsatellites in Norway spruce (*Picea abies* K.). *Genome* 40: 491-498.

Powell, W., Morgante, M., McDevitt, R., Vendramin, G.G., and Rafalski, J.A. (1995). Polymorphic simple sequence repeat regions in chloroplast genomes: Applications to the population genetics of pines. *Proc. Natl. Acad. Sci. USA* 92: 7759-7763.

Primmer, C.R., Moeller, A.P., and Ellegren, H. (1996). A wide-range survey of cross-species microsatellite amplification in birds. *Mol. Ecol.* 5: 365-378.

Primmer, C.R., Saino, N., Moeller, A.P., and Ellegren, H. (1998). Unravelling the processes of microsatellite evolution through analysis of germline mutations in barn swallows *Hirundo rustica. Mol. Biol. Evol.* 15: 1047-1054.

Provan, J., Powell, W., and Hollingsworth, P.P. (2001). Chloroplast microsatellites: New tools for studies in plant ecology and evolution. *Trends Ecol. Evol.* 16: 142-147.

Provan, J., Soranzo, N., Wilson, N.J., Goldstein, D.B., and Powell, W. (1999). A low mutation rate for chloroplast microsatellites. *Genetics* 153: 943-947.

Provan, J., Soranzo, N., Wilson, N.J., McNicol, J.W., Forrest, G.I., Cottrell, J., and Powell, W. (1998). Gene-pool variation in Caledonian and European Scots pine (*Pinus sylvestris* L.) revealed by chloroplast simple-sequence repeats. *P. Roy. Soc. Lond. Ser. B* 265: 1-9.

Rahman, M.H. and Rajora, O.P. (2001). Microsatellite DNA somaclonal variation in micropropagated trembling aspen *(Populus tremuloides). Plant Cell Rep.* 20: 531-536.

Ribeiro, M.M., LePrevost, G., Gerber, S., Vendramin, G.G., Anzidei, M., Decroocq, S., Marpeau, A., Mariette, S., and Plomion, C. (2002). Origin identification of maritime pine stands in France using chloroplast simple-sequence repeats. *Ann. Sci. For.* 59: 53-62.

Ribeiro, M.M., Plomion, C., Petit, R., Vendramin, G.G., and Szmidt, A.E. (2001). Variation in chloroplast-single-sequence repeats in Portuguese maritime pine (*Pinus pinaster* Ait.). *Theor. Appl. Genet.* 102: 97-103.

Richardson, B.A., Brunsfeld, S.J., and Klopfenstein, N.B. (2002). DNA from bird-dispersed seed and wind-disseminated pollen provides insights into postglacial colonization and population genetic structure of whitebark pine *(Pinus albicaulis). Mol. Ecol.* 11: 215-227.

Rienzo, A.D., Peterson, A.C., Garza, J.C., Valdes, A.M., Slatkin, M., and Freimer, N.B. (1994). Mutational processes of simple-sequence repeat loci in human populations. *Proc. Nat. Acad. Sci. USA* 91: 3166-3170.

Ritter, E., Aragones, A., Markussen, T., Achere, V., Espinel, S., Fladung, M., Wrobel, S., Faivre-Rampant, P., Jeandroz, S., and Favre, J.M. (2002). Towards construction of an ultra high density linkage map for *Pinus pinaster. Ann. For. Sci.* 59: 637-643.

Rousset, F. (1996). Equilibrium values of measures of population subdivision for stepwise mutation processes. *Genetics* 142: 1357-1362.

Salvini, D., Anzidei, M., Fineschi, S., Malvolti, M.E., Taurchini, D., and Vendramin, G.G. (2001). Low genetic differentiation among Italian populations of *Populus tremula* L. estimated using chloroplast PCR-RFLP and microsatellite markers. *For. Gen.* 8: 81-87.

Samadi, S., Erard, F., Estoup, A., and Jarne, P. (1998). The influence of mutation, selection and reproductive systems on microsatellite variability: A simulation approach. *Genet. Res.* 71: 213-222.

Schlotterer, C., Ritter, R., Harr, B., and Brem, G. (1998). High mutation rate of a long microsatellite allele in *Drosophila melanogaster* provides evidence for allel-specific mutation rates. *Mol. Biol. Evol.* 15: 1269-1274.

Schmid, K. and Tautz, D. (2000). Evolutionäre Genomforschung: Welche Rolle spielen schnell evolvierende Gene? [Evolutionary genome research: What is the role of fast evolving genes?] *Biospektrum* 3: 175-179.

Schmidt, A., Doudrick, R.L., Heslop-Harrison, J.S., and Schmidt, T. (2000). The contribution of short repeats of low sequence complexity to large conifer genomes. *Theor. Appl. Genet.* 101: 7-14.

Schueler, S., Tusch, A., Schuster, M., and Ziegenhagen, B. (2003). Characterization of microsatellites in wild and sweet cherry (*Prunus avium* L.)—Markers for individual identification and reproductive processes. *Genome* 46: 95-102.

Scotti, I., Magni, F., Fink, R., Powell, W., Binelli, G., and Hedley, P.E. (2000). Microsatellite repeats are not randomly distributed within Norway spruce (*Picea abies* K.) expressed sequences. *Genome* 43: 41-46.

Scotti, I., Paglia, G.P., Magni, F., and Morgante, M. (2002). Efficient development of dinucleotide microsatellite markers in Norway spruce (*Picea abies* Karst.) through dot-blot selection. *Theor. Appl. Genet.* 104: 1035-1041.

Shepherd, M., Cross, M., Maguire, T.L., Dieters, M.J., Williams, C.G., and Henry, R.J. (2002). Transpecific microsatellites for hard pines. *Theor. Appl. Genet.* 104: 1007-1112.

Shriver, M.D., Jin, L., Boerwinkle, L.E., Deka, R., Ferrel, R.E., and Chakraborty, R. (1995). A novel measure of genetic distance for highly polymorphic tandem repeat loci. *Mol. Biol. Evol.* 12: 914-920.

Sia, E.A., Kokoska, R.J., Dominska, M., Greenwell, P., and Petes, T.D. (1997). Microsatellite instability in yeast: Dependence on repeat unit size and mismatch repair genes. *Mol. Cell. Biol.* 17: 2851-2858.

Slatkin, M. (1995). A measure of population subdivision based on microsatellite allele frequencies. *Genetics* 139: 457-462.

Smith, G.P. (1976). Evolution of repeated DNA sequences by unequal crossover. *Science* 191: 528-535.

Soranzo, N., Provan, J., and Powell, W. (1999). An example of microsatellite length variation in the mitochondrial genome of conifers. *Genome* 42: 158-161.

Sperisen, C, Büchler, U., Gugerli, F., Matyas, G., Geburek, T., and Vendramin, G.G. (2001). Tandem repeats in plant mitochondrial genomes: Application to the analysis of population differentiation in the conifer Norway spruce. *Mol. Ecol.* 10: 257-263.

Steinkellner, H., Fluch, S., Turtschek, E., Lexer, C., Streiff, R., Kremer, A., Burg, K., and Glössl, J. (1997). Identification and characterization of (GA/CT)$_n$-microsatellite loci from *Quercus petraea*. *Plant Mol. Biol.* 33: 1093-1096.

Steinkellner, H., Lexer, C., Turetschek, E., and Glössl, J. (1997). Conservation of (GA$_n$) microsatellite loci between *Quercus* species. *Mol. Ecol.* 6: 1189-1194.

Stephan, W. and Kim, Y. (1998). Persistence of microsatellite arrays in finite population. *Mol. Biol. Evol.* 15: 1332-1336.

Streiff, R., Ducousso, A., Lexer, C., Steinkellner, H., Gloessl, J., and Kremer, A. (1999). Pollen dispersal inferred from paternity analysis in a mixed oak stand of *Quercus* robur L. and *Quercus petraea* (Matt.) Liebl. *Mol. Ecol.* 8: 831-842.

Taberlet, P. and Liukart, G. (1999). Non-invasive genetic sampling and individual identification. *Mol. Gent. Animal Ecol* 68: 41-55.

Takezaki, N. and Nei, M. (1996). Genetic distance and reconstruction of phylogenetic trees from microsatellite DNA. *Genetics* 144: 389-399.

Taramino, G., Tarchini, R., Ferrario, S., Lee, M., and Pe, M.E. (1997). Characterization and mapping of simple sequence repeats (SSRs) in *Sorghum bicolor*. *Theor. Appl. Genet.* 95: 66-72.

Taramino, G. and Tingey, S. (1996). Simple sequence repeats for germplasm analysis and mapping in maize. *Genome* 39: 277-287.

Taylor, J.S., Sanny, P., and Breden, F. (1999). Microsatellite alleles size homoplasy in the guppy *(Poecilia reticulate)*. *J. Mol. Evol.* 48: 245-247.

Tautz, D. (1989). Hypervariability of simple sequences as a general source for polymorphic DNA markers. *Nucleic Acids Research* 16: 6463-6471.

Tenzer, I., degli Ivanissevich, S., Morgante, M., and Gessler, C. (1999). Identification of microsatellite markers and their application to population genetics of *Venturia inaequalis*. *Phytopathology* 89: 748-753.

Van Treuren, R., Kuittinen, H., Karkkainen, K., Baena-Gonzalez, E., and Savolainen, O. (1997). Evolution of microsatellites in *Arabis petraia* and *Arabis lyrata,* outcrossing relatives of *Arabidopsis thaliana. Mol. Biol. Evol.* 14: 220-229.

Vendramin, G.G., Anzidei, M., Madaghiele, A., Sperisen, C., and Bucci, G. (2000). Chloroplast microsatellite analysis reveals the presence of population subdivision in Norway spruce *(Picea abies* K.). *Genome* 43: 68-78.

Vendramin, G.G., Degen, B., Petit, R.J., Anzidei, M., Madaghiele, A., and Ziegenhagen, B. (1999). High level of variation at *Abies alba* chloroplast microsatellite loci in Europe. *Mol. Ecol.* 8: 1117-1126.

Vendramin, G.G., Lelli, L., Rossi, P., and Morgante, M. (1996). A set of primers for the amplification of 20 chloroplast microsatellites in *Pinaceae. Mol. Ecol.* 5: 111-114.

Vendramin, G.G. and Ziegenhagen, B. (1997). Characterization and inheritance of polymorphic plastid microsatellites in *Abies. Genome* 40: 857-864.

Viard, F., Franck, P., Dupois, M.P., Estoup, A., and Jarne, P. (1998). Variation of microsatellite size homoplasy across electromorphs, loci, and populations in three invertebrate species. *J. Mol. Evol.* 47: 42-51.

Wagner, D.B. (1992). Nuclear, chloroplast, and mitochondrial DNA polymorphisms as biochemical markers in population genetic analyses of forest trees. *New Forests* 6: 373-390.

Weber, J.L. (1990). Informativeness of human (dC-dA)n(dG-dT)n polymorphisms. *Genomics* 7: 524-530.

Weber, J.L. and Wong, C. (1993). Mutation of human short tandem repeats. *Hum. Mol. Genet.* 2: 1123-1128.

Weising, K. and Gardner, R.C. (1999). A set of conserved PCR primers for the analysis of simple sequence repeat polymorphisms in chloroplast genomes of dicotyledonous angiosperms. *Genome* 42: 9-19.

Young, E.T., Sloan, J.S., and Van Riper, K. (2000). Trinucleotide repeats are clustered in regulatory genes in *Saccharomyces cerevisiae*. *Genetics* 154: 1053-1068.

Ziegenhagen, B., Scholz, F., Madaghiele, A., and Vendramin, G.G. (1998). Chloroplast microsatellites as markers for paternity analysis in *Abies alba*. *Can. J. For. Res.* 28: 317-321.

Chapter 16

Genome Mapping in *Populus*

María Teresa Cervera
Mitchell M. Sewell
Patricia Faivre-Rampant
Véronique Storme
Wout Boerjan

INTRODUCTION

Genetic linkage maps have significantly contributed to both the genetic dissection of complex inherited traits and positional cloning of genes of interest, and have become a valuable tool in molecular breeding (Yano, 2001). The development of mapping strategies particularly designed for species with long generation intervals and high levels of heterozygosity, together with the development of new marker technologies, has paved the way for constructing genetic maps of forest tree species (e.g., Barreneche et al., 1998; Butcher and Moran, 2000).

Poplars, cottonwoods, and aspens (*Populus* spp.) have a number of characteristics that make them attractive both as model trees and as commercial crops (Bradshaw et al., 2000). They are fast-growing trees that can be propagated vegetatively by cuttings, they have a small genome size (550 Mb), and several species can be efficiently transformed. Furthermore, the *Populus* genus consists of approximately 30 species, many of which can be sexually crossed to generate new superior hybrids, and several well-studied pedigrees exist (Stettler et al., 1996; Klopfenstein et al., 1997). Breeding institutes around the world have identified elite genotypes for use in breeding programs to improve performance for traits such as fast growth, wood quality, disease resistance, and environmental adaptation. The challenge today is to accelerate selection by using markers associated to the traits of interest (Bradshaw, 1996).

The aim of this chapter is to present an overview of all the existing genetic maps of the *Populus* genome. First, the different marker technologies and mapping strategies that have been used for generating genetic maps of *Populus* genomes are briefly described. Second, the individual poplar genome maps generated from different research groups are presented. Information regarding the design and construction of these maps is presented in Table 16.1. Additional information, mostly reflecting the efficiency of the different marker systems, is presented both in the text and Table 16.1. Third, the results from comparative mapping between several linkage maps, based on the map position of common codominant markers, are discussed in the text and presented in Table 16.2. Finally, the potential of integrating the *Populus* genome and expressed sequence tag (EST) information with the current genetic maps as well as the use of this information in a candidate-gene approach is discussed.

MARKER SYSTEMS

The development of a large number of marker technologies has made it possible to construct genetic maps for nondomesticated species. These markers can be grouped in two classes based on their dominant/codominant and multiallelic nature (for a review, see O'Malley and Whetten, 1997; Karp and Edwards, 1998). Dominant markers, such as random amplified polymorphic DNAs (RAPDs), amplified fragment length polymorphisms (AFLP is a registered trademark of Keygene N.V. in The Netherlands), selective amplification of microsatellite polymorphic loci (SAMPLs), inter-simple sequence repeats (ISSRs), and sequence-specific amplification polymorphisms (S-SAPs), require no previous sequence information and can, therefore, quickly generate a high number of anonymous markers for construction of a saturated genetic map (for acronyms used in genome research and their references, see <http://www.dpw.wageningen-ur.nl/pv/aflp/acronyms.html>). Codominant multiallelic marker systems, such as restriction fragment-length polymorphisms (RFLPs), microsatellites (SSRs), sequence-tagged sites (STSs), cleaved amplified products (CAPs), expressed sequence tags (ESTs), single-nucleotide polymorphisms (SNPs), and sequence characterized amplified regions (SCARs), are sequence dependent and therefore more laborious to identify than dominant markers. However, codominant markers are orthologous (i.e., descending from the same gene) among species and can be used in conjunction with dominant-marker-based maps to determine synteny among homoeologous linkage groups. Over the past six years, international consortia have developed large sets of codominant markers such as SSRs and ESTs. As genomic saturation of codominant markers increases, and as comparative mapping progresses, markers for future genetic maps can be chosen directly from the more informative codominant marker systems.

MAPPING STRATEGIES

Different strategies have been followed to construct genetic linkage maps of *Populus:* the "pseudo-testcross strategy," the "F2 inbred model," and the "three-generation outbred model." The pseudo-testcross strategy (Grattapaglia and Sederoff, 1994) takes advantage of the high level of heterozygosity of most forest-tree species. It is based on the linkage analysis of dominant markers that are heterozygous in one parent and null in the other and, thus, segregate 1:1 in their F1 progeny as in a testcross configuration. As a consequence, two linkage maps are generated (i.e., two-way pseudo-testcross), one for each parent. The F2 inbred model is based on a three-generation pedigree for which the grandparents are treated as inbred lines (represented by AA and BB). In the F2 generation, three genotypes occur at any locus: AA, AB, and BB, segregating 1:2:1. Software programs, such as MAPMAKER (Lander et al., 1987) can assemble a combined parental map from the F2 progeny data using the intercross mating type. The three-generation outbred model (Sewell et al., 1999) is an extension of the pseudo-testcross strategy. Within a single outbred pedigree, any given codominant marker will segregate in one of three different ways. When one parent is heterozygous and the other is homozygous, segregation will be 1:1 (i.e., testcross mating type). When both parents are heterozygous, segregation will be either 1:2:1 if both parents have the same genotype (i.e., intercross mating type), or 1:1:1:1 if they have different genotypes (i.e., fully informative mating type). This segregation data is then subdivided into two independent data sets that separately contain the meiotic segregation data from each parent, and independent maps are constructed for each parent. A sex-average map is then constructed using an outbred mapping program, such as JoinMap (Stam, 1993), which uses fully informative and intercross markers to serve as common anchor points between each parental data set. This integrated sex-average map is one component of mapping quantitative trait loci (QTLs) using the four-allele outbred model, which allows the simultaneous analysis of segregation data from both parents of an outbred species (Knott et al., 1997).

GENETIC LINKAGE MAPS
OF THE POPULUS *GENOME*

Genetic linkage maps have been constructed for species belonging to three main *Populus* sections (Eckenwalder, 1996): Leuce (*P. adenopoda* Maxim., *P. alba* L., *P. tremuloides* Michaux), Tacamahaca (*P. trichocarpa* T. and G., *P. cathayana* Rehder), and Aigeiros (*P. deltoides* Marshall, *P. nigra* L.). Maps were generated by analyzing marker segregation in intraspecific and inter-specific crosses between species of the same section, although hybrid progenies between *P. trichocarpa* and *P. deltoides* have also been used for genetic map-ping. Even though these two species have been classified in two different sec-tions, they nevertheless show close genetic relationships based on nuclear mo-lecular markers (M.-T. Cervera, unpublished data).

Liu and Furnier (1993) were the first to make a map of the *Populus* ge-nome. This linkage map of trembling aspen *(P. tremuloides)* was based on the segregation analysis of four allozymes and 75 RFLPs, using 49 *PstI* genomic DNA probes, in 93 individuals derived from the full-sib family GRS7 × 347. No segregation distortion was observed for any of these markers. The essentials of this map are presented in Table 16.1.

Bradshaw and colleagues (1994) published a genetic linkage map of *Populus* F2 hybrid family 331. Family 331 consists of 350 individuals from a cross between two F1 trees (53-246 × 53-242) from F1 hybrid family 53, which was generated from an interspecific cross between *P. trichocarpa* (93-968) and *P. deltoides* (ILL-129). A subset of 90 progeny was used for mapping. A total of 232 RFLP probes (generated from *PstI* and *XbaI* genomic DNAs and cDNAs) were tested on the parents. Of these RFLP probes, 11 percent hybridized to more than one locus, revealing genome duplication, and 215 of these RFLPs were used for mapping. Of the 44 STSs developed from *P. trichocarpa* (93-968), 29 could be amplified using genomic DNA from *P. trichocarpa,* and 26 of these 29 could be amplified from *P. deltoides.* Of these STSs, 17 were informa-tive among both parents of the mapping pedigree and were introduced into the mapping data set. Sixty-three RAPD primers (35 percent of those attempted) detected segregation for 111 dominant markers. The map was constructed us-ing the F2 inbred model, as previously described. This approach was followed because the map was to be used in a QTL analysis, in which the alleles among each grandparent of this interspecific cross were genetically highly divergent, while the alleles within each grandparent were expected to be relatively similar. Therefore, Bradshaw and Stettler (1995) made the assumption that the segrega-tion of QTLs would closely follow an F2 inbred model. The use of both domi-nant and codominant markers ensured the construction of a framework map in-cluding anchors, which can be used to identify homoeologous linkage groups when building new genetic linkage maps. Moreover, codominant markers were useful for identifying aneuploids in the progeny, which were excluded from the linkage analysis. Segregation distortion of markers (6.1 percent of RFLPs, 13.8 percent of STS, and 14 percent of RAPDs) resulted from a lethal allele at a sin-gle locus affecting embryo development (Bradshaw and Stettler, 1994). Final map characteristics are presented in Table 16.1.

The map for family 331 was used to identify genomic regions controlling quantitative traits that segregate under the assumptions of the F2 inbred model

TABLE 16.1. Genetic linkage maps of *Populus*—Overview

Mapping family	Mapping strategy	Number of markers analyzed	Mapped species	Mapped markers	Number of linkage groups (linkage criteria)	Average distance between 2 markers (cM)	Coverage (cM)	Reference
93 F1 *P. tremuloides* (GRS7 × 347)	F2 inbred model	Allozymes: 4 RFLP: 75	*P. tremuloides*	Allozymes: 3 RFLP: 54	14 (θ ⌐0.30) 22 unlinked	15.4 (K)	664 (K)	Liu and Furnier (1993)
90 progeny of the F2 hybrid family 331: [*P. trichocarpa* (93-968) × *P. deltoides* (ILL-129)] × [*P. trichocarpa* (93-968) × *P. deltoides* (ILL-129)]	F2 inbred model	RFLP: 232 probes tested, 215 used for mapping STS: 44 tested, 17 used for mapping RAPD: 180 primers screened, 63 used, 111 loci	*P. trichocarpa* × *P. deltoides* hybrid	RFLP: 203 STS: 17 RAPD: 92	35 (LOD>3.0, map distance ⌐0.30 cM (H) 31 unlinked	11.2 (H)	1261 (H)[a]	Bradshaw et al. (1994)[b]
F2 hybrid family [*P. deltoides* × *P. cathayana*] × [*P. deltoides* × *P. cathayana*] number of progeny ni	F2 inbred model	RAPD: 300 primers screened, 100 used, nr of loci ni	*P. deltoides* × *P. cathayana* hybrid	RAPD: 110	20	17.3	1899	Su et al. (1998)

391

TABLE 16.1 (continued)

346 progeny of the interspecific F2 hybrid 822 family: [P. trichocarpa (93-968) × P. deltoides (S7C4)] × [P. trichocarpa (93-968) × P. deltoides (S7C4)]	pseudo-testcross strategy	AFLP: 29 pc: 305 dominant loci, 10 codominant loci	P. trichocarpa	AFLP: nl SSR[c]: 8 genes: 5	26 (LOD ≥5.0, map distance ≤40 cM)	13.6	2002	Frewen et al. (2000)
		SSR: 9		AFLP: ni SSR[c]: 8				
		Candidate genes: 5	P. deltoids	Genes: 5	24 (LOD ≥5.0, map distance ≤40 cM)	12.3	1778	
93 BC1 trees: [P. deltoides I-69 × P. deltoides I-63] × I-63	pseudo-testcross strategy	22 AFLP pc: 523 loci, 154 from I-63: used for mapping I-63, 77 from C-135: used for mapping C-135, 105 intercross markers: used for both maps, 187 not used for mapping	P. deltoides I-63	AFLP: 137	19 (LOD ≥4.0, θ ≤0.25) 5 triplets, 19 doublets, 69 unlinked	23.3 (K)[d]	2927 (K)[d]	Wu et al. (2000)
			P. deltoides BC1 C-135	ni	ni			

Progeny/family	Strategy	Marker details	Parent	Mapped	Linkage groups			Reference
121 progeny of the interspecific F1 hybrid family 87001: P. deltoides (S9-2) x P. nigra 'Ghoy'	pseudo-testcross strategy	AFLP: 50 pc, 438 loci, 403 used for mapping	P. deltoides	AFLP: 394	21 (19[e]) (LOD≥5, θ≤0.30) 1 doublet, 16 unlinked	10.0 (K)[f]	2304 (K) Z (2178[f])	Cervera et al. (2001)
		SSR: 153 tested, 61 used for mapping		SSR: 53				
		STS: 25 tested, 1 used for mapping		STS: 0				
		Resistance marker: 1		Resistance marker: 1				
		genes: 2		genes: 2 (ur)				
		AFLP: 50 pc, 383 loci, 355 used for mapping	P. nigra	AFLP: 329	34 (28[e]) (LOD≥5, θ≤0.30) 4 triplets, 2 doublets, 19 unlinked	12.5 (K)[f]	2791 (K) (2356[f])	
		SSR: 153 tested, 49 used for mapping		SSR: 40				
		STS: 25 tested, 0 used for mapping		STS: 0				
101 progeny of the interspecific F1 hybrid family 87001: P. deltoides (S9-2) x P. trichocarpa (V24)	pseudo-testcross strategy	AFLP: 41 pc, 321 loci, 309 used for mapping	P. deltoides	AFLP: 305	23 (19[e]) (LOD≥5, θ≤0.30) 3 doublets, 13 unlinked	10.4 (K)[f]	1838 (K) (1626[f])	Cervera et al. (2001)
		SSR: 153 tested, 63 used for mapping		SSR: 51				
		STS: 25 tested, 1 used for mapping		STS: 0				
		Resistance marker: 1		Resistance marker: 1				
		genes: 4		genes: 2 (ur)				
		AFLP: 41 pc, 314 loci, 287 used for mapping	P. trichocarpa	AFLP: 278	23 (19[e]) (LOD≥5, θ≤0.30) 3 triplets, 2 doublets, 14 unlinked	11.2 (K)[f]	2326 (K) (1920[f])	
		SSR: 153 tested, 76 used for mapping		SSR: 60				
		STS: 25 tested, 1 used for mapping		STS: 1				
		genes: 6		genes: 4 (ur)				

393

TABLE 16.1 (continued)

80 progeny of the interspecific F1 hybrid family P. adenopoda Maxim. × P. alba L.	pseudo-testcross strategy	RAPD: 640 primers screened, 142 used: 12 intercross markers: 2 used for mapping, 92 dominant markers: 80 used for mapping	P. adenopoda	RAPD: 36	7 (LOD ≥4.0, θ ≤0.30) 2 triplets, 10 doublets, 20 unlinked	8.9 (K)[f]	553 (K)[f]	Yin et al. (2001)
		RAPD: 640 primers screened, 142 used: 12 intercross markers: 2 used for mapping, 256 dominant markers: 210 used for mapping	P. alba	RAPD: 186	19 (LOD) 4.0, θ ≤0.30) 1 triplet, 4 doublets, 15 unlinked	11.4 (K)[f]	2300 (K)[f]	
93 interspecific F1 hybrid family: P. deltoides (I-69) × P. euramericana (I-45) [P. deltoides × P. nigra]	pseudo-testcross strategy	RAPD: 1040 primers screened, 127 used: 94 dominant markers, 39 intercross markers, and 19 non-parental markers	P. deltoides	RAPD: 85				Yin et al. (2002)
		AFLP: 256 pc screened, 48 pc used for mapping: 214 dominant markers, 154 intercross markers, and 54 codominant markers		AFLP: 219	31 (LOD ≥4.0, θ ≤0.30)	13.6 (K)	3801 (K)	
		ISSR: 75 pc screened, 29 pc used: 13 dominant markers, 13 intercross markers		ISSR: 6				

		Markers		Map			Reference
93 interspecific F1 hybrid family: *P. deltoides* (I-69) x *P. euramericana* (I-45) [*P. deltoides* x *P. nigra*]	pseudo-testcross strategy	RAPD: 1040 primers screened, 127 used: 78 dominant markers, 39 intercross markers, and 19 nonparental markers	*P. euramericana*	RAPD: 68			Yin et al. (2002)
		AFLP: 256 pc screened, 48 pc used for mapping: 142 dominant markers, 154 intercross markers, and 54 codominant markers		AFLP: 162	34 (LOD ≥4.0, θ ≤0.30)	16 (K)	3452 (K)
		ISSR: 75 pc screened, 29 pc used: 19 dominant markers, 13 intercross markers		ISSR: 11			
44 trees of F2 hybrid family 331 (see above)	3-generation outbred model	SSR: 180	*P. trichocarpa* x *P. deltoides* hybrid	ni	23 (LOD ≥5)	ni	850	M.M. Sewell ur
325 trees of F2 hybrid family 331	3-generation outbred model	SSR: 80	*P. trichocarpa* x *P. deltoides* hybrid	ni	23 (LOD ≥5)	ni	850	M.M. Sewell ur

TABLE 16.1 *(continued)*

							P. Faivre-ampant ur
91 interspecific F1 hybrid family: *P. deltoides* (73028-62) x *P. trichocarpa* (101-74)	pseudo-test cross strategy	AFLP: 24 pc: 158 markers RAPD: 125 primers:135 markers SCAR: 1 STS: 17 tested, 0 used for mapping RFLP: 200 tested, 29 used for mapping SSR: 45 tested, 21 used for mapping genes: 6 tested	*P. deltoides*	AFLP: 139 RAPD: 107 SCAR: 1 STS: 0 RFLP: 29 SSR: 16 genes: 5	26 with more than 3 markers (LOD 4.0, 0.30) 5 triplets, 4 doublets, 17 unlinked	12 (K)	2845 (K)
		AFLP: 24 pc: 164 markers RAPD: 125 primers: 139 markers STS: 17 tested, 1 used for mapping RFLP: 200 tested, 27 used for mapping SSR: 45 tested, 27 used for mapping genes: 8 tested	*P. trichocarpa*	AFLP: 92 RAPD: 98 STS: 1 RFLP: 24 SSR: 19 genes: 5	27 with more than 3 markers (LOD 4.0, 0.30) 10 triplets, 11 doublets, 5 unlinked	14 (K)	2095 (K)

a coverage based on 19 major linkage groups
b the map has been extended to 512 markers (H.D. Bradshaw, unpublished data)
c not indicated on which map
d based on 19 major linkage groups obtained by alignment of linkage groups with intercross heteroduplex markers
e after alignment of linkage groups based on SSR and STS
f based on framework
pc: primer combinations
ur: unpublished results
ni: data not indicated
K: Kosambi units
H: Haldane units

(Bradshaw, 1996). QTLs have been mapped for commercial traits such as stem proportion and height, basal area and volume growth (Bradshaw and Stettler, 1995), tree growth and architecture (Wu, 1998; Wu et al., 1998), leaf variation (Wu et al., 1997), resistance to *Septoria populicola* (Newcombe and Bradshaw, 1996), *Melampsora medusae* (Newcombe et al., 1996; Newcombe, 1998), and *M.* × *columbiana* (Newcombe et al., 2001), as well as for adaptive traits such as spring leaf phenology (Bradshaw and Stettler, 1995).

Su and colleagues (1998) constructed a genetic linkage map from an F2 hybrid family derived from a three-generation pedigree (Table 16.1). The F1 resulted from a controlled cross between *P. deltoides* × *P. cathayana*. One hundred RAPD primers (33 percent of those attempted) yielded at least one polymorphic marker. The final map data are presented in Table 16.1. This map is being used to study quantitative traits related to pest and pathogen resistance.

Genetic linkage maps were constructed by Frewen and colleagues (2000) using segregation data from AFLPs, SSRs, and candidate genes and the two-way pseudo-testcross strategy. The maps were based on 346 progeny from the F2 hybrid family 822, which was derived from two F1 individuals from a cross between *P. trichocarpa* (93-968) and *P. deltoides* (S7C4). A total of 29 AFLP primer combinations were used to analyze 315 segregating loci, of which 305 were scored as dominant and ten as codominant. Nine SSRs provided by the Poplar Molecular Genetics Cooperative (PMGC; Tuskan et al., manuscript in preparation) were used for mapping. Ten aneuploids, identified by two SSRs, were excluded from the linkage analysis. In addition, five genes were mapped using a sequence-based polymerase chain reaction (PCR) assay; the genes *PtPHYB1* and *PtPHYB2* (encoding phytochrome B1 and B2, respectively), and *PtABI1B, PtABI1D,* and *PtABI3* (*Populus* homologs of the *Arabidopsis thaliana ABSCISIC ACID-INSENSITIVE* genes) were amplified and polymorphisms revealed by either length variation (*PtPHYB1*) or amplified-fragment digestion with restriction enzymes (*PtPHYB2, PtABI1B, PtABI1D,* and *PtABI3*). These same candidate genes and microsatellite markers were also mapped on the previously published map of family 331, which shared the same *P. trichocarpa* grandparent (Bradshaw et al., 1994), and some homoeologous linkage groups were identified (Table 16.2; see section on comparative mapping). These maps were used to analyze the genetic control of bud phenology.

Wu and colleagues (2000) published genetic linkage maps of two clones of *P. deltoides*. The maps were generated by combining the two-way pseudo-testcross strategy with AFLP markers. The intraspecific backcross pedigree consisted of 93 progeny from the cross C-135 × I-63, in which C-135 resulted from a cross between natural "elite" clones I-69 and I-63. Genetic maps for I-63 and C-135 were constructed based on linkage analysis of 523 AFLPs obtained with 22 primer combinations. Seventy-eight percent of the AFLPs segregated 1:1, 15 percent segregated 3:1, and 7 percent showed segregation distortion.

TABLE 16.2. Comparative mapping: Alignment of homoeologous groups

P. deltoides (S9-2) (Cervera et al. 2001) group	marker	P. nigra (Ghoy) (Cervera et al. 2001) group	marker	P. trichocarpa (V24) (Cervera et al. 2001) group	marker	Family 331 (Bradshaw et al. 1994) group	marker	Family 331 (M.M. Sewell, unpublished) group	marker	P. trichocarpa (93-968) (Frewen et al. 2000) group	marker	P. deltoides (ST24) (Frewen et al. 2000) group	marker	P. deltoides (73028-62) (P. Faivre-Rampant, unpublished) group	marker	P. trichocarpa (101-74) (P. Faivre-Rampant, unpublished) group	marker
I	PMGC2852							LG09	PMGC2852								
I	PMGC93	I	PMGC93			A	PMGC93	LG09	PMGC93							XXV	PMGC93
I	PMGC2098			I	PMGC2098	A	PMGC2098							I	PMGC2098		
II	PMGC2815	II	PMGC2618	II	PMGC667	M	PMGC667	LG01	PMGC667					XIII	PMGC667	V	PMGC667
II	PMGC667	II	PMGC667	II	PMGC684	M	PMGC684	LG01	PMGC422							V	PMGC684
III	PMGC422			II	PMGC456			LG01	PMGC684	BB1	PMGC223	BBb	PMGC223				
III	PMGC684			II	PMGC2650			LG01	PMGC456	BB1	PMGC456	BBb	PMGC456				
				II	PMGC2088			LG01	PMGC2660								
II	PMGC2088			II	PMGC223			LG01	PMGC2098	BB1	PMGC683	BBb	PMGC683	IV	PMGC2088	XXI	PMGC2088
II	PMGC2418			II	PMGC2418			LG01	PMGC223							XXI	PMGC223
								LG01	PMGC2418								
III	PMGC486	III	PMGC486	III	PMGC486					F1	PMGC486	F2	PMGC486				
III	PMGC2611	III	wpms10	III	PMGC2611			LG12	PMGC2611								
III	PMGC2501			III	PMGC2501			LG12	PMGC2501								
III	PMGC2858			III	PMGC2858			LG12	PMGC2858								
III	PMGC2879	III	PMGC2879	III	PMGC2879			LG12	PMGC2879								
				IV	PMGC2209											IV	PMGC2020
IV	PMGC2235	IV	PMGC2235	IV	PMGC2826	I	PMGC2235	LG17	PMGC2235					XIX	PMGC2235	IV	PMGC2215
IV	PMGC2826	IV	PMGC2826	IV	PMGC2826			LG17	PMGC2826								
IV	PMGC2881	IV	PMGC2881	IV	PMGC2881		PMGC2270	LG17	PMGC2881								
		IV	PMGC2270														
V	PMGC639	V		V	PMGC639									XXIII	PMGC639		
V		V	PMGC2838	V	PMGC2838			LG07	PMGC2838								
V	PMGC2156	V	PMGC2873	V	PMGC2156	B	PMGC2156	LG07	PMGC2873								
		V	PMGC2156	V	PMGC576	B	PMGC576	LG07	PMGC2155							II	PMGC576
				V	PMGC2606			LG07	PMGC576								
				V	PMGC2536			LG07	PMGC2606								
				V	PMGC2558			LG07	PMGC2536								
		V	PMGC2839	V	PMGC2839			LG07	PMGC2558								
				V	wpms15			LG07	PMGC2839								
VI	PMGC2557	VI		VI	PMGC2557												
VI	PMGC2556	VI	wpms9	VI	PMGC2556												
		VI	wpms9														

TABLE 16.2 (continued)

	wpms5	XII	wpms5	XII	wpms5 / PMGC108			LG			
XIII	PMGC2420							LG05	PMGC2420		
								LG05	PMGC649		
XIII	PMGC2599			XIII	PMGC2599			LG05	PMGC2599		
XIII	PMGC2658			XIII				LG05	PMGC2658		
XIII	PMGC14									XIV	PMGC14
XIV	PMGC420		PMGC420	XIV		X		LG10	PMGC420	XI	PMGC420
				XIV	PMGC2055	X		LG10	PMGC2055	XI	PMGC2055
XIV	PMGC571		PMGC14b	XIV	PMGC571			LG10	PMGC571		
				XIV	PMGC2515			LG10	PMGC2515		
XV	PMGC520			XV	PMGC520			LG08	PMGC520		
				XV	PMGC2105			LG08	PMGC2105		
				XV	PMGC2408			LG08	PMGC2408		
XV	PMGC690			XV	PMGC690						
XVI	PMGC2804							LG1	PMGC2804		
XVI	PMGC2403c										
XVI	PMGC2143c		PMGC433	IX						IX	PMGC433
XVIII	PMGC2525							LG03	PMGC2525		
		B		B	PMGC2289	O	PMGC2289				
		C	PMGC451					LG16	PMGC461		
		O	PMGC2885					LG13+18	PMGC2885		
						X		LG13+18	PMGC2274	X	PMGC2274
						G	P1273				P1273

Note: Markers with prefix PMGC are SSR markers developed by the PMGC. Markers with prefix WPMS are SSR markers developed at PRI. Markers with prefix STS and P are STS and RFLP markers, respectively (Bradshaw et al., 1994). The relative marker order of the RFLP markers to the SSR markers remains unpublished.

aDiscrepancy found in the map of P. Faivre-Rampant (unpublished results).

bDiscrepancies are explained in Cervera et al. (2001).

cThe markers PMGC433 and PMGC2143 are located near PMGC61 in linkage group LG02 (M.M. Sewell, unpublished results); for this marker a chromosomal rearrangement in *P. deltoides* has been suggested (Cervera et al., 2001).

The genetic map of the male I-63 was based on 154 AFLPs segregating 1:1 and 105 heteroduplex markers (nonparental markers as described by Wu et al. [2000]). The genetic map of the female C-135 was based on 77 AFLPs segregating 1:1 (only eight of them were placed on the framework map) and the same 105 heteroduplex markers. These nonparental heteroduplex markers were used as "allelic bridges" to identify homoeologous linkage groups between both maps, thereby merging the genetic linkage maps of both progenitors (Wu et al., 2000). The final map data is presented in Table 16.1. These maps will be saturated with markers using bulked segregant analysis (Michelmore et al., 1991) to increase the marker density at QTL regions associated with traits of interest.

Cervera and colleagues (2001) published genetic linkage maps of three clones from three *Populus* species: *P. deltoides, P. nigra,* and *P. trichocarpa.* Four genetic maps were constructed from two F1 hybrid pedigrees using the two-way pseudo-testcross strategy (Table 16.1). Family 87001 consisted of 127 individuals derived from a full-sib cross between *P. deltoides* (S9-2) and *P. nigra* 'Ghoy'. Family 87002 consisted of 105 individuals resulting from a full-sib cross between the same *P. deltoides* female (S9-2) and *P. trichocarpa* (V24). Three different molecular marker systems were used: high-throughput markers (AFLPs) to create the scaffold and SSRs and STSs that were located on this scaffold as codominant markers. Seven illegitimate progeny and three aneuploids were identified in these pedigrees by AFLP and SSR analysis, respectively, and were rejected from further linkage analyses. Fifty AFLP primer combinations produced 821 polymorphic markers in family 87001, and 41 combinations produced 635 markers in family 87002. In both pedigrees, 153 SSRs, provided by PMGC and Plant Research International (PRI; Wageningen, The Netherlands) were analyzed. Only three of the 25 STSs published by Bradshaw and colleagues (1994) were polymorphic in these pedigrees. The genetic map of *P. deltoides* (87001) was constructed from 403 AFLPs (1:1), 61 SSRs, the resistance phenotype against *M. larici-populina* (MER) and one STS. The *P. nigra* map was constructed from 355 AFLPs (1:1) and 49 SSRs; the map of *P. deltoides* (87002) from 309 AFLPs (1:1), 63 SSRs, the MER phenotype, and one STS; and, finally, the map of *P. trichocarpa* from 287 AFLPs (1:1), 76 SSRs, and one STS. Several known genes have also been mapped on at least one of the parental maps: genes involved in lignin biosynthesis (*CCoAOMT* and *CCR*), a gene involved as a positive regulator of photomorphogenesis *(PtHY5)*, a gene putatively involved in the first committed and rate-limiting step of the abscisic acid biosynthesis *(PtVP14)*, as well as *PtABI1* and *PtABI3*. The final map data for all four maps are presented in Table 16.1. Segregation distortion was detected for 8 percent of the AFLP markers in family 87001, and 6 percent in family 87002. In family 87002, markers segregating with the MER phenotype were distorted due to death of susceptible trees.

Because both pedigrees shared the female progenitor (S9-2), two genetic maps were constructed for *P. deltoides*. From a total of 237 markers (193 AFLPs, 43 SSRs, and the MER phenotype) found in common, the locus ordering was the same for 96 percent of the framework markers. The maps for families 87001 and 87002 were compared by aligning the location of common codominant markers. This alignment allowed a reduction of linkage groups to 28 for *P. nigra* and 19 for both *P. deltoides* and *P. trichocarpa* ($n = 19$). These homoeologous linkage groups are described at <http://www.psb.rug.ac.be/~vesto>. Furthermore, 12 linkage groups from family 331 were aligned to the maps described by Cervera and colleagues (2001; Table 16.2).

These maps are used to study quantitative traits related to wood production (stem height, basal area, and volume growth) and disease resistance (to *M. larici-populina* and *Xanthomonas populi*). The locus for the MER phenotype has been fine-mapped to a 2.2 cM interval as a starting point for positional cloning of the gene conferring resistance to *M. larici-populina* races E1, E2, and E3 (Zhang et al., 2001).

Yin and colleagues (2001) constructed genetic maps for *P. adenopoda* and *P. alba* from an interspecific F1 pedigree, using the two-way pseudo-testcross strategy and RAPD markers. One hundred forty-two RAPD primers (22 percent of those attempted) yielded polymorphic bands between the two parents. Twelve intercross markers (including four nonparental markers) were identified and used to align linkage groups between both maps. Eighteen percent of the testcross markers showed segregation distortion in *P. alba,* and 13 percent in *P. adenopoda*. The final map data are presented in Table 16.1. These maps will be used to identify genomic regions associated with species differentiation during long evolutionary divergences.

The same group recently published the genetic linkage maps for *P. deltoides* and *P. euramericana*. The maps were generated by combining the two-way pseudo-testcross strategy with RAPD, AFLP, and ISSR markers (Yin et al., 2002). The interspecific full-sib family consisted of 93 progeny resulting from the cross between *P. deltoides* (I-69) and *P. euramericana* (I-45), in which I-45 was selected from natural hybrids between *P. deltoides* and *P. nigra*. One hundred twenty-seven RAPD primers (12 percent of those attempted) detected segregation for 230 markers including 172 testcross markers (segregating 1:1), 39 dominant intercross markers (segregating 3:1), and 19 nonparental markers (segregating 3:1). Forty-eight AFLP primer combinations (19 percent of those attempted) produced 564 segregating markers including 356 testcross markers, 154 intercross dominant markers, and 54 fully informative codominant markers (segregating 1:1:1:1). Twenty-nine ISSR primer combinations (27 percent of those attempted) detected segregation for 45 markers, including 32 testcross markers and 13 dominant intercross markers. Segregation distortion of testcross markers was found in 9 percent of the RAPDs, 6 percent of the AFLPs, and

5 percent of the ISSRs in *P. deltoides,* and 8 percent of the RAPDs, 4 percent of the AFLPs, and 9 percent of the ISSRs in *P. ×euramericana.* Dominant intercross markers were excluded from the map construction to avoid biased estimates of recombination rates. Both maps were compared by aligning the location of nonparental RAPDs and codominant AFLPs. This alignment allowed a reduction of linkage groups to 30 for each parent. The final map data are presented in Table 16.1. These maps will be used to identify genomic regions controlling economically important traits and to study genome structure and function relationships.

Both the Oak Ridge National Laboratory (ORNL; Oak Ridge, Tennessee) and the INRA/Université Nancy (Vandoeuvre, France) are currently constructing genetic maps for elite *Populus* clones to dissect complex inherited traits. At ORNL, family 331 is being reanalyzed (M.M. Sewell, unpublished data) with the four-allele three-generation outbred model. Two approaches are employed to construct SSR maps for different objectives. First, from a subpopulation of 44 progeny, new and previously available SSR primers are used to construct a *comparative* map. This map can be used both in future mapping projects to select SSR markers from throughout the genome, and in comparative studies to relate this map with other published maps. Second, evenly spaced and fully informative markers are being selected from the comparative map to construct a *framework* map using the full 325-plus progeny population from family 331. The framework map will then be used in QTL analyses to identify genomic regions associated with wood quality and growth traits using an outbred QTL model (Knott et al., 1997). The comparative map is currently being constructed from approximately 180 SSR markers, and the framework map from approximately 80 (Table 16.1).

At INRA, genetic maps are constructed for the elite clones *P. deltoides* (73028-62) and *P. trichocarpa* (101-74) with 91 F1 progeny and the two-way pseudo-testcross strategy (P. Faivre-Rampant, unpublished data). The maps are based on the linkage analysis of RAPDs, AFLPs, STSs, SCARs, RFLPs, and SSRs (Table 16.1). Markers previously developed by Bradshaw and colleagues (1994) and used to map family 331 (also a *P. deltoides* × *P. trichocarpa* hybrid pedigree) have been tested in order to compare the linkage maps. From the 45 SSRs available from PMGC, 21 were included in the *P. deltoides* data set and 27 in *P. trichocarpa*. From the 17 STS primer pairs mapped to the family 331, only a single polymorphism was detected (in *P. trichocarpa* using a restriction digestion). Also, from the 200 RFLP probes mapped to family 331, only 29 were mapped to *P. deltoides* and 27 to *P. trichocarpa*. Four genes involved in the phenylpropanoid pathway (*PAL, COMT, CCR,* and *CHS*) and a peroxidase were also placed on the genetic maps as RFLP markers. The genetic map of *P. deltoides* (73028-62) was constructed from 158 AFLPs, 135 RAPDs, 29 RFLPs, 21 SSRs, and one SCAR, and that for *P. trichocarpa* (101-74) was constructed from 164 AFLPs, 139 RAPDs, 27 RFLPs, 27 SSRs, and one STS. Segregation distortion occurred for 14 (3.9 percent) markers in *P. deltoides* and 30 (8.2 percent) markers for *P. trichocarpa.* The final map data are presented in Table 16.1. These maps are being used to dissect the genetic architecture of quantitative resistance to *M. larici-populina, X. populi, Marssonina brunnea,* and *Chrysomela tremulae,* as well as leaf shape and wood quality.

COMPARISON OF LINKAGE MAPS

Comparative mapping based on the coalignment of common molecular markers among genetic linkage maps enables one to correlate linkage information from different genetic maps and to validate the accuracy of locus ordering from the different mapping strategies. With the coalignment of genetic maps from different experiments, it is possible to compare QTLs from different genetic and environmental backgrounds. Comparative mapping also allows the comparison of genomic structure within the genus and, thus, helps to study chromosomal evolution by detecting chromosome rearrangements.

Codominant, multiallelic marker systems are ideal for comparative mapping studies. O'Brien (1991) classified markers in two groups based on their usefulness for comparative mapping: type I markers are evolutionary conserved and based on coding sequences (e.g., isozymes and ESTs), whereas type II markers are species- or genus-specific DNA markers, exhibiting a high degree of polymorphism (for instance, SSRs). Isozymes are codominant and multiallelic but display relatively low heterozygosity levels and are limited in number. RFLPs, based on probes directed at single or low-copy number sequences, are also well suited for comparative studies; heterozygosity levels are intermediate to high. However, RFLP analysis is a laborious task: high amounts of high-quality DNA are required and the banding patterns are often complex. SSRs have a highly variable number of repeats and, consequently, result in high heterozygosity levels. The limiting factor for SSRs is the high cost of development. Once the SSRs are developed, only a simple PCR reaction is required prior to resolution on a polyacrylamide gel. Scoring is easy and can be automated. To date, 175 SSR primer combinations have been developed for *Populus* and are available to the public: 132 have been developed by PMGC (Tuskan et al., manuscript in preparation) and are derived from *P. trichocharpa,* 21 by PRI derived from *P. nigra* (van der Schoot et al., 2000; Smulders et al., 2001), and 22 derived from *P. tremuloides* (Dayanandan et al., 1998; Rahman et al., 2000). Currently, ORNL is identifying additional SSR primers from the PMGC pedigree (Tuskan et al., manuscript in preparation).

The most extensive comparative study between linkage maps of different poplar species was published by Cervera and colleagues (2001). Based on the SSR markers developed by PMGC and PRI, and the 17 informative STS markers mapped by Bradshaw and colleagues (1994), homologies were established between 12 linkage groups from *P. deltoides* and 18 from *P. nigra,* and between 15 linkage groups from *P. deltoides* and 19 from *P. trichocarpa.* The success rate for PCR amplification of the 153 SSRs was 66 percent for the *P. deltoides,* 53 percent for *P. nigra,* and 66 percent for *P. trichocarpa.* However, only 42 percent, 32 percent, and 50 percent were polymorphic in these clones, respectively, and only 38 percent, 26 percent, and 39 percent were successfully mapped in *P. deltoides, P. nigra,* and *P. trichocarpa,* respectively. Twenty-seven SSR markers were common among the *P. deltoides* and *P. nigra* maps, 34 among *P. deltoides* and *P. trichocarpa,* 16 among *P. nigra* and *P. trichocarpa,* and only 14 (~10 percent) among all three maps. The success rate for mapping the STS primers was very low, in which only one of the STS analyzed could be mapped in *P. deltoides.* A large number of highly polymorphic markers is thus necessary for a successful comparative analysis.

RAPD and AFLP markers are not well suited for comparative studies. Cervera and colleagues (2001) have tried to use AFLP markers that were heterozygous in both parents (segregating 3:1) to align the parental maps. However, for AFLP markers, sequence data are needed to prove that markers of the same size represent the same locus. Therefore, the AFLP markers segregating 3:1 were cloned and sequenced. Coamplification of contaminating bands appeared to be a major problem. Most of the markers also remained unlinked due to the low information content between marker pairs segregating 1:1 and 3:1 (Ritter et al., 1990). Therefore, the efficiency and reliability of these markers to align genetic maps is low. Furthermore, to align linkage groups between different genetic maps, RAPD or AFLP markers need to be transformed into SCAR or AFLP-STS markers, which is a laborious task.

For the *Populus* maps constructed at INRA (P. Faivre-Rampant, unpublished data), 17 STS primer pairs already mapped by Bradshaw and colleagues (1994) were used in PCR reactions of the F1 progeny. From these, only one marker segregated using the restriction enzymes described by Bradshaw and colleagues (1994) and could be mapped on the *P. trichocarpa* linkage map. From the 200 RFLP probes developed and mapped by Bradshaw and colleagues (1994), only three resulted in heterozygous markers in both parents and could thus be used for comparative mapping. Finally, only one RFLP marker and five SSRs were mapped on both genetic maps.

Frewen and colleagues (2000) mapped nine PMGC SSR markers and found seven homoeologous linkage groups between the maps of *P. trichocarpa* (93-968) and *P. deltoides* (S7C4). Both maps showed homology with eight linkage groups from family 331 (Bradshaw et al., 1994). These maps were used to analyze the genetic control of bud set and bud flush, two quantitative traits. The results from both family 331 (Bradshaw and Stettler, 1995) and family 822 were examined using comparative mapping, indicating that three QTLs for bud flush were detected in both studies. The map position of *PtPHYB2* and *PtABI1B* coincided with QTLs affecting bud set and bud flush.

In conclusion, only highly polymorphic markers are suitable for efficient comparative mapping. An overview of the current alignment for many of the maps described here is presented in Table 16.2. Except for a possible translocation of a part of linkage group VIII in *P. deltoides* (S9-2) (Cervera et al., 2001), there is a complete agreement in linkage grouping and marker order among the three *Populus* species compared. Table 16.2 demonstrates that SSR markers are useful for comparative mapping between different species of poplar and that they validate the different map constructions. Based on the analysis of chromosomal distribution of ribosomal sequences by fluorescence in situ hybridization, no variation was found between Tacamahaca and Aigeiros sections, although differences between *P. alba* and the other species of Leuce section were observed (Prado et al., 1996).

PERSPECTIVES

Recently, the U.S. Department of Energy has committed itself to create a 6 × draft sequence of *Populus trichocarpa* clone "383-2499" (Tuskan et al., 2002). The entire genome will be sequenced by the Joint Genome Institute (<http://www.jgi.doe.gov/>). Approximately 2,000 contig-scaffolds, available by mid-2003, will create the first unabridged catalog of genes in a tree. This will allow intensive comparison of genome structure and gene order, and ultimately comparison of orthogonal gene expression and function. Large EST sequencing projects on different tissues of poplar are also ongoing (Sterky et al., 1998; <http://www.pierroton.inra.fr/Lignome/>, <http://mycor.nancy.inra.fr/poplardb/>) and will be extremely valuable for the correct annotation of the poplar genome sequence. It will now be essential to align the genetic maps with the poplar genome sequence. This will allow the location of large numbers of candidate genes on the genetic maps and to compare their map positions with QTLs. The function of selected candidate genes can then be studied by genetic transformation or association studies. The availability of the genome sequence will also allow the development of SNPs. These markers will be extremely valuable to saturate genetic linkage maps for more accurate localization of QTLs controlling traits of interest. SNP markers will also aid in the development of a "consensus" map for each *Populus* species or for "comparative mapping" within the *Populus* genus to identify common QTLs in different genetic backgrounds (Sewell and Neale, 2000).

Furthermore, the construction of bacterial artificial chromosome (BAC) libraries of different poplar species (Stirling et al., 2001; W. Boerjan, unpublished data) will contribute to build physical maps. Anchoring molecular markers to overlapping BACs will help in aligning the physical map with the genetic linkage maps and thus facilitate candidate gene approaches and positional cloning of genes.

REFERENCES

Barreneche, T., Bodenes, C., Lexer, C., Trontin, J.-F., Fluch, S., Streiff, R., Plomion, C., Roussel, G., Steinkellner, H., Burg, K., et al. (1998). A genetic linkage map of *Quercus robur* L. (pedunculate oak) based on RAPD, SCAR, microsatellite, minisatellite, isozyme and 5S rDNA markers. *Theor. Appl. Genet.* 97: 1090-1103.

Bradshaw, H.D. Jr. (1996). Molecular genetics of *Populus*. In R.F. Stettler, H.D. Bradshaw Jr., P.E. Heilman, and T.M. Hinckley (Eds.), *Biology of Populus and Its Implications for Management and Conservation* (pp. 183-199). Ottawa, Canada: NRC Research Press.

Bradshaw, H.D. Jr., Ceulemans, R., Davis, J., and Stettler, R. (2000). Emerging model systems in plant biology: Poplar *(Populus)* as a model forest tree. *J. Plant Growth Regul.* 19: 306-313.

Bradshaw, H.D. Jr. and Stettler, R.F. (1994). Molecular genetics of growth and development in *Populus*. II. Segregation distortion due to genetic load. *Theor. Appl. Genet.* 89: 551-558.

Bradshaw, H.D. Jr. and Stettler, R.F. (1995). Molecular genetics of growth and development in *Populus*. IV. Mapping QTLs with large effects on growth, form, and phenology traits in a forest tree. *Genetics* 139: 963-973.

Bradshaw, H.D. Jr., Villar, M., Watson, B.D., Otto, K.G., Stewart, S., and Stettler, R.F. (1994). Molecular genetics of growth and development in *Populus*. III. A genetic linkage map of a hybrid poplar composed of RFLP, STS, and RAPD markers. *Theor. Appl. Genet.* 89: 167-178.

Butcher, P.A. and Moran, G.F. (2000). Genetic linkage mapping in *Acacia mangium*. 2. Development of an integrated map from two outbred pedigrees using RFLP and microsatellite loci. *Theor. Appl. Genet.* 101: 594-605.

Cervera, M.-T., Storme, V., Ivens, B., Gusmão, J., Liu, B.H., Hostyn, V., Van Slycken, J., Van Montagu, M., and Boerjan, W. (2001). Dense genetic linkage maps of three Populus species (*Populus deltoides, P. nigra* and *P. trichocarpa*) based on AFLP and microsatellite markers. *Genetics* 158: 787-809.

Dayanandan, S., Rajora, O.P., and Bawa, K.S. (1998). Isolation and characterization of microsatellites in trembling aspen *(Populus tremuloides)*. *Theor. Appl. Genet.* 96: 950-956.

Eckenwalder, J.E. (1996). Systematics and evolution of *Populus*. In R.F. Stettler, H.D. Bradshaw Jr., P.E. Heilman, and T.M. Hinckley (Eds.), *Biology of Populus and Its Implications for Management and Conservation* (pp. 7-32). Ottawa, Canada: NRC Research Press.

Frewen, B.E., Chen, T.H.H., Howe, G.T., Davis, J., Rohde, A., Boerjan, W., and Bradshaw, H.D. Jr. (2000). Quantitative trait loci and candidate gene mapping of bud set and bud flush in *Populus*. *Genetics* 154: 837-845.

Grattapaglia, D. and Sederoff, R. (1994). Genetic linkage maps of *Eucalyptus grandis* and *Eucalyptus urophylla* using a pseudo-testcross mapping strategy and RAPD markers. *Genetics* 137: 1121-1137.

Karp, A. and Edwards, K.J. (1998). DNA markers: A global overview. In G. Caetano-Anollés and P.M. Gresshoff (Eds.), *DNA Markers: Protocols, Applications and Overviews* (pp. 1-13). New York: John Wiley and Sons.

Klopfenstein, N.B., Chun, Y.W., Kim, M.-S., and Ahuja, M.R. (1997). *Micropropagation, Genetic Engineering, and Molecular Biology of* Populus (General Technical Report RM-GTR-297). Fort Collins, CO: Rocky Mountain Forest and Range Experiment Station.

Knott, S.A., Neale, D.B., Sewell, M.M., and Haley, C.S. (1997). Multiple marker mapping of quantitative trait loci in an outbred pedigree of loblolly pine. *Theor. Appl. Genet.* 94: 810-820.

Lander, E.S., Green, P., Abrahamson, J., Barlow, A., Daly, M.J., Lincoln, S.E., and Newburg, L. (1987). MAPMAKER: An interactive computer package for constructing primary genetic linkage maps of experimental and natural populations. *Genomics* 1: 174-181.

Liu, Z. and Furnier, G.R. (1993). Comparison of allozyme, RFLP, and RAPD markers for revealing genetic variation within and between trembling aspen and bigtooth aspen. *Theor. Appl. Genet.* 87: 97-105.

Michelmore, R.W., Paran, I., and Kesseli, R.V. (1991). Identification of markers linked to disease-resistance genes by bulked segregant analysis: A rapid method to detect markers in specific genomic regions by using segregating populations. *Proc. Natl. Acad. Sci. USA* 88: 9828-9832.

Newcombe, G. (1998). Association of *Mmd1*, a major gene for resistance to *Melampsora medusae* f. sp. *deltoidae*, with quantitative traits in poplar rust. *Phytopathology* 88: 114-121.

Newcombe, G. and Bradshaw, H.D. Jr. (1996). Quantitative trait loci conferring resistance in hybrid poplar to *Septoria populicola*, the cause of leaf spot. *Can. J. For. Res.* 26: 1943-1950.

Newcombe, G., Bradshaw, H.D. Jr., Chastagner, G.A., and Stettler, R.F. (1996). A major gene for resistance to *Melampsora medusae* f. sp. *deltoidae* in a hybrid poplar pedigree. *Phytopathology* 86: 87-94.

Newcombe, G., Stirling, B., and Bradshaw, H.D. Jr. (2001). Abundant pathogenic variation in the new hybrid rust *Melampsora × columbiana* on hybrid poplar. *Phytopathology* 91: 981-985.

O'Brien, S.J. (1991). Mammalian genome mapping: Lessons and prospects. *Curr. Opin. Genet. Dev.* 1: 105-111.

O'Malley, D.M. and Whetten, R. (1997). Molecular markers and forest trees. In G. Caetano-Anollés and P.M. Gresshoff (Eds.), *DNA Markers: Protocols, Applications and Overviews* (pp. 237-257). New York: John Wiley and Sons.

Prado, E.A., Faivre-Rampant, P., Schneider, C., and Darmency, M.A. (1996). Detection of a variable number of ribosomal DNA loci by fluorescent in situ hybridization in *Populus* species. *Genome* 39: 1020-1026.

Rahman, M.H., Dayanandan, S., and Rajora, O.P. (2000). Microsatellite DNA markers in *Populus tremuloides*. *Genome* 43: 293-297.

Ritter, E., Gebhardt, C., and Salamini, F. (1990). Estimation of recombination frequencies and construction of RFLP linkage maps in plants from crosses between heterozygous parents. *Genetics* 125: 645-654.

Sewell, M. and Neale, D. (2000). Mapping quantitative traits in forest trees. In S.M. Jain and S.C. Minocha (Eds.), *Molecular Biology of Woody Plants,* Volume 1, (Forestry Sciences, Volume 64) (pp. 407-423). Dordrecht, The Netherlands: Kluwer Academic Publishers.

Sewell, M.M., Sherman, B.K., and Neale, D.B. (1999). A consensus map for loblolly pine (*Pinus taeda* L.). I. Construction and integration of individual linkage maps from two outbred three-generation pedigrees. *Genetics* 151: 321-330.

Smulders, M.J.M., Van der Schoot, J., Arens, P., and Vosman, B. (2001). Trinucleotide repeat microsatellite markers for black poplar (*Populus nigra* L.). *Mol. Ecol. Notes* 1: 188-190.

Stam, P. (1993). Construction of integrated genetic linkage maps by means of a new computer package: JoinMap. *Plant J.* 3: 739-744.

Sterky, F., Regan, S., Karlsson, J., Hertzberg, M., Rohde, A., Holmberg, A., Amini, B., Bhalerao, R., Larsson, M., and Villarroel, R. (1998). Gene discovery in the wood-forming tissues of poplar: Analysis of 5,692 expressed sequence tags. *Proc. Natl. Acad. Sci. USA* 95: 13330-13335.

Stettler, R.F., Bradshaw, H.D. Jr., Heilman, P.E., and Hinckley, T.M. (1996). *Biology of* Populus *and Its Implications for Management and Conservation*. Ottawa, Canada: National Research Council of Canada.

Stirling, B., Newcombe, G., Vrebalov, J., Bosdet, I., and Bradshaw, H.D. Jr. (2001). Suppressed recombination around the *MXC3* locus, a major gene for resistance to poplar leaf rust. *Theor. Appl. Genet.* 103: 1129-1137.

Su, X.H., Zhang, Q.W., Zheng, X.W., Zhang, X.H., and Harris, S. (1998). Construction of *Populus deltoides* Marsh x *P. cathayana* Rehd. molecular linkage map. *Sci. Silvae Sin.* 34: 29-37.

Tuskan, G.A., Wullschleger, S.D., Bradshaw, H.D., and Dalhman, R.C. (2002). Sequencing the *Populus* genome: Applications to the energy-related missions of DOE. Abstract presented at the Plant, Animal and Microbe Genomes X Conference, San Diego, CA, January 12-16, 2002.

van der Schoot, J., Pospíšková, M., Vosman, B., and Smulders, M.J.M. (2000). Development and characterization of microsatellite markers in black poplar (*Populus nigra* L.). *Theor. Appl. Genet.* 101: 317-322.

Wu, R.L. (1998). Genetic mapping of QTLs affecting tree growth and architecture in *Populus:* Implication for ideotype breeding. *Theor. Appl. Genet.* 96: 447-457.

Wu, R., Bradshaw, H.D. Jr., and Stettler, R.F. (1997). Molecular genetics of growth and development in *Populus* (Salicaceae). V. Mapping quantitative trait loci affecting leaf variation. *Am. J. Bot.* 84: 143-153.

Wu, R., Bradshaw, H.D. Jr., and Stettler, R.F. (1998). Developmental quantitative genetics of growth in *Populus*. *Theor. Appl. Genet.* 97: 1110-1119.

Wu, R.L., Han, Y.F., Hu, J.J., Fang, J.J., Li, L., Li, M.L., and Zeng, Z.-B. (2000). An integrated genetic map of *Populus deltoides* based on amplified fragment length polymorphisms. *Theor. Appl. Genet.* 100: 1249-1256.

Yano, M. (2001). Genetic and molecular dissection of naturally occurring variation. *Curr. Opin. Plant Biol.* 4: 130-135.

Yin, T., Huang, M., Wang, M., Zhu, L.-H., Zeng, Z.-B., and Wu, R. (2001). Preliminary interspecific genetic maps of the *Populus* genome constructed from RAPD markers. *Genome* 44: 602-609.

Yin, T., Zang, X., Huang, M., Wang, M., Zhuge, Q., Tu, S., Zhu, L.-H., and Wu, R. (2002). Molecular linkage maps of the *Populus* genome. *Genome* 45: 541-555.

Zhang, J., Steenackers, M., Storme, V., Neyrinck, S., Schamp, K., Van Montagu, M., Gerats, T., and Boerjan, W. (2001). Fine mapping and identification of nucleotide-binding site/leucine-rich repeat sequences at the *MER* locus in *Populus deltoides* "S9-2." *Phytopathology* 91: 1069-1073.

Chapter 17

Genetic Mapping in Acacias

Penelope A. Butcher

INTRODUCTION

Acacia is a widespread genus that occurs naturally on all continents except Europe and Antarctica (Maslin, 2001). It includes three subgenera: *Acacia* and *Aculeiferum* which are pantropical, occurring from Central and South America, through Africa, to Southeast Asia, and *Phyllodineae* which occurs primarily in Australia (Ross, 1981). Over 1,200 species have been described, 955 of which are confined to Australia (Maslin, 2001). Commercial plantations have been developed for the tropical species *Acacia mangium* Willd., *A. auriculiformis* A. Cunn ex Benth., and *A. crassicarpa* A. Cunn ex Benth., and the temperate species *A. melanoxylon* R. Br., *A. saligna* (Labill.) H.L. Wendl., and *A. mearnsii* De Wild. (reviewed in Doran and Turnbull, 1997), from the subgenus *Phyllodineae*. *Acacia nilotica* and *A. senegal*, from subgenus *Acacia*, have also been established in plantations (CAB International, 2000). Acacias are legumes (subfamily Mimosoideae) that form symbiotic relationships with nitrogen-fixing bacteria of the genera *Rhizobium* and *Bradyrhizobium* (Doyle, 1994). They therefore play an important role in soil improvement and are widely planted for land rehabilitation, amenity plantings, and fuel wood. The seeds of several species are used as a source of vegetable protein (Maslin et al., 1998).

Genetic mapping in the genus has been limited to *A. mangium*, a diploid (2n = 26), predominantly outcrossing tree species that is in the early stages of domestication. The species occurs naturally in Australia, New Guinea, and the Maluku islands. Its fast growth rate, good pulping qualities, and tolerance of a range of soil types have led to rapid expansion of the area of plantations in the humid and subhumid tropics (over 800,000 ha in 2000) (Barr, 2000). The first plantations of the species were established using seed from seed production stands that were established from a narrow genetic base (Butcher et al., 1996) and that had inferior growth rates when compared to natural populations in New Guinea and Cape York, Australia (Harwood and Williams, 1992; Tuomela et al., 1996). More recently, efforts have been directed toward broadening the genetic base of breeding populations for recurrent selection and the development of advanced generation breeding programs with clonal propagation. This has opened opportunities to further improve the efficiency of breeding programs using marker-assisted selection.

ASPECTS OF THE REPRODUCTIVE BIOLOGY OF ACACIAS THAT AFFECT GENETIC MAPPING STRATEGIES

The mapping strategy for acacias differs from most forest trees due to two aspects of their reproductive biology. First, the difficulty of producing large progeny arrays from controlled crosses due, in part, to the small size of flowers, compound inflorescences with variable proportions of male flowers, and low seed set (Sedgley et al., 1992). Second, acacias produce compound pollen grains (polyads), and seeds within a pod may not be produced from independent meiotic events. The polyad is considered to be a mechanism for ensuring full pod set following a single pollination event (Kenrick and Knox, 1982; see also Muona et al., 1991 for review). Cytological studies in acacias have shown that ovules in an ovary are derived from independent meiotic events (Buttrose et al., 1981). All ovules within a flower can be fertilized by a single polyad because the number of pollen grains in the polyad (16 in *A. mangium*) is the same as the maximum number of ovules within a flower (Sedgley et al., 1992). The polyads are derived from two rounds of mitosis preceding meiosis in the sporogenous cell (Newman, 1933; Kenrick and Knox, 1979). This implies that up to four seeds in a pod can be produced from a single meiosis. Because tests for linkage are based on the assumption that all individuals in a mapping pedigree are derived from independent meioses, ideally only one seed per pod would be used for mapping. When the number of pods produced is limited, the need for independence must be balanced with that of maximizing the precision with which genetic linkage is measured. If more than one individual per pod is sampled, the position of crossovers on a linkage group can be used to test for independence. If the position of a crossover differs among individuals from the same pod, it can be assumed that different recombination events were sampled (see Butcher and Moran, 2000).

Breeding System of Acacias and Development of Mapping Pedigrees

Genetic linkage maps for forest trees are usually constructed for outbred pedigrees in which segregation is the result of meiotic recombination from both parents. Inbreeding depression has generally precluded the development of inbred lines (Griffin and Cotterill, 1988; Williams and Savolainen, 1996)—for exceptions see Bradshaw and colleagues (1994) and Plomion and colleagues (1995). However, outcrossing rates and self-compatibility vary widely among *Acacia* species. Outcrossing rates, estimated from allozyme analysis of open-pollinated progeny arrays, range from highly outcrossing in the tropical species *A. auriculiformis* ($t_m = 0.93$) and *A. crassicarpa* ($t_m = 0.96$) (Moran et al., 1989b) to significant selfing in the polyploid African acacias *A. nilotica* (L.) Willd. ex Del. ssp. *leiocarpa* ($t_m = 0.38$) (Mandal et al., 1994) and *A. tortilis* (Forssk.) Hayne ($t_m = 0.35$) (Olng'otie, 1991). Outcrossing rates can also vary widely among populations within species, for example natural populations of *A. mangium* range from complete selfing to complete outcrossing (Butcher et al., 2003). *Acacia mangium* flowers have a short female phase that reduces the effectiveness of the protogynous outcrossing mechanism reported in other acacias (Sedgley et al., 1992).

Quantitative estimates of self-incompatibility, based on results from hand pollination using self and outcross pollen, also vary widely among acacia species (Kenrick and Knox, 1989). Species range from highly self-incompatible (index of self-incompatibility [ISI] = 0 in *A. mearnsii*) to fully self-compatible (ISI = 0.96 in *A. ulicifolia* [Salisb.] Court). In *A. mangium*, cross-pollination experiments revealed no difference in ovule penetration with self and outcross pollen, indicating that selection for outcross pollen in this species is not strong (Sedgley et al., 1992). In crosses made to produce mapping pedigrees from *A. mangium*, 30 percent of the pods produced from flowers that were not emasculated prior to pollination contained selfs (Butcher and Moran, 2000). In contrast, less than 1 percent of pods produced from emasculated flowers contained selfs (Butcher and Moran, 2000). No selfing was detected in progeny arrays from the same trees following open-pollination (Butcher et al., 2003). During controlled pollinations, a limited number of flowers within an inflorescence are pollinated as flowers mature sequentially along the spike. Bagging allows self-pollen from flowers that have not been emasculated to fertilize later maturing flowers in the inflorescence. The fact that no selfs were detected from open-pollinated flowers on the same trees suggests some form of maternal resource allocation favoring outcrossed embryos. Evidence of self-compatibility, in addition to the wide variation in selfing rates in natural populations, suggests the species could best be described as a facultative outcrosser. If offspring produced from self-pollination are fertile, selfing could be used as a breeding tool for acacias to increase selection efficiency and increase uniformity within breeding lines (Williams and Savolainen, 1996). The short generation time of tropical acacias (*A. mangium* flowers at two years of age) means inbred lines could be developed more quickly than in most other forest trees.

LINKAGE MAPPING IN ACACIA MANGIUM

Mapping Pedigrees

No direct comparisons of the performance of inbred and outcrossed progeny of *A. mangium* have been reported. However, the observed decline in performance of individuals derived from a single mother tree in successive generations (Sim, 1984) provides some evidence of inbreeding depression associated with high genetic load. This is supported by estimates of the relative fitness of selfed compared with outcrossed progeny of 11 to 36 percent in populations of *A. mangium* with significant selfing (Butcher et al., 2003) inferred from inbreeding coefficients (Ritland, 1990). Mapping pedigrees in *A. mangium* have therefore been developed using outcrossed full-sib families.

Linkage mapping relies on the ability to detect allelic differences in full-sib pedigrees—a task made more difficult in species with low genetic diversity. Allozyme variation in *A. mangium* is among the lowest detected in forest trees (Moran et al., 1989a), and the species has an unusual pattern of diversity across a disjunct geographic range (Butcher et al., 1998). Even the more diverse populations from New Guinea have approximately half the level of restriction fragment-length polymorphism (RFLP) variation (H_E = 0.21) (Butcher et al., 1998) recorded in eucalypts (*Eucalyptus nitens* mean H_E = 0.37, Byrne et al., 1998; mallee eucalypts H_E = 0.49, Byrne, 1999; *E. camaldulensis* H_E = 0.49, Butcher, Otero, et al., 2002). To maximize the number of segregating loci, two independent two-generation mapping pedigrees were developed in *A. mangium* using parents from four populations in Western Province, Papua New Guinea (cross A: Bura × Bensbach and cross B: Boite × Bimadebun) (Butcher, Moran, and Bell, 2000). The pedigrees consisted of 108 progeny from cross A and 123 progeny from cross B. Populations from Western Province have superior growth rates in provenance trials on a range of sites (Harwood and Williams, 1992; Tuomela et al., 1996) and are the preferred source of breeding material in current tree improvement programs throughout Southeast Asia.

Marker Selection

The utility of markers for mapping depends, in part, on whether polymorphism can be detected in other mapping populations and in other species. Aspects of the taxonomy of the *Acacia* genus remain unresolved (Pedley, 1986; Maslin and Stirton, 1997; Robinson and Harris, 2000; Miller and Bayer, 2001). Markers that could be used across the genus would provide a useful tool for comparative studies of genome organization and phylogenetic studies. To develop a generic map that can be used in other commercially planted acacia species, it is also important that loci are orthologous across as wide a group of species as possible.

Marker selection is also influenced by the fact that quantitative trait dissection analysis in full-sib pedigrees under an outbred model requires multiallelic, codominant markers to follow the inheritance of alleles from both parents to the progeny. The alternatives are to convert dominant markers to codominant markers by a laborious process of cloning, sequencing, and polymerase chain reaction (PCR), or to develop trans-dominant linked marker pairs from arbitrarily primed PCR markers (see Plomion et al., 1996). However, the latter requires prior information on marker arrangement.

Based on findings that codominant RFLPs were highly conserved across the genus *Acacia* (Butcher, Moran, and Bell, 2000), these markers were used to develop a baseline map in *A. mangium*. Screening of 20 RFLPs over a range of *Acacia* species revealed all probes hybridized with the commercially important species *A. crassicarpa* and *A. auriculiformis* in section Juliflorae and *A. melanoxylon* in the Plurinerves, and 95 percent of probes hybridized with *A. mearnsii* in section Botrycephalae (Butcher, Moran, and Bell, 2000).

In contrast to RFLPs, the proportion of *A. mangium* microsatellite (simple sequence repeat [SSR]) primers that amplified alleles in other species in the genus *Acacia* was relatively low. One third of primers amplified alleles in species within the same subgenus *Phyllodineae,* but only one out of 12 amplified DNA from species in subgenus *Acacia* (Butcher, Decroocq, et al., 2000). Intergeneric comparisons have also shown higher homology of RFLPs when compared to SSRs. DNA sequenced from 51 out of 68 RFLP clones (72 percent) in soybean had significant homology with *Arabidopsis* genomic or cDNA sequence. In contrast, only one out of 23 sequences surrounding soybean SSRs had any detectable homology with *Arabidopsis* (Grant et al., 2000).

Cross-species amplification of *A. mangium* SSRs was low compared to cited rates of 50 to 100 percent within other genera (Peakall et al., 1998).

This may reflect the high evolutionary divergence among species within *Acacia*. The *Phyllodineae* is thought to have evolved independently from subgenus *Acacia* (Maslin and Stirton, 1997; Murphy et al., 2000; Robinson and Harris, 2000; Miller and Bayer, 2001) and was treated by Pedley (1986) as a distinct genus. The only other report of cross-species amplification of SSRs in *Acacia* is for six pairs of primers developed in *Pithecellobium elegans* (tribe Ingeae) and tested on other Leguminosae, which failed to amplify *Acacia collincii* (subgenus *Acacia*) (Dayanandan et al., 1997). The low cross-species amplification of *Acacia* microsatellites suggests that regions flanking the microsatellites are not highly conserved, in nucleotide sequence or relative position. Microsatellites are, however, approximately three times more polymorphic than RFLPs in *A. mangium* (Butcher, Decroocq, et al., 2000) and were therefore used to increase the number of fully informative loci. Fully informative loci are particularly powerful for quantitative trait dissection (see, for example, Byrne et al., 1997).

Linkage Mapping

Linkage maps have been developed for two outcrossed two-generation pedigrees of *A. mangium*. A disadvantage of using two-generation pedigrees for mapping is that the linkage phase between pairs of loci is not known a priori. Most mapping packages are not optimal for outcrossed pedigrees, because they either cannot order phase-ambiguous data or use only pairwise information when ordering loci within linkage groups. Orders determined using pairwise information cannot be expected to be as accurate as those determined using multilocus likelihood (reviewed by Ott, 1992). Linkage maps were developed for the two *A. mangium* pedigrees using pairwise recombination frequencies (Butcher and Moran, 2000) and the mapping package JoinMap (Stam and van Oiijen, 1995). The marker order within linkage groups was later compared with that produced using multilocus likelihood and the package OutMap (Butcher, Williams, et al., 2002). The OutMap package was specifically designed for analyzing segregation data from codominant loci in outcrossed pedigrees and deals effectively with phase-ambiguous data. Comparison of marker orders revealed that those produced using multilocus likelihood (OutMap) were consistently of higher likelihood than the order produced using only pairwise information (JoinMap) (see Butcher, Williams, et al., 2002). Map distances estimated using multilocus likelihood were approximately 10 percent greater than when using pairwise data.

The use of two pedigrees allowed mapping of a larger number of loci than with a single pedigree, thereby increasing genomic coverage and marker density in specific genomic regions. Increased marker density increases the probability of identifying closely linked polymorphic loci for marker-assisted selection or for map-based cloning (see, for example, *Phaseolus vulgaris* integrated map, Freyre et al., 1998). The use of two pedigrees also provides a means of evaluating the repeatability of map construction and may provide an efficient means of validating quantitative trait loci (QTL). Simulation studies indicate that use of more than one full-sib pedigree increases the power to detect QTL, especially when the QTL explains more than 10 percent of the phenotypic variance (Muranty, 1996). In other genera, mapping with multiple populations has provided evidence for chromosomal rearrangements (Beavis and Grant, 1991; Kianian and Quiros, 1992), gene duplication (Gentzbittel et al., 1995), and differences in recombination rates between sexes and genotypes (Fatmi et al., 1993; Sewell et al., 1999). In pooling information from multiple pedigrees, if heterogeneity of recombination is common, then assembling a consensus map for a species will be difficult (Beavis and Grant, 1991; Plomion and O'Malley, 1996).

In *A. mangium,* a high percentage of markers segregated in both sexes (33 percent) and in both pedigrees (41 percent), allowing the comparison of recombination rates in male and female meiosis and between genotypes. Genome-wide trends were not evident. Differences in meiotic recombination rates between the sexes were concentrated in a single linkage group in one pedigree and did not result in major differences in the ordering of loci between pedigrees. There was little evidence of segregation distortion, with the number of loci with distorted segregation ratios no greater than the 5 percent expected by chance alone (Butcher and Moran, 2000). These loci were randomly distributed throughout the genome (Figure 17.1); the absence of any clustering suggests there is no biological basis for the observed distortion. This contrasts with findings in several leguminous crop species in which high levels of segregation distortion have been reported, for example, in diploid alfalfa *(Medicago sativa)* (Echt et al., 1994) and *Vigna* species (Kaga et al., 2000). It has since been demonstrated in alfalfa that genetic linkage of markers displaying distorted segregation can lead to problems of artificial linkage of genome regions (Kaló et al., 2000).

Consistency in the ordering of loci in the two *A. mangium* pedigrees facilitated the construction of an integrated map. The pairwise recombination frequencies estimated for the two pedigrees were combined into a single data set that was reanalyzed for linkage using the JoinMap package (Butcher and Moran, 2000). The additional linkage information provided by using two pedigrees enabled several of the smaller linkage groups on the individual pedigree maps to be combined. The high percentage of common markers (41 percent), together with the consistency of gene order between the two individual maps, also provided a measure of reliability for the inte-

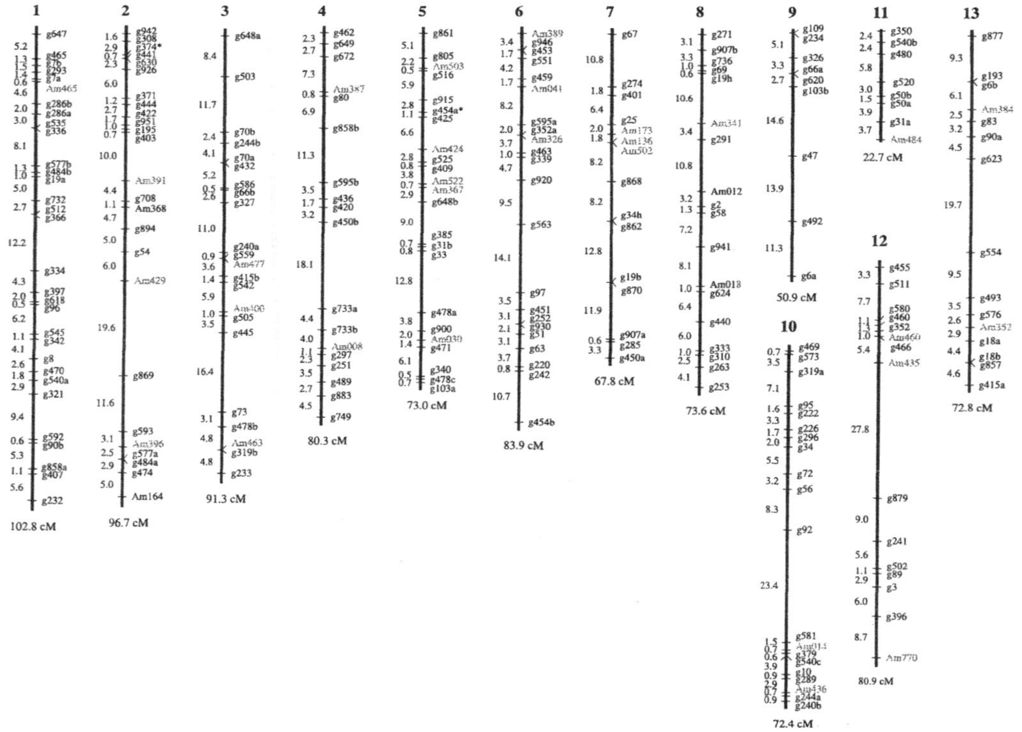

FIGURE 17.1 Genetic linkage map for *Acacia mangium* constructed from two outcrossed, unrelated pedigrees. Loci are listed on the right and recombination distances (cM, Kosambi) on the left of each linkage group. RFLP markers are denoted by the prefix g and microsatellites by the prefix *Am*. Loci with distorted segregation ratios (*P* ≤ 0.05) are marked with an asterisk.

grated map. The map consisted of 219 RFLP and 33 microsatellite loci in 13 linkage groups with an average distance between markers of 4.6 cM (Figure 17.1). The linkage groups ranged in size from 23 to 103 cM, and the total map length was 966 cM.

The distribution of markers in the 13 linkage groups of the integrated map was reasonably uniform, with only three regions of the map in which the distance between two adjacent markers was greater than 20 cM. Similar gaps have been reported on most plant RFLP maps and may represent either regions of high recombination or genomic regions that were not sampled with probe isolation techniques. Based on estimates that the power to detect QTL does not significantly increase for marker densities greater than one every 10 to 20 cM (Darvasi et al., 1993), the map should provide a firm basis for location of QTL, assuming 100 percent coverage of the genome.

The 966 cM covered by the *A. mangium* map is shorter than most maps reported for forest trees, reflecting either incomplete genome coverage or a lower rate of recombination. Low levels of recombination have been reported in other Leguminosae (reviewed in Young et al., 1996). *Acacia mangium* has relatively low RFLP (Butcher et al., 1998), and a relationship between the level of DNA diversity and the overall rate of recombination would not be unexpected. Positive correlations between RFLP diversity and rates of recombination have been reported in *Drosophila melanogaster* (Begun and Aquadro, 1992), *Aegilops* (Dvorak et al., 1998), and *Beta vulgaris* (Kraft et al., 1998).

Estimated Map Length and Genome Coverage

The expected length of the *A. mangium* genome, based on the methods of Chakravarti and colleagues (1991), was 1550 cM, and the estimated genome coverage was 85 to 90 percent (Butcher and Moran, 2000). This is within the range of estimates for other forest trees, 1160 cM in *Eucalyptus urophylla* (Grattapaglia and Sederoff, 1994) to 2840 cM in *Picea abies* (Paglia et al., 1998). The observed length of the *A. mangium* map calculated was, however, less than expected (62 percent based on distances estimated from pairwise recombination frequencies; 70 percent using distances estimated using multilocus likelihood), indicating the need for more markers to be mapped (Butcher and Moran, 2000; Butcher, Williams, et al., 2002).

Based on an estimated genome length of 1550 cM, DNA content of 2.277 pg/2C Mbp (Mukherjee and Sharma, 1995), and that 1 pg of DNA = 0.965×10^9 bp (Bennett and Smith, 1976), the average physical equivalent per unit of genetic distance is 710 kbp/cM. This value is within the range reported for other legumes, for example, 530 to 758 kbp/cM for *Phaseolus vulgaris* (Vallejos et al., 1992; Nodari et al., 1993); 517 kbp/cM for *Glycine max* (Diers et al., 1992); and 1000 to 1300 kbp/cM in *Medicago sativa* (Kaló et al., 2000), but the value is at the upper end of the range reported for angiosperm trees (350 kbp/cM for *Eucalyptus nitens;* Byrne et al., 1995, and 400 to 600 kbp/cM for *E. grandis* and *E. urophylla;* Grattapaglia and Sederoff, 1994; Verhaegen and Plomion, 1996). Higher average physical distance estimates reduce the possibility of map-based cloning of genes, assuming the genes are randomly distributed in the genome.

One source of error in these calculations is the reported variability in the nuclear DNA concentrations in acacias. Not only is there significant variation in the amount of nuclear DNA in different acacia species, associated with their evolution and divergence (Mukherjee and Sharma, 1995), but values also differ among authors. The 4C values reported for diploid Australian acacias range from 4.4 pg in *A. auriculiformis* to 8.4 pg in *A. simsii* (Mukherjee and Sharma, 1995). The values often differed from those reported by Bennett and Leitch (1997), which ranged from 4C = 2.4 pg for *A. modesta,* 4C = 3.3 pg for *A. auriculiformis,* to 8.2 pg in *A. victoriae.*

APPLICATIONS OF THE ACACIA MAP

The *A. mangium* map is based on four parents from the Western Province in Papua New Guinea which is the preferred source of breeding material in current tree improvement programs for the species. It will therefore provide a sound basis on which to carry out molecular breeding of this species. The linkage relationships in the *A. mangium* map are being used to locate genes affecting quantitative traits of economic importance. Replicated clonal trials have been established from the mapping populations, and the quantitative data are being used to locate markers linked to genes controlling resistance to the phyllode rust *Atelocauda digitata.* This disease is emerging as a potentially serious problem in *A. mangium* nurseries and plantations in Southeast Asia (Old et al., 2000). Differences have been identified at the population, family, and individual level in rust resistance (Old et al., 1999). In the replicated clonal trials, one pedigree was fully resistant, while segregation for resistance was recorded in the second pedigree. Sixteen of the 17 loci with significant associations with high rust scores contained an allele in the susceptible pedigree that was not present in the resistant pedigree. Using a single-factor analysis of variance of rust resistance scores, four putative QTL regions were identified, on linkage groups 5 (6 linked markers), 9 (3 linked markers), significance threshold of $P < 0.001$), 10 (six linked markers, 3 loci at $P < 0.05$, 3 loci at $P < 0.01$), and 13 (two linked markers, $P < 0.01$) (Butcher et al., 2003). Interval mapping will be used to clarify whether only one QTL is present per region. Individual loci on linkage groups 5, 9, 10, and 13 accounted for up to 20 percent, 19 percent, 7 percent, and 3 percent of variance in resistance, respectively. Use of codominant markers in the mapping pedigree enabled the mode of action of QTL to be inferred. In all cases, the QTL effect segregated from only one parent; for loci on linkage groups 5, 9, and 13 the QTL effect was from the male parent, while the QTL affect on linkage group 10 was from the female parent. Loci within these QTL regions that segregated from only the alternate parent were not associated with resistance. The identification of specific alleles associated with susceptibility to rust opens opportunities for marker-assisted selection.

Knowledge of linkage relationships between loci in *Acacia* provides a baseline for selecting markers spanning the genome for use in population studies. Mapped loci at different levels of linkage are being used to develop new statistical approaches to assess the demographic history of a species. For example, the pattern of variation across unlinked microsatellite loci has been used to test whether population size has been constant or increasing (Goldstein et al., 1999). Variation among closely linked microsatellite loci has been used to measure gene flow from introduced populations to native populations (Estoup et al., 1999).

The *A. mangium* map provides a useful reference map for comparative studies of genome organization in other acacias. Comparative genomics is being used to determine the mechanisms and pathways by which plant genes and genomes have diverged. A high degree of linkage conservation and colinearity of marker order has been reported between other legumes, for example, large conserved sequence blocks of 37 to 104 cM between mung bean *(Vigna radiata)* and common bean *(Phaseolus vulgaris)* (Boutin et al., 1995) and between mung bean and cowpea *(Vigna unguiculata)* (Menancio-Hautea et al., 1993). By contrast, soybean contained much shorter conserved linkage blocks of 12 to 13 cM length, possibly associated with the frequent presence of duplicated sequences in the soybean genome (Boutin et al., 1995). Conservation of linkage relationships has been reported between pea *(Pisum sativum)* and lentil *(Lens ervoides* × *L. culinaris)* (Weeden et al., 1992). QTL mapping has shown the same marker loci to be associated with seed weight in three different genera: pea, mung bean, and cowpea (Timmerman Vaughan et al., 1996). Comparison of sequences from soybean genomic DNA with the model species *Arabidopsis* also revealed intergenomic synteny (Grant et al., 2000), indicating the potential to extrapolate genomic information from *Arapidopsis* to other dicotyledonous species, including legumes. These results are encouraging for comparative mapping in acacias and other legumes that should help clarify steps in their evolutionary divergence.

CONCLUSIONS

Linkage analysis in outbred pedigrees of *A. mangium* has provided the first genetic map for *Acacia*. The map is based on two independent pedigrees from the Western Province in New Guinea which is the preferred source of breeding material in current tree improvement programs. It therefore provides a sound basis on which to carry out molecular breeding. Replicated clonal trials established from the mapping pedigrees have been used to locate QTLs for resistance to the phyllode rust *Atelocauda digitata*. Mapping of resistance gene analogs to these QTL regions would allow variation in the gene sequence to be explored between resistant and susceptible phenotypes in open-pollinated populations. Differences in gene sequences or alleles could then be used for marker-assisted selection. Given the importance of the species for pulp production in Southeast Asia, future research should be directed toward identifying genes linked to pulp yield and other wood-quality traits. The genetic map provides a useful tool for transferring information on the location of genes in *A. mangium* to other *Acacia* species.

REFERENCES

Barr, C. (2000). *Profits on Paper: The Political Economy of Fibre, Finance and the Debt in Indonesia's Pulp and Paper Industries*. Bogor, Indonesia: Center for International Forestry Research.

Beavis, W. D. and Grant, D. (1991). A linkage map based on information from four F_2 populations of maize (*Zea mays* L.). *Theor. Appl. Genet.* 82: 636-644.

Begun, D. J. and Aquadro, C. F. (1992). Levels of naturally occurring DNA polymorphism correlate with recombination rates in *D. melanogaster*. *Nature* 356: 519-520.

Bennett, M. D. and Leitch, I. J. (1997). Nuclear DNA amounts in angiosperms— 583 new estimates. *Annals of Botany* 80: 169-196. <http://www.rbgkew.org.uk/cval/homepage.html>.

Bennett, M. D. and Smith, J. B. (1976). Nuclear DNA amounts in angiosperms. *Philos. T. Roy. Soc. B* 274: 227-274.

Boutin, S. R., Young, N. D., Olsen, T. C., Yu, Z.-H., Shoemaker, R. C., and Vallejos, C. E. (1995). Genome conservation among three legume genera detected with DNA markers. *Genome* 38: 928-937.

Bradshaw, H. D., Villar, M., Watson, B. D., Otto, K. G., Stewart, S., and Stettler, R. F. (1994). Molecular genetics of growth and development in Populus. III. A genetic linkage map of a hybrid poplar composed of RFLP, STS, and RAPD markers. *Theor. Appl. Genet* 89: 167-178.

Butcher, P. A., Decroocq, S., Gray, Y., and Moran, G. F. (2000). Development, inheritance and cross-species amplification of microsatellite markers. *Theor. Appl. Genet.* 101: 594-605.

Butcher, P. A., Harwood, C. E., and Quang, T. H. (2003). Studies of mating systems in seed stands suggest possible causes of variable outcrossing rates in natural populations of *Acacia mangium.* IUFRO Symposium Population and Evolutionary Genetics of Forest Trees, Stará Lesná, Slovakia, August 25-29, 2002 (in press).

Butcher, P. A. and Moran, G. F. (2000). Genetic linkage mapping in *Acacia mangium.* 2. Development of an integrated map from two outbred pedigrees using RFLP and microsatellite loci. *Theor. Appl. Genet.* 101: 594-605.

Butcher, P. A., Moran, G. F., and Bell, R (2000). Genetic linkage mapping in *Acacia mangium.* 1. Evaluation of restriction endonucleases, inheritance of RFLP loci and their conservation across species. *Theor. Appl. Genet.* 100: 576-583.

Butcher, P. A., Moran, G. F., and Perkins, H. D. (1996). Genetic resources and domestication of *Acacia mangium.* In M. J. Dieters., A. C. Matheson, D. G. Nikles, C. E. Harwood, and S. M. Walker (Eds.), *Tree Improvement for Sustainable Tropical Forestry* (pp. 467-471). Proceedings QFRI-IUFRO Conference, Caloundra, Queensland, October 27-November 1, 1996. Glympie: 1 Queensland Forestry Research Institute.

Butcher, P. A., Moran, G. F., and Perkins, H. D. (1998). RFLP diversity in the nuclear genome of *Acacia mangium. Heredity* 81: 205-213.

Butcher, P. A., Otero, A., McDonald, M. W., and Moran, G. F. (2002). Nuclear RFLP variation in *Eucalyptus camaldulensis* Dehnh. from northern Australia. *Heredity* 88: 402-412.

Butcher, P. A., Williams, E. R., Whitaker, D., Ling, S., Speed, T. P., and Moran, G. F. (2002). Improving linkage analysis in outcrossed forest trees—An example from *Acacia mangium. Theor. Appl. Genet.* 104: 1185-1191.

Buttrose, M. S., Grant, W. J. R., and Sedgley, M. (1981). Floral development in *Acacia pycnantha* Benth. in Hook. *Aust. J. Bot.* 29: 385-395.

Byrne, M. (1999). High genetic identities between three oil mallee taxa, *Eucalyptus kochii* ssp. *kochii,* ssp. *plenissima* and *E. horistes,* based on nuclear RFLP analysis. *Heredity* 82: 205-211.

Byrne, M., Murrell, J. C., Allen, B., and Moran, G. F. (1995). An integrated genetic linkage map for eucalypts using RFLP, RAPD and isozyme markers. *Theor. Appl. Genet.* 91: 869-875.

Byrne, M., Murrell, J. C., Owen, J. V., Kriedemann, P., Williams, E. R., and Moran, G. F. (1997). Identification and mode of action of quantitative trait loci affecting seedling height and leaf area in *Eucalyptus nitens. Theor. Appl. Genet.* 94: 674-681.

Byrne, M., Parrish, T. L., and Moran, G. F. (1998). Nuclear RFLP diversity in *Eucalyptus nitens. Heredity* 81: 225-233.

CAB International (2000). *Forestry Compendium Global Module.* Wallingford, UK: CAB International.

Chakravarti, A., Lasher, L. K., and Reefer, J. E. (1991). A maximum likelihood method for estimating genome length using genetic linkage data. *Genetics* 128: 175-182.

Darvasi, A., Weinreb, A., Minke, V., Weller, J. I., and Soller, M. (1993). Detecting marker-QTL linkage and estimating QTL gene effect and map location using a saturated genetic map. *Genetics* 134: 943-951.

Dayanandan, S., Bawa, K. S., and Kesseli, R. (1997). Conservation of microsatellites among tropical trees (Leguminosae). *Am. J. Bot.* 84: 1658-1663.

Diers, B. W., Keim, P., Fehr, W. R., and Shoemaker, R. C. (1992). RFLP analysis of soybean seed protein and oil content. *Theor. Appl. Genet.* 83: 13-26.

Doran, J. C. and Turnbull, J. W. (1997). *Australian Trees and Shrubs: Species for Land Rehabilitation and Farm Planting in the Tropics.* Australian Centre for International Agricultural Research (ACIAR) Monograph No. 24. Canberra, Australia: ACIAR.

Doyle, J. J. (1994). Phylogeny of the legume family: An approach to understanding the origins of nodulation. *Annu. Rev. Ecol. Syst.* 25: 325-349.

Dvorak, J., Luo, M.-C., and Yang, Z.-L. (1998). Restriction fragment length polymorphism and divergence in the genomic regions of high and low recombination in self-fertilizing and cross-fertilizing *Aegilops* species. *Genetics* 148: 423-434.

Echt, C. S., Kidwell, K. K., Knapp, S. J., Osborn, T. C., and McCoy, T. J. (1994). Linkage mapping in diploid alfalfa *(Medicago sativa). Genome* 37: 61-71.

Estoup, A., Cornuet, J.-M., Rousset, F., and Guyomard, R. (1999). Juxtaposed microsatellite systems as diagnostic markers for admixture: Theoretical aspects. *Mol. Biol. Evol.* 16: 898-908.

Fatmi, A. C., Poneleit, G., and Pfeiffer, T. W. (1993). Variability of recombination frequencies in the Iowa Stiff Stalk synthetic (*Zea mays* L.). *Theor. Appl. Genet.* 86: 859-866.

Freyre, R., Skroch, P. W., Geffroy, V., Adam-Blondon, A.-F., Shirmohamadali, A., Johnson, W. C., Llaca, V., Nodari, R. O., Pereira, P. A., Tsai, S.-M., et al. (1998). Towards an integrated linkage map of common bean. 4. Development of a core linkage map and alignment of RFLP maps. *Theor. Appl. Genet.* 97: 847-856.

Gentzbittel, L., Vear, F., Zhang, Y.-Z., Bervillé, A., and Nicolas, P. (1995). Development of a consensus linkage RFLP map of cultivated sunflower (*Helianthus annuus* L.). *Theor. Appl. Genet.* 90: 1079-1086.

Goldstein, D. B., Roemer, G. W., Smith, D. A., Reich, D. E., Bergman, A., and Wayne, R. K. (1999). The use of microsatellite variation to infer population structure and demographic history in a natural model system. *Genetics* 151: 797-801.

Grant, D., Cregan, P., and Shoemaker, R. C. (2000). Genome organisation in dicots: Genome duplication in *Arabidopsis* and synteny between soybean and *Arabidopsis. P. Natl. Acad. Sci. USA* 97: 4168-4173.

Grattapaglia, D. and Sederoff, R. (1994). Genetic linkage maps of *Eucalyptus grandis* and *Eucalyptus urophylla* using a pseudo-testcross mapping strategy and RAPD markers. *Genetics* 137: 1121-1137.

Griffin, A. R. and Cotterill, P. P. (1988). Genetic variation in growth of outcrossed, selfed and open-pollinated progenies of *Eucalyptus regnans* and some implications for breeding strategy. *Silvae Genetica* 37: 124-131.

Harwood, C. E. and Williams, E. R. (1992). A review of provenance variation in growth of *Acacia mangium*. In L. T. Carron and K. M. Aken (Eds.), *Breeding Technologies for Tropical Acacias* (pp. 22-30). ACIAR Proceedings No. 37. Canberra, Australia: ACIAR.

Kaga, A., Ishii, T., Tsukimoto, K., Tokoro, E., and Kamijima, O. (2000). Comparative molecular mapping in *Ceratotropis* species using an interspecific cross between azuki bean *(Vigna angularis)* and rice bean *(V. umbellata). Theor. Appl. Genet.* 100: 207-213.

Kaló, P., Endre, G., Zimányi, L., Csanádi, G., and Kiss, G. B. (2000). Construction of an improved linkage map of diploid alfalfa *(Medicago sativa). Theor. Appl. Genet.* 100: 641-657.

Kenrick, J. and Knox, R. B. (1979). Pollen development and cytology in some Australian species of *Acacia. Aust. J. Bot.* 27: 413-427.

Kenrick, J. and Knox R. B. (1982). Function of the polyad in reproduction of *Acacia. Ann. Bot.* 50: 721-727.

Kenrick, J. and Knox, R. B. (1989). Quantitative analysis of self-incompatibility in trees of seven species of Acacia. *J. Hered.* 80: 241-245.

Kianian, S. F. and Quiros, C. F. (1992). Generation of a *Brassica oleracea* composite RFLP map: Linkage arrangements among various populations and evolutionary implications. *Theor. Appl. Genet.* 84: 544-554.

Kraft, T., Säll, T., Magnusson-Rading, I., Nilsson, N.-O., and Halldén, C. (1998). Positive correlation between recombination rates and levels of genetic variation in natural populations of sea beet *(Beta vulgaris* subsp. *maritima). Genetics* 150: 1239-1244.

Mandal, A. K., Ennos, R. A., and Fagg, C. W. (1994). Mating system analysis in a natural population of *Acacia nilotica* subspecies *Leiocarpa. Theor. Appl. Genet.* 89: 931-935.

Maslin, B. R. (2001). *Introduction to Acacia Flora of Australia 11A Mimosaceae, Acacia.* Part 1 (pp. 3-13). Melbourne: Australian Biological Resources Study (ABRS)/Commonwealth Scientific and Industrial Research Organization (CSIRO).

Maslin, B. R. and Stirton, C. H. (1997). Generic and infrageneric classification in *Acacia* (Leguminosae: Mimosoideae): A list of critical species on which to build a comparative data set. *Bulletin of the International Group for the Study of Mimosoideae* 20: 22-44.

Maslin, B. R., Thomson, L. A. J., McDonald, M. W., and Hamilton-Brown, S. (1998). *Edible Wattle Seeds of Southern Australia.* Melbourne: CSIRO.

Menancio-Hautea, D., Fatokun, C. A., Kumar, L., Danesh, D., and Young, N. D. (1993). Comparative genome analysis of mungbean *(Vigna radiat* L. Wilczek) and cowpea *(V. unfuiculata* L Walpers). *Theor. Appl. Genet.* 86: 797-810.

Miller, J. T. and Bayer, R. J. (2001). Molecular phylogenetics of *Acacia* (Fabaceae: Mimosoideae) based on the chloroplast *MATK* coding sequence and flanking *TRNK* intron spacer regions. *Am. J. Bot.* 88: 697-705.

Moran, G. F., Muona, O., and Bell, J. C. (1989a). *Acacia mangium:* A tropical forest tree of the coastal lowlands with low genetic diversity. *Evolution* 43: 231-235.

Moran, G. F., Muona, O., and Bell, J. C. (1989b). Breeding systems and genetic diversity in *Acacia auriculiformis* and *A. crassicarpa. Biotroprica* 21: 250-256.

Mukherjee, W. and Sharma, A. K. (1995). *In situ* nuclear DNA variation in Australian species of *Acacia. Cytobios* 83: 59-64.

Muona, O., Moran, G. F., and Bell, J. C. (1991). Hierarchical patterns of correlated mating in *Acacia melanoxylon. Genetics* 127: 619-626.

Muranty, H. (1996). Power of tests for quantitative trait loci detection using full-sib families in different schemes. *Heredity* 76: 156-165.

Murphy, D. J., Udovicic, F., and Ladiges, P. Y. (2000). Phylogenetic analysis of Australian *Acacia* (Leguminosae: Mimosoideae) by using sequence variation of an intron and two intergenic spacers of chloroplast DNA. *Aust. Syst. Bot.* 13: 745-754.

Newman, I. V. (1933). Studies in the Australian Acacias. II. The life history of *Acacia baileyana* (F.v.M.). Part I. Some ecological and vegetative features, spore production and chromosome number. *J. Linn. Soc. London, Bot.* 49: 145-171.

Nodari, R. O., Tsai, S. M., Gilbertson, R. L., and Gepts, P. (1993). Towards an integrated map of common bean 2. Development of an RFLP-based linkage map. *Theor. Appl. Genet.* 85: 513-520.

Old, K. M., Butcher, P. A., Harwood, C. E., and Ivory, M. I. (1999). *Atelocauda digitata,* a rust disease of tropical plantation acacias. Poster Abstract, Twelfth Australasian Plant Pathology Conference, Canberra, September 27-31, 1999.

Old, K. M., See, L. S., Sharma, J. K., and Yuan, Z. Q. (2000). *A Manual of Diseases of Tropical Acacias in Australia, Southeast Asia and India* (pp. 13-20). Jakarta, Indonesia: Center for International Forestry Research (CIFOR).

Olng'otie, P. A. S. (1991). *Acacia tortilis:* A study of genetic structure and breeding systems. PhD Thesis, Oxford University, Oxford, UK.

Ott, J. (1992). *Analysis of Human Genetic Linkage,* Revised Edition (pp. 116-117). Baltimore, MD: The Johns Hopkins University Press.

Paglia, G. P., Olivieri, A. M., and Morgante, M. (1998). Towards second-generation STS (sequence-tagged sites) linkage maps in conifers: A genetic map of Norway spruce (*Picea abies* K.). *Mol. Gen. Genet.* 258: 466-478.

Peakall, R., Gilmore, S., Keys, W., Morgante, M., and Rafalski, A. (1998). Cross-species amplification of soybean *(Glycine max)* simple-sequence-repeats (SSRs) within the genus and other legume genera. *Mol. Biol. Evol.* 15: 1275-1287.

Pedley, L. (1986). Derivation and dispersal of *Acacia* (Leguminosae), with particular reference to Australia, and the recognition of *Senegalia* and *Racosperma. Bot. J. Linn. Soc.* 92: 219-254.

Plomion, C., Liu, B.-H., and O'Malley, D. M. (1996). Genetic analysis using trans-dominant linked markers in an F2 family. *Theor. Appl. Genet.* 93: 1083-1089.

Plomion, C. and O'Malley, D. M. (1996). Recombination rate differences for pollen parents and seed parents in *Pinus pinaster. Heredity* 77: 341-350.

Plomion, C., O'Malley, D. M., and Durel, C. E. (1995). Genomic analysis in maritime pine *(Pinus pinaster).* Comparison of two RAPD maps using selfed and open-pollinated seeds of the same individual. *Theor. Appl. Genet.* 90: 1028-1034.

Ritland, K. (1990). Inferences about inbreeding depression based on changes of the inbreeding coefficient. *Evolution* 44: 1230-1241.

Robinson, J. and Harris, S. A. (2000). A plastid DNA phylogeny of the genus *Acacia* Miller (Acacieae, Leguminoseae). *Bot. J. Linn. Soc.* 132: 195-222.

Ross, J. H. (1981). An analysis of the African *Acacia* species: Their distribution, possible origins and relationships. *Bothalia* 13: 389-413.

Sedgley, M., Harbard, J., Smith, R. M. M., Wickneswari, R., and Griffin, A. R. (1992). Reproductive biology and interspecific hybridisation of *Acacia mangium* and *Acacia auriculiformis* A. Cunn. ex Benth. (Leguminosae: Mimosoideae). *Aust. J. Bot.* 40: 37-48.

Sewell, M. M., Sherman, B. K., and Neale, D. B. (1999). A consensus map for loblolly pine (*Pinus taeda* L.). I. Construction and integration of individual linkage maps from two outbred three-generation pedigrees. *Genetics* 151: 321-330.

Sim, B. L. (1984). The genetic base of *Acacia mangium* Willd. in Sabah. In R. D. Barnes and G. L. Gibson (Eds.), *Provenance and Genetic Improvement Strategies in Tropical Forest Trees* (pp. 597-603). Harare, Zimbabwe: Commonwealth Forestry Institute Oxford and Forest Research Centre.

Stam, P. and van Oiijen, J. W. (1995). JoinMap Version 2.0: Software for the calculation of genetic linkage maps. Wageningen, The Netherlands: Centre for Plant Breeding and Reproduction Research (CPRO-DLO).

Timmerman Vaughan, G. M., McCallum, J. A., Frew, T. J., Weeden, N. F., and Russell, A. C. (1996). Linkage mapping of quantitative trait loci controlling seed weight in pea (*Pisum sativum* L.) *Theor. Appl. Genet.* 93: 431-439.

Tuomela, K., Otsamo, A., Kuusipalo, J., Vuokko, R., and Nikles, G. (1996). Effect of provenance variation and singling and pruning on early growth of *Acacia mangium* Willd. plantation on *Imperata cylindrica* (L.) Beauv. dominated grassland. *Forest Ecol. and Manag.* 84: 241-249.

Vallejos, C. E., Sakiyama, N. S., and Chase, C. D. (1992). A molecular marker-based linkage map of *Phaseolus vulgaris* L. *Genetics* 131: 733-740.

Verhaegen, D. and Plomion, C. (1996). Genetic mapping in *Eucalyptus urophylla* and *Eucalyptus grandis* using RAPD markers. *Genome* 39: 1051-1061.

Weeden, N. F., Muehlbauer, F. J., and Ladizinsky, G. (1992). Extensive conservation of linkage relationships between pea and lentil genetic maps. *J. Hered.* 83: 123-129.

Williams, C.G. and Savolainen, O. (1996). Inbreeding depression in conifers: Implications for breeding strategy. *Forest Sci.* 42: 102-117.

Young, N. D., Weeden, N. F., and Kochert, G. (1996). Genome mapping in legumes (Fam. Fabaceae). In A. H. Paterson (Ed.), *Genome Mapping in Plants* (pp. 211-228). San Diego, CA: Academic Press.

Index

Page numbers followed by the letter "f" indicate figures; those followed by the letter "t" indicate tables.

T - #0027 - 101024 - C8 - 229/152/25 [27] - CB - 9781560229582 - Gloss Lamination